T0336086

Frontiers in
Nanobiomedical
Research

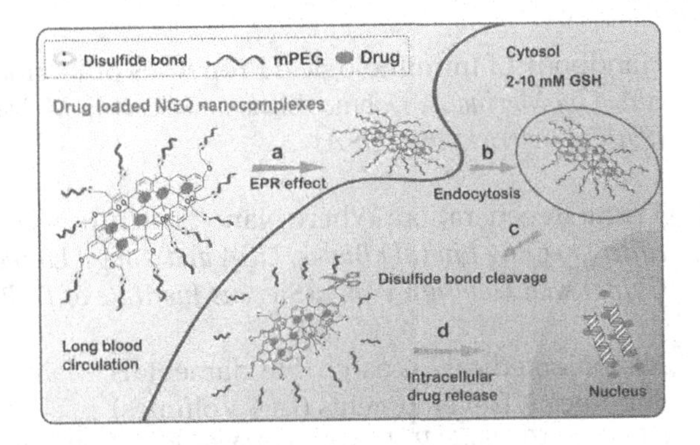


# BIO-INSPIRED NANOMATERIALS AND APPLICATIONS

## Nano Detection, Drug/Gene Delivery, Medical Diagnosis and Therapy

## Frontiers in Nanobiomedical Research

**ISSN: 2251-3965**



---

*Published*

Vol. 1: Handbook of Immunological Properties of Engineered Nanomaterials
*edited by Marina A. Dobrovolskaia and Scott E. McNeil (SAIC-Frederick, Inc., USA)*

Vol. 2: Tissue Regeneration: Where Nano-Structure Meets Biology
*edited by Qing Liu (3D Biotek, USA and Tongji University, China) and Hongjun Wang (Stevens Institute of Technology, USA)*

Vol. 3: Nanobiomedical Research: Fundamentals, Main Applications and Recent Developments (In 4 Volumes)
*edited by Vladimir P. Torchilin (Northeastern University, USA)*

Vol. 4 Bio-Inspired Nanomaterials and Applications: Nano Detection, Drug/Gene Delivery, Medical Diagnosis and Therapy
*edited by Donglu Shi (Tongji University School of Medicine, China and University of Cincinnati, USA)*

*Forthcoming titles*

Cancer Therapeutics and Imaging: Molecular and Cellular Engineering and Nanobiomedicine
*edited by Kaushal Rege (Arizona State University, USA)*

Nano Vaccines
*edited by Balaji Narasimhan (Iowa State University, USA)*

Nano Pharmaceuticals
*edited by Rajesh N. Dave (New Jersey Institute of Technology, USA)*

Thermal Aspects in Nanobiomedical Systems and Devices
*by Dong Cai (Boston College, USA)*

Nano Mechanochemistry in Biology
*edited by Jeffrey Ruberti (Northeastern University, USA)*

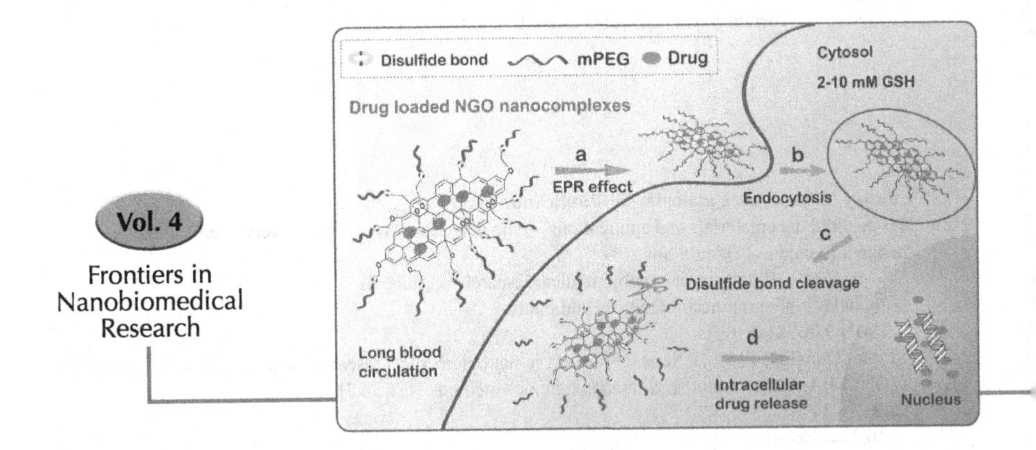


Vol. 4

Frontiers in Nanobiomedical Research

# BIO-INSPIRED NANOMATERIALS AND APPLICATIONS

## Nano Detection, Drug/Gene Delivery, Medical Diagnosis and Therapy

editor

**Donglu Shi**

Tongji University School of Medicine, China

University of Cincinnati, USA

NEW JERSEY • LONDON • SINGAPORE • BEIJING • SHANGHAI • HONG KONG • TAIPEI • CHENNAI

*Published by*

World Scientific Publishing Co. Pte. Ltd.
5 Toh Tuck Link, Singapore 596224
*USA office:* 27 Warren Street, Suite 401-402, Hackensack, NJ 07601
*UK office:* 57 Shelton Street, Covent Garden, London WC2H 9HE

**Library of Congress Cataloging-in-Publication Data**
Bio-inspired nanomaterials and applications : nano detection, drug/gene delivery, medical diagnosis and therapy / [edited by] Donglu Shi.
p. ; cm. -- (Frontiers in nanobiomedical research ; volume 4)
Includes bibliographical references and index.
ISBN 978-9814616911
I. Shi, Donglu, editor. II. Series: Frontiers in nanobiomedical research ; v. 4. 2251-3965
[DNLM: 1. Nanostructures. 2. Nanomedicine--methods. QT 36.5]
R857.N34
610.28'4--dc23

2014031062

**British Library Cataloguing-in-Publication Data**
A catalogue record for this book is available from the British Library.



In-house Editor: Rhaimie Wahap

Typeset by Stallion Press
Email: enquiries@stallionpress.com

Printed in Singapore

# Preface

Materials science and engineering has been a field that is primarily concerned with the studies of the structure, property, and processing of all bulk solids with considerable rigidities. These materials are often known as metals, ceramics, polymers, and composites. Based on their distinctively different properties, the materials can also be divided into structural and functional materials. The former is widely employed, with large volumes, in mechanical systems where high strengths are required, such as building constructions, automobiles, aviation, and naval vessels. The latter, for its unique electromagnetic properties, may be characterized as the electronic materials. Sensors and semiconductor thin films fall into this category that makes up the main components of today's electronic devices and electrical instruments. Another important group, for its main applications in medical implants, is called biomaterials. Typical biomaterials include hydroxyapatite bone graft and titanium alloy hip joint. These materials are designed, selected, processed, and developed according to specific applications where important materials properties are correlated to the (micro)structures and optimized for both engineering and economic considerations.

A common characteristic among all traditional materials is its bulk form. Even for most of the materials used in the powder form assume several micrometers in diameter. For semiconducting films, the substrates of integrated circuitry extend into the millimeter range. As such, the materials behaviors are characterized as the "bulk property." It is easy to understand the bulk strength of a structural material as its stress-strain behavior must reply upon appreciable dimension of the testing bar. The electronic properties such as electrical resistivity, thermal conductivity, energy bands, magnetic domains, and dielectricity are all established based on the lattice structures of the bulk materials. The corresponding theories are developed under the boundary conditions such that the material surfaces and interfaces are not interplayed in the calculations of the intrinsic parameters.

However, the situation is complicated when the dimension of a material is reduced to nanoscale, say a few nanometers. As is well known, the surface of a particle is classically defined, in materials science, a two-dimension defect (or surface defect) where the atomic arrangement is no long ordered compared to the bulk. Therefore, the properties of the surfaces must be altered significantly. For instance, for a particle of Palladium with a diameter of 0.77 cm, the surface area is 3.6 $cm^2$. If the diameter is reduced to 0.01 cm, the surface area increases to 260 $cm^2$. The surface area will be stunningly $2.8 \times 10^7$ $cm^2$ for a particle of 1 nm. Therefore, for nanoparticles, the volume fraction of a given mass will be primarily the surfaces. In other words, the bulk volume fraction is negligibly small. If this is the case, application of theories that are developed based on the bulk structures may not be appropriate.

The typical nanomaterials that are being extensively investigated include nanoscale particles of polymeric micelles, oxides ($Fe_3O_4$, $SiO_2$, ZnO...), gold, and carbon-based structures (graphene and nanotubes). There has been intense current research on their new properties that have risen from the "nano effects." New physical properties have been discovered from these nanomaterials. For instance, the quantum dots are developed based quantum confinement which gives size-dependent emissions. Superparamagnetism is a unique behavior resulting from the single magnetic domains of $Fe_3O_4$ nanoparticles in a size range of a few nanometers. Novel synthesis routs have been developed for a variety of nanomaterials that can be used for applications in efficient energy devices, biotechnology, medical diagnosis, and cancer therapeutics.

Ever since the beginning of nanoscience, the research communities soon found the real possibilities of using nanotechnology to solve the key biomedical problems that cannot be easily dealt with by conventional approaches. Collaborations quickly began among researchers from the biomedical and physical sciences. On the other hand, the materials scientists found medical, especially clinical problems extremely inspiring to their innovative design of the nanostructures and development of multifunctional carriers for biomedical applications. However, new materials issues arise concerning the interfaces between the nanoparticles and cells, nano surface properties that affect the biological behaviors such as endocytosis and toxicity, and nano size effect on transport kinetics in a biological system. These are the new challenges rarely faced by chemists and materials scientists in conventional materials research.

For instance, the current clinical magnetic resonance imaging (MRI) scan can readily detect lesions as small as a few millimeters. Dynamic MRI of human cancer in mice has reached a high spatial resolution. But MRI is costly and time consuming to implement. Compared to MRI, optical imaging is easier to use and cost-effective, and can be applied to cancer diagnosis with high resolution in a straightforward fashion. Recently, for diagnosing breast cancer, MRI and near-infrared optics have been combined as a potentially more accurate method. By combining these two techniques, MRI produces the basic image of the breast, while the near-infrared optical means provides the functional information of the tissue, such as the oxygen consumption rate at a particular region, which could indicate early cancer development. Thus, multimodality will enable imaging from different perspectives, giving much higher accuracy in medical diagnosis. In optical imaging, quantum dots (QD) have superior properties, including higher quantum yield and much sharper emission spectra. Quantum dots can be "tuned" based on their size, resulting from quantum confinement effects. They offer great advantages as a contrasting agent for *in vivo* imaging.

There is an increasing need for the early diagnosis of cancer, prior to the detection of anatomic anomalies. A great challenge in cancer diagnosis is to locally biomark cancer cells for maximum therapeutic benefit. Tumor cell targeting has been a complex issue in medical diagnosis using nanotechnologies. Although there have been extensive efforts in this area, a few have reported successful attempts that enabled effective targeting with high specificity. Several reasons may have contributed to this difficulty, and the major one is due to the lack of biomarkers and non-specific binding. The success of targeted treatments will depend on the expression of specific proteins or genes present in cancer cells. Targeting strategies may be divided into those utilizing specific binding to ligands or receptors and those based on nonspecific adsorptive mechanism. If the cancer cells exhibit cell-specific carbohydrates, they may serve as binding sites to colloidal drug carriers containing appropriate ligands. However, none of these strategies will guarantee absolute cell targeting with high specificity as they primarily reply on common antibodies of the tumor cells.

There have been many effective approaches developed in tumor therapy by nanotechnology. But selective delivery of drugs into tumor lesions has been a key challenge for successful management of cancers. In current drug delivery, the release of the drug has mainly relied on the natural diffusion process and bio degradation of the carrier systems. However, the tumor treatment needs well-developed procedures whereby, anti-cancer drug can be delivered in a selective, localized, and timely controlled fashion. There has been a critical need to develop novel approaches that could provide precision spatial and temporal drug delivery at the tumor site. New delivery mechanisms must be identified and developed from the perspectives of materials science and biomedicine. Novel nanostructures need to be designed and developed with multi functionalities which can enable drug delivery in a selective, predictable, and controlled manner. Drugs are well protected by special coatings during transport, but released via different but intelligent mechanisms. The nanocarriers are also rendered strongly fluorescent and surface functionalized with tumor specific antibodies. Therefore they are targeted on the tumor cells and observable by *in vivo* fluorescent imaging. In medical therapy, delivery of siRNAs to target cells has been a novel approach. Using small interfering RNA for studying gene function and drug target validation has received great attention in the medical communities. The key challenge is on the design of the nanocarriers for siRNAs delivery. In the delivery systems, the chemical modification of siRNA duplex is critical for increased stability and sensitive detection of siRNA.

Recently, a new trend is becoming prevalent in the development of the nanosystems for simultaneous tumor diagnosis and therapy. This requires high versatility of the nanocarrier system with multiple functionalities of cell targeting, drug storage, optical imaging, and effective means of treatment such as magnetic and photothermal hyperthermia, photodynamic therapy, and drug release via various intelligent mechanisms (pH, temperature, and biochemical variations in the tumor environment). A new terminology "theranosics" has been frequently used and applied in medical communities, which implies simultaneous therapy and diagnosis. For instance, a nanosystem can simultaneously achieve both cell targeted *in vivo* imaging and photothermal treatment of cancer by nanotechnology. While achieving concurrent high spatial and temporal resolution of the lesions via cell targeting; special non-evasive treatments are implemented at the same time by various means, such as localized drug release, hyperthermia, and photo-thermal therapy.

Inspired by these challenging problems in biomedical fields, the development of the nanotechnologies will be the key in addressing some of the critical issues in medicine, especially in early cancer diagnosis and treatment. This book, *Bio-inspired Nanomaterials and Devices,* summarizes the most recent developments in nanomaterials, biotechnology, and medical diagnosis and therapy in a comprehensive fashion for researchers from diverse fields of chemistry, materials science, physics, engineering, biology, and medicine. Not only does the book touch up on the most fundamental topics of nanoscience, but also deal with critical clinical issues of translational medicine.

The book is written in a straightforward and tutorial fashion, typically suitable for technical non-specialists. All chapters are written by active researchers in frontier research of nanobiomedicine. We hope this book will provide timely and useful information for the progress of nanomaterials and biomedical applications. We are grateful to all invited authors for their excellent contributions to this book. We would like to thank Drs. Haiqing Dong and Yong Yong Li for their great talent and efforts in developing the schematic diagram for the book cover.

*Prof. Donglu Shi*
Shanghai East Hospital
The Institute for Biomedical Engineering & Nano Science
Tongji University School of Medicine
Shanghai 200120, China

and

The Materials Science and Engineering Program
College of Engineering and Applied Science
University of Cincinnati, USA

September, 2014

# List of Contributors

**Bingdi Chen**
Research Center for Translational Medicine, East Hospital, Tongji University School of Medicine, No. 150 Jimo Road, Shanghai 200120;
The Institute for Biomedical Engineering & Nano Science Tongji University, No. 1239 Siping Road, Shanghai, 200092, P. R.China
*Chapter 8*

**Yao Chen**
Institute for Biomedical Engineering and Nano Science,East Hospital, Tongji University School of Medicine, No. 150 Jimo Road, Shanghai 200120, P. R. China
and
Shanghai Tongji Hospital, Tongji University School of Medicine, Shanghai 200065, P. R. China
*Chapter 9*

**Liang Cheng**
Jiangsu Key Laboratory for Carbon-Based Functional Materials & Devices, Institute of Functional Nano & Soft Materials (FUNSOM), Soochow University, Suzhou, Jiangsu 215123, P. R. China
*Chapter 1*

**Maoquan Chu**
School of Life Science and Technology, Tongji University, 1239 Siping Road, Shanghai 200092, P. R. China
and
Research Center for Translational Medicine at Shanghai East Hospital, Tongji University, 150 Jimo Road, Shanghai 200120, P. R. China
*Chapter 4*

**Jia Huang**
School of Materials Science and Engineering, Tongji University, Shanghai 201804, P. R. China
and
The Institute for Biomedical Engineering and Nano Science, Tongji University School of Medicine, Shanghai 200092, P. R. China
*Chapter 10*

**Yongyong Li**
The Institute for Biomedical Engineering and Nano Science, Tongji University, Shanghai, 200092, P. R. China
*Chapter 2*

**Chao Lin**
East Hospital, The Institute for Biomedical Engineering and Nano Science, Tongji University School of Medicine, Shanghai 200092, P. R. China
*Chapter 3*

**Zhuang Liu**
Jiangsu Key Laboratory for Carbon-Based Functional Materials & Devices, Institute of Functional Nano & Soft Materials (FUNSOM), Soochow University, Suzhou, Jiangsu 215123, P. R. China
*Chapter 1*

**Kun Lu**
Institute for Biomedical Engineering and Nano Science, East Hospital, Tongji University School of Medicine, No. 150 Jimo Road, Shanghai 200120, P. R. China
and
Shanghai Tenth People's Hospital, Tongji University School of Medicine, Shanghai 200072, P. R. China
*Chapter 9*

**Lifeng Qi**
Institute for Biomedical Engineering and Nano Science, East Hospital, Tongji University School of Medicine, No. 150 Jimo Road, Shanghai 200120, P. R. China
*Chapter 9*

**Yao Qin**
East Hospital, The Institute of Biomedical Engineering and Nano Science, Tongji University School of Medicine, Shanghai 201203, P. R. China
*Chapter 7*

**Ting Shi**
East Hospital, The Institute for Biomedical Engineering and Nano Science, Tongji University School of Medicine, Shanghai 200092, P. R. China
*Chapter 3*

**Xiaohui Tan**
Institute for Biomedical Engineering and Nano Science, East Hospital, Tongji University School of Medicine, No. 150 Jimo Road, Shanghai 200120, P. R. China
and
Shanghai Shuguang Hospital, Shanghai University of Traditional Chinese Medicine, Shanghai 201203, P. R. China
*Chapter 9*

**Changhui Wang**
Shanghai Tenth People's Hospital, Tongji University School of Medicine, Shanghai 200072, P. R. China
*Chapter 9*

**Feng Wang**
Department of Physics and TcSUH, University of Houston, Houston, TX77204, USA
*Chapter 6*

**Lei Wang**
Institute for Biomedical Engineering and Nano Science, East Hospital, Tongji University School of Medicine, No. 150 Jimo Road, Shanghai 200120, P. R. China
and
Shanghai Tongji Hospital, Tongji University School of Medicine, Shanghai 200065, P. R. China
*Chapter 9*

**Yilong Wang**
Research Center for Translational Medicine, East Hospital, The Institute for Biomedical Engineering & Nano Science, Tongji University School of Medicine
Shanghai, 200092, P. R. China
*Chapter 6*

**Huiyun Wen**
School of Chemical Engineering, Northwest University, Xi'an, 710069, P. R. China
*Chapter 2*

**Xiaohan Wu**
School of Materials Science and Engineering, Tongji University, Shanghai 201804, P. R. China
and
The Institute for Biomedical Engineering and Nano Science, Tongji University School of Medicine, Shanghai 200092, P. R. China
*Chapter 10*

**Jinhu Yang**
Department of Chemistry, Tongji University, Shanghai 200092, P. R. China
and
Research Center for Translational Medicine, East Hospital The Institute for Biomedical Engineering and Nano Science, Tongji University School of

Medicine, Shanghai 200092,
P. R. China
*Chapter 5*

**Bingbo Zhang**
Research Center for Translational Medicine, East Hospital, The Institute for Biomedical Engineering and Nano Science, Tongji University School of Medicine, Shanghai 200092, P. R. China
*Chapter 1*

**Peng Zhao**
East Hospital, The Institute for Biomedical Engineering and Nano Science, Tongji University School of Medicine, Shanghai 200092, P. R. China
*Chapter 3*

**Xin Zhao**
Harvard Medical School, MA, Cambridge, 02139, USA
*Chapter 2*

# Contents

# Chapter 1

# Up-Conversion Nanoparticles for Early Cancer Diagnosis

*Liang Cheng**, *Bingbo Zhang*† *and Zhuang Liu**

**Jiangsu Key Laboratory for Carbon-Based Functional Materials & Devices, Institute of Functional Nano & Soft Materials (FUNSOM), Soochow University, Suzhou, Jiangsu 215123, P. R. China*
†*Research Center for Translational Medicine, East Hospital, The Institute for Biomedical Engineering and Nano Science, Tongji University School of Medicine, Shanghai 200092, P. R. China*

Up-conversion nanoparticles (UCNPs), particularly lanthanide-doped nanocrystals, which emit high energy photons under excitation by the near-infrared (NIR) light, have found potential applications in many different fields including biomedicine. Compared with traditional down-conversion fluorescence imaging, the NIR light excited up-conversion luminescence (UCL) imaging, relying on UCNPs, exhibits improved tissue penetration depth, higher photochemical stability and is free of auto-fluorescence background. UCL shows promise in biomedical imaging with high sensitivities. On the other hand, the unique up-conversion (UC) process of UCNPs may be utilized to photosensitive therapeutic agents for applications in cancer treatment. This chapter will focus on the biomedical imaging and cancer therapy applications of UCNPs and their nanocomposites, and discuss recent advances and future prospects in the imaging field.

## 1. Introduction of Up-Conversion Nanoparticles (UCNPs) in Nanomedicine

Rare earth (RE)-containing UCNPs are able to emit high-energy photons under near-infrared (NIR) excitation, referring to a nonlinear optical up-conversion (UC) process in which the sequential absorption of two or more photons leads to the emission of a single photon at the shorter wavelength. UC processes are mainly divided into three classes of mechanisms: excited state absorption (ESA), energy transfer up-conversion (ETU), and photo avalanche (PA).[1] Since its discovery in the mid-1960s,[2] it has attracted broad interests because of its potential applications in many different areas including solid state lasers,[3] 3D flat-panel displays,[4] low-intensity IR imaging,[5] and bio-probes.[6–14] With the development of nanotechnology, a variety of methods have been developed to synthesize various kinds of UCNPs, many of which are lanthanide-doped nanocrystals with fine crystalline structures and controlled sizes.[7,15–18] Compared with the traditionally used down-conversion fluorescent organic dyes and quantum dots (QDs), optical probes based on UCNPs show a number of advantages, such as sharp emission bandwidths, long lifetime, tunable emission, high photostability, low cytotoxicity, and importantly, little background auto-fluorescence, making them attractive contrast agents in optical biomedical imaging.[15,19–22]

In the past few years, a large number of groups have investigated the potential applications of UCNPs in biomedicine.[19,20,23] Prasad *et al.* reported *in vivo* whole body up-conversion luminescence (UCL) imaging of mice injected with UCNPs for the first time.[24] Using targeting ligand conjugated UCNPs, Li *et al.* achieved efficient *in vivo* tumor targeting and UCL imaging.[25] UCNPs could be further engineered to be contrast agents in magnetic resonance (MR) imaging, positron emission tomography (PET), and computer tomography (CT), for *in vitro* and *in vivo* multimodal imaging as demonstrated by a number of groups.[26–30] UCNPs may also be combined with anti-cancer drugs,[31] photosensitizers (PS)[32,33] or gold nanostructures[34] for potential therapeutic applications including chemo-therapy, photodynamic therapy (PDT), and photothermal therapy (PTT), respectively. In this chapter, we thus summarize recent progress regarding the use of UCNPs and their nanocomposites for *in vivo* biomedical imaging, and discuss the future challenges and promises of using UCNPs as novel theranostic nano-platforms.

## 2. Synthesis and Functionalization of UCNPs and Their Nanocomposites

### 2.1. *Synthetic approaches for UCNPs*

Generally speaking, UCNPs are composed of three components: a host matrix, a sensitizer, and an activator. An ideal host matrix needs to have low lattice photon energies, a requirement to minimize non-radiative losses and maximize the radiative emission. Among the available types of up-conversion host materials, fluorides such as $NaYF_4$,[7,15–18,35–40] $NaGdF_4$,[27,41–45] $NaLuF_4$,[46–49]

$KYF_4$,[50] $NaYbF_4$,[51] $LaF_3$,[52] $CaF_2$,[53] $KMnF_3$,[54] $YF_3$,[55] $KGdF_4$[10] and a few others, have proven to be the ideal host candidates due to their low photon energies and high chemical stability. Several approaches have been used to synthesize RE-doped UCNPs, including thermal decomposition method, hydrothermal method, co-precipitation method, and several others.

Thermal decomposition is an efficient and convenient method to synthesize highly monodispersed UCNPs. In this process, $Ln^{3+}$ trifluoroacetates are decomposed in the presence of octadecene (ODE)/oleic acid (OA) or oleic amine (OM). The ratio of trifluoroacetate precursors, reaction time, temperatures, and coordination properties of the solvents play important roles in controlling the size, phase, morphology, and the UCL efficiency of the yielded nanocrystals during the thermal decomposition process.[15–17] Yan *et al.* reported a general method to synthesize high quality fluoride-based nanocrystals *via* the co-thermolysis of the $Ln^{3+}$ trifluoroacetate precursors.[18,52,56,57] The $Er^{3+}$, $Tm^{3+}$, and $Ho^{3+}$ ion-doped $NaYbF_4$ UCNPs could generate red, blue, and green light under the 980 nm excitation.[51] The thermal decomposition method has been proven to be an effective approach to fabricate monodispersed, single crystalline and phase-pure UCNPs.

Hydrothermal/solvothermal method refers to a chemical synthesis procedure within a sealed environment under high temperature and pressure. It has been widely used to synthesize various kinds of inorganic nanomaterials,[58,59] including UCNPs. There are several examples for the hydrothermal synthesis of high quality UCNPs using ethylenediaminetetraacetic acid (EDTA), citrate, polyvinylpyrrolidone (PVP), polyethylene glycol (PEG), or polyacrylic acid (PAA) as capping reagents.[60–63] The advantages of the hydrothermal approach include the easy control of the reaction conditions, high synthesis yields and relatively low costs.

The co-precipitation method is one of the most convenient approaches to synthesize UCNPs. In a typical procedure, capping ligands (such as EDTA) are used to tune and manipulate the size of nanoparticles.[64] This method is very simple considering its mild reaction conditions, low costs, simple protocols and short reaction time; however, it fails to obtain nanoparticles with single crystalline structures.

There are some other methods to synthesize UCNPs, such as sol-gel processing,[65,66] combustion synthesis, and flame synthesis. Prasad *et al.* developed an interesting variation of the sol-gel method that produces $Er^{3+}$-doped $ZrO_2$ nanocrystals.[65] Combustion synthesis is a time-saving method that can afford reaction products, which usually are oxide and oxysulfide UCNPs ($Y_2O_3$, $La_2O_2S$, and $Gd_2O_3$), within minutes.[67–69] Flame synthesis is another technique of synthesizing UCNPs. The particle size, morphology, and photoluminescence intensity are strongly affected by the flame temperature in this method.[70]

Despite the variety of reported synthetic methods, thermal decomposition and solvothermal methods are still the two most widely adopted approaches to synthesize UCNPs of high qualities and bright UCL emissions.

## 2.2. *Surface modifications of UCNPs*

As-synthesized UCNPs by the above-mentioned protocols usually are capped with organic ligands and thus are not soluble in water. Surface modification is the critical step to fabricate UCNP-based bio-probes for various biomedical applications. To date, several approaches have been used to modify UCNPs including ligand exchange, ligand oxidation reaction, layer-by-layer (LBL) self-assembly method, and coating with amphiphilic polymers.

Hydrophobic ligands (such as OA, ODE, or OM) on the surface of the synthesized UCNPs can be replaced through ligand exchange reactions with a variety of hydrophilic organic molecules including PAA,[12,34,40,71–77] polyethylene glycol (PEG),[15,78] hexanedioic acid (HDA),[79] 6-aminohexanoic acid (AHA),[80] Thioglycolic acid (TGA),[81] 3-mercaptopropionic acid (3-MA),[24,42,82] Dimercaptosuccinic acid (DMSA),[83] citrate,[84]1, 10-decanedicarboxylic acid (DDA),[85] and 11-mercaptoundecanoic acid (MUA).[85]After these modifications, the obtained UCNPs would become water-soluble.

Ligand oxidation reaction is another method to render nanoparticles water-solubility and to provide reactive functional groups for subsequent bioconjugation. The carbon–carbon double bond of the ligand (such as OA) on the surface of UCNPs can be oxidized directly to form two carboxyl groups with the aid of an oxidation reagent such as $KMnO_4$, $NaIO_4$, and ozone.[86–88] The morphology, composition, and UCL of the UCNPs should not be influenced after the ligand oxidation. However, this method is applicable only to a specific class of UCNPs capped with ligand molecules containing unsaturated carbon–carbon bonds in their long alkyl chains.

Apart from ligand engineering, LBL assembly is other widely used surface modification strategy to functionalize UCNPs. In such a method, several layers of counter-charged polymers are successively deposited on the nanoparticle surface, utilizing the electrostatic attractions between the oppositely charged species.[7] Poly allylamine hydrochloride (PAH), PAA and poly sodium 4-styrenesulfonate (PSS) are among the most commonly used charged polymers to coat nanoparticles by this approach.[89] The LBL assembly technique has several advantages, including simplicity, universality, and coating thickness control in nanoscale.

Coating with amphiphilic polymers is another way to modify UCNPs. Hydrophobic UCNPs coated by OA/OM/ODE could be further coated with amphiphilic polymers *via* hydrophobic interactions and then transferred into the aqueous phase. A number of PEG grafted amphiphilic polymers, such as PEG/octylamine co-modified PAA (OA–PAA–PEG), PEG grafted poly maleic anhydride-alt-1-octadecene (C18PMH-PEG), and PEGylated phospholipids, have been developed by our and other groups to functionalize UCNPs.[31,78,90] PEGylated UCNPs usually offer improved stability in physiological solutions and better biocompatibility.

## 2.3. *Fabrication of UCNP-based nanocomposites*

Recently, much attention has been paid to the development of UCNP-based composite nanostructures with multiple functions by coating or integrating UCNPs with other inorganic nanoscale components. Silica coating on UCNPs is one of the most popular techniques for the fabrication of nanocomposites,[32,91,92] not only to facilitate surface coating with biocompatible polymers and bioconjugation,[91] but also to render the obtained nanocomposites more functionalities. The size of the coated silica layer can be controlled by adjusting the concentration of added tetraethoxysilane (TEOS), a precursor for silica formation by hydrolysis and polycondensation. Silica is biocompatible and its surface can be easily modified with amines, thiols, and carboxyl groups, allowing the linking of biomolecules. Zhang *et al.* prepared core/shell structured UCNP nanospheres with a thin and uniform silica coating,[92] into which organic dyes or QDs are encapsulated to tune the UCL spectra by fluorescence resonance energy transfer (FRET). Therapeutic agents such as PS can also be encapsulated within the silica shell grown on UCNPs for applications in PDT.[32,93]

UCNPs may also be coupled with other functional nanomaterials.[94] Kim *et al.* were able to couple QDs with UCNPs, and used the obtained nanocomposites for multiplexed NIR excited *in vivo* imaging.[95] Duan *et al.* reported a new approach to modulate the UC emission through plasmonic interactions between UCNPs and the attached gold nanostructures, and observed enhanced UCL emission by coating UCNPs with a gold shell with appropriate thickness.[96]Stucky *et al.* reported a facile method to fabricate Ag@$SiO_2$@UCNP nanomaterials based on a core/spacer/shell approach.[97] The UC fluorescence enhancements and quenching could be tuned by the thickness of the dielectric silica spacer. Song *et al.* synthesized highly crystalline, monosized core/shell UCNP@Ag nanocomposites with multi-functionalities and used them as a useful PTT agent as well as an UC imaging label.[81]

Super-paramagnetic iron oxide nanoparticles (IONP) have been widely applied in MR imaging.[98,99] Yan *et al.* have developed a cross-linker anchoring process to synthesize $Fe_3O_4$/UCNPs hetero-nanoparticles using DDA or MUA as the cross-linker.[85] Mi *et al.* synthesized magnetic/UCNPs multi-functional nanoparticles (MFNPs) by covalently linking multiple carboxyl-functionalized IONPs onto individual amino-functionalized silica-coated UCNPs.[100] $Fe_3O_4$ could also be directly grown on the surface of $NaFY_4$-based UCNPs as reported by Li and coworkers for dual model UCL and MR imaging.[101] Very recently, several groups have developed a new strategy to synthesize MFNPs based on UCNPs and IONPs using silica coating method,[101] such as core/shell $Fe_3O_4$@n$SiO_2$@m$SiO_2$@UCNPs nanostructures by Lin *et al.*,[28] $Fe_3O_4$@$SiO_2$@UCNPs nanocomposite by Chen group,[102] and $Fe_3O_4$&UCNPs@$SiO_2$ by Shi *et al.*[103] The nanocomposites based on silica modified UCNPs, showed excellent super-paramagnetic properties and strong UCL emissions useful for *in vitro* and *in vivo* imaging. Beyond these, our group used a LBL method to synthesize a novel class of MFNPs containing a UCNP as the core, closely packed IONPs as the inter-layer, and a thin layer of gold as the shell. [34] The obtained UCNP@IONP@Au MFNPs were then coated with PEG *via* the Au-thiol bond and used for *in vitro* and *in vivo* multimodal imaging and therapy.[34,104]

## 3. UCNPs and Their Nanocomposites for *in vivo* Biomedical Imaging

Starting from 1999, UCNPs have been developed as a new class of fluorescence bio-probes owing to their many intriguing advantages; several groups have started using UCNPs as optical nano-probes for *in vitro* cell imaging.[22,87,105–108] In 2006, Lim *et al.* studied the use of UCNPs in live organism imaging[109](**Figure 1(a)**). In their study, $Y_2O_3$:Yb/Er nanoparticles in the size range of 50–150 nm were inoculated into live nematode *Caenorhabditis elegans* worms and subsequently imaged in the digestive system of the worms. Upon excitation at 980 nm, the statistical distribution of the nanoparticles in the intestines can be clearly visualized. Importantly, the nanoparticles have shown good biocompatibility as the worms do not exhibit unusual behavior in feeding. In 2008, Nyk *et al.* for the first time demonstrated *in vivo* imaging of Balb/c mouse injected with UCNPs,[24] which showed high uptake in the liver and spleen after intravenous injection (**Figure 1(b)**). Compared with traditional down-conversion fluorescence imaging using QDs or organic dyes, the UCNP-based UCL imaging has no autofluorescence. Recently, Cheng *et al.* compared the *in vivo* imaging sensitivity of the UCNPs and QDs

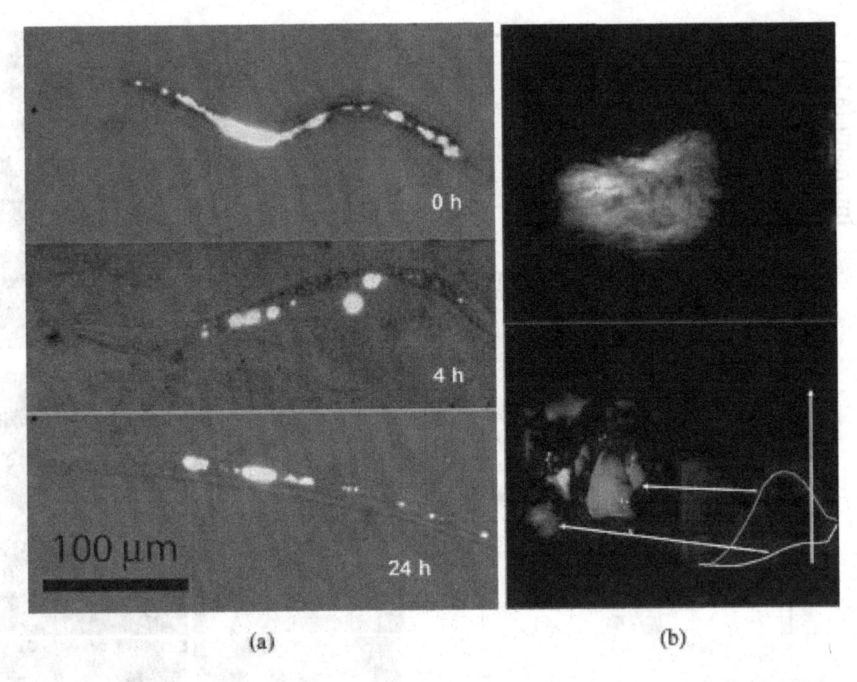


**Figure 1.** *In vivo* organism and small animal imaging with UC nanoparticles (**a**) False color images of *C. elegans* after being deprived of food over various periods of time: the red color represents the bright field and green for the UC emission.[24] (**b**) Whole body images of intact mouse (above) and the same mouse after injection with UC nanoparticles and dissection (below): the red color indicates emission from UC nanoparticles and green and black colors show the imaging background; inserted graphics represent the photoluminescence spectra from three different areas as indicated by the arrows.[109]

(QD545 and QD625) in a side-by-side experiment.[78] Because of the auto-fluorescence free nature of the UCL imaging, a long exposure time was employed in UCL imaging without introducing any obvious background, allowing the *in vivo* detection limit of UCNPs to be at least one order of magnitude lower than that of QDs by using our imaging system. In the recent few years, UCNPs have thus been widely used for *in vivo* multicolor imaging, tumor-targeted imaging, multimodal imaging, and *in vivo* cell tracking.

## 3.1. *Multicolor imaging*

By varying the $Ln^{3+}$ dopants during UCNP synthesis, the UCL emission spectra of nanoparticles could be well tuned, enabling multicolor UCL imaging in biological systems.[16,51,110–112] In a study by Liu's group, a series of PEGylated UCNPs ($NaY_{0.78}Yb_{0.2}Er_{0.02}F_4$, $NaY_{0.69}Yb_{0.3}Er_{0.01}F_4$, $NaY_{0.78}Yb_{0.2}Tm_{0.02}F_4$) with

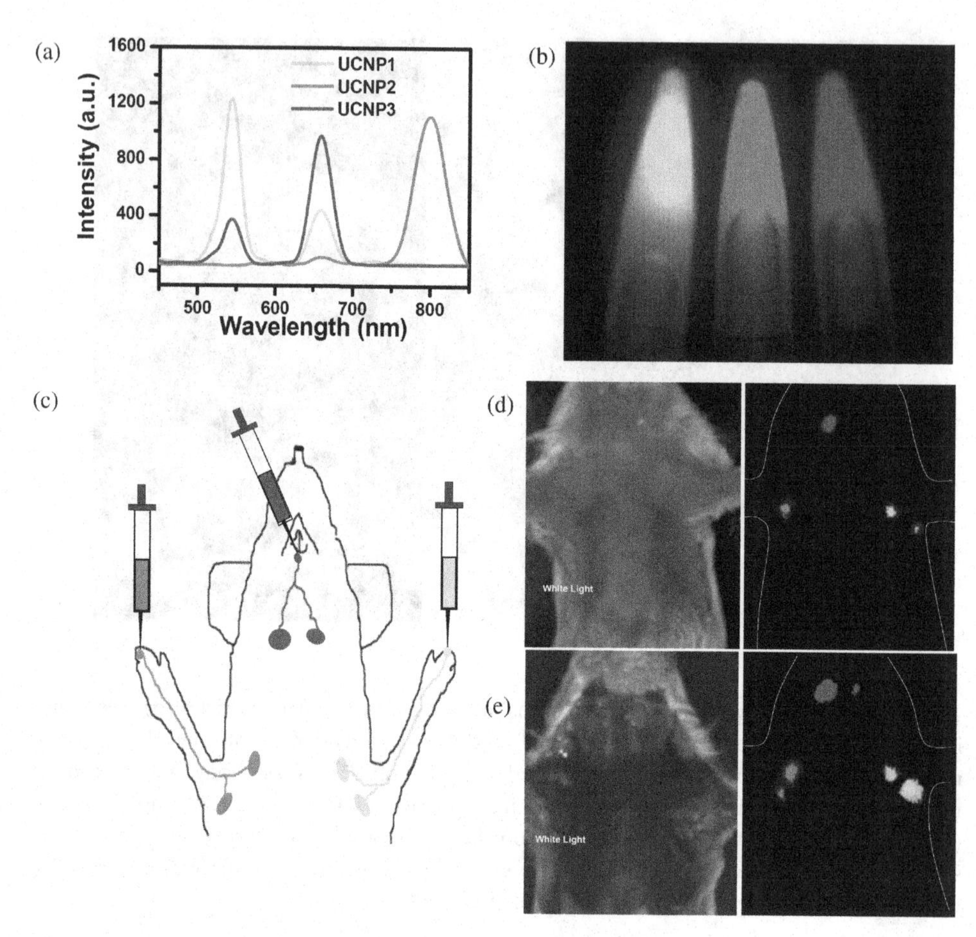


**Figure 2.** Multicolor UCL imaging **(a)** UCL emission spectra of three UCNP solutions under 980 nm NIR laser excitation. **(b)** A multicolor UCL image of three solutions obtained by *in vivo* imaging system (CRi, Inc.) after spectral unmixing. **(c)** A schematic illustration of UCNP based multicolor lymph node mapping. **(d)** White light and *in vivo* UCL images of a mouse injected with UCNPs. **(e)** White light and UCL images of the same mouse after dissection.[78]

different UCL emission spectra[78] were synthesized (**Figure 2**). Three types of UCNPs upon 980 nm laser excitation were easily differentiated after spectral deconvolution. Multicolor *in vivo* UCL imaging was demonstrated by imaging subcutaneously injected UCNPs and applied in multiplexed *in vivo* lymph node mapping as well as multicolor *in vivo* cell labeling and tracking.

FRET is another approach to modulate UCL spectra of UCNPs for multicolor imaging. In 2008, Zhang *et al.* fabricated the UCNP/silica core/shell structure by encapsulating organic dyes or QDs into the silica shell.[92] The up-conversion emission spectra under 980 nm excitation were tuned by FRET from the UCNP core to organic dyes or QDs inside the silica shell, generating UCNP-composites with different UCL 'colors'. Recently, Cheng *et al.* showed that PEGylated UCNPs physically loaded with organic dyes *via* hydrophobic force could also serve as a FRET nanosystem for *in vivo* multicolor UCL imaging.[113] As many as five UCL 'colors' could be clearly separated after spectral deconvolution in the imaging of mice.

## 3.2. *Tumor-targeted molecular UCL imaging*

Tumor-targeted molecular imaging plays an important role in tumor diagnosis and prognosis. Recently, much attention has been paid to investigate *in vivo* targeted imaging using UCNP-based nano-probes.[25,114–116] In 2009, Li and co-workers for the first time realized *in vivo* tumor targeting using folic acid (FA) modified UCNPs and arginine–glycine–asparatic acid (RGD)-peptide conjugated UCNPs, in their two separate studies.[25,115]After intravenous injection of FA-modified UCNPs into HeLa tumor-bearing nude mouse for a day, obvious UCL signals were observed in the tumor, while no significant luminescence was observed in the control group. The RGD-peptide has a high binding affinity with the $\alpha_{\upsilon}\beta_3$ integrin receptor which plays a pivotal role in tumor angiogenesis. Li *et al.* showed that PEGylated UCNPs conjugated with RGD could effectively target U87 MG human glioblastoma tumors in mice as revealed by *in vivo* UCL imaging[25](**Figure 3**). In a later work, Yu *et al.* reported the development of neurotoxin-mediated UCNPs nano-probes for tumor targeting and visualization in living animals.[116] The nano-probes were synthesized by conjugating polyethyleneimine (PEI) coated UCNPs with recombinant chlorotoxin, a typical peptide neurotoxin that could bind to many types of cancer cells with high specificity. Those UCNP-based nano-probes with high imaging sensitivity and specific tumor-binding ability may have unique potential in molecular cancer imaging in future explorations.

Apart from molecular targeting, magnetic targeting can also be used for enhanced tumor uptake and *in vivo* UCL imaging. Cheng *et al.* developed a novel class of MFNPs based on UCNPs with combined optical and magnetic properties useful in multimodality imaging and therapy[34] *via* a LBL self-assembly strategy followed by seed-induced gold shell growth. These UCNP@IONP@Au MFNPs were successfully used for *in vivo* dual-modal imaging guided by magnetically targeted PTT.[104] By placing a magnet close to the tumor, MFNPs intravenously injected into mice would accumulate in the tumor region by magnetic attraction as revealed by dual modal UCL & MR imaging. Owing to the strong NIR optical absorption of these Au shelled MFNPs and the highly efficient magnetic tumor targeting, tumors on mice

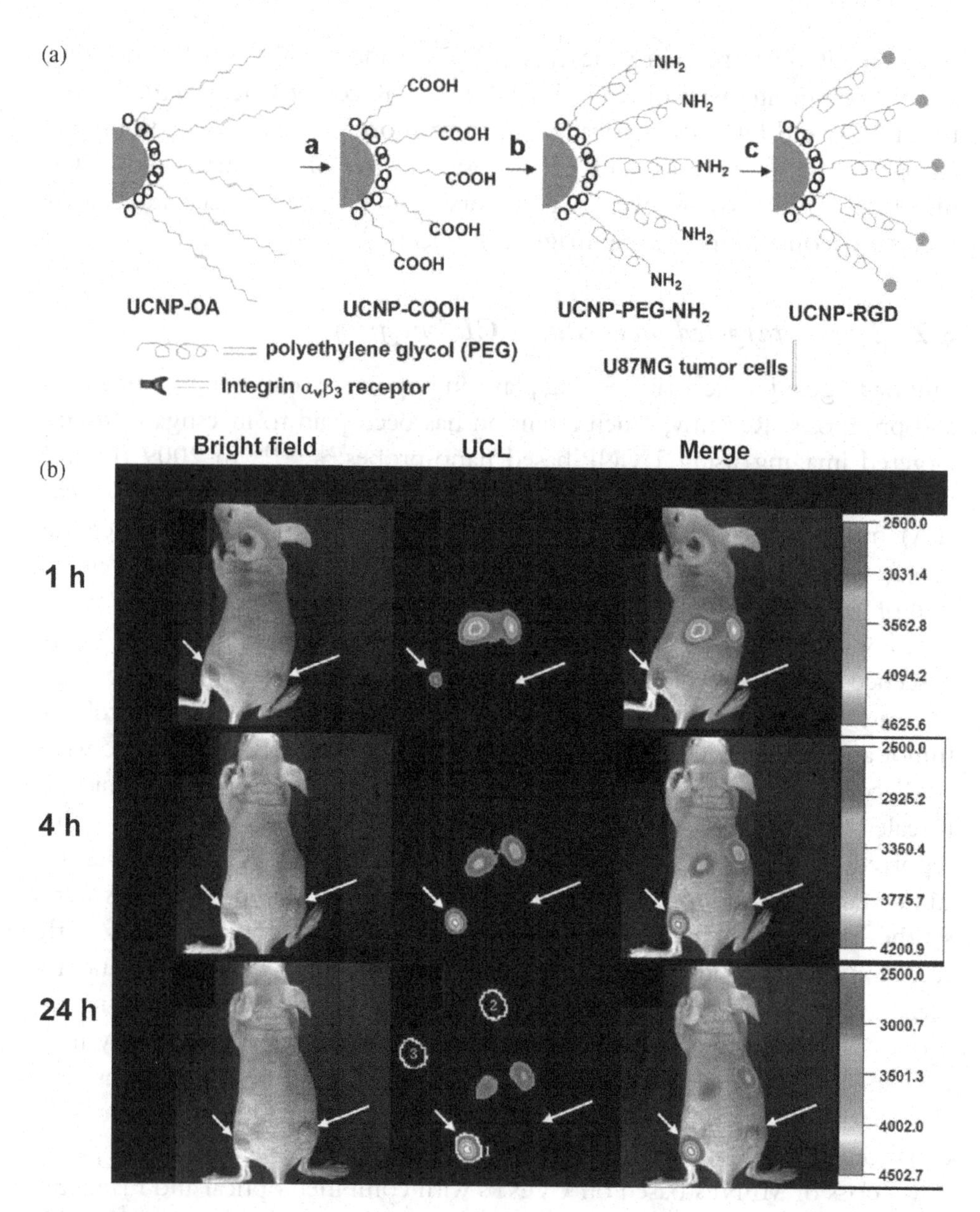


**Figure 3.** *In vivo* tumor-targeted UCL imaging **(a)** A scheme showing synthesis of UCNP-RGD. **(b)** Time-dependent *in vivo* UCL images of a mouse bearing a U87 MG tumor (left hind leg, indicated by short arrows) and a MCF-7 tumor (right hind leg, indicated by long arrows) after intravenous injection of UCNP-RGD.[25]

after MFNP-injection and magnetic tumor-targeting were completely eliminated upon exposure to the 808 nm NIR laser. This work highlights the promise of using UCNP-based multifunctional nanostructures for novel cancer imaging-guided therapy.

### 3.3. *Multimodal imaging of UCNPs*

Multimodal imaging that combines several different imaging modalities to overcome limitations of each single imaging approach, has received tremendous attention in the area of *in vivo* biomedical imaging.[117,118] Recently, great efforts have been made to obtain multimodal imaging probes based on UCNPs and their nanocomposites.[26,27,30,34,41–43,119,120]

Gadolinium ($Gd^{3+}$) is a paramagnetic relaxation agent extensively used in MR imaging. By doping light-emitting RE ions into a $Gd^{3+}$ containing host matrix such as $NaGdY_4$, many groups synthesized Gd-based UCNPs that offered both UCL emission and T1-weighted MR contrast. In a representative example, Hyeon *et al.* demonstrated the use of PEG-phospholipid functionalized $NaGdF_4$:Yb/Er nanoparticles for optical and MR imaging in breast cancer cells (SK-BR3) for the first time.[41] Combined optical and MR imaging of cells using $NaGdF_4$:Yb/Er nanocrystals were also reported by Prasad *et al.*[42] However, the $NaGdF_4$: Yb/ Er nanoparticles were used only for *in vitro* multimodal imaging of cells in these two studies. Starting from 2011, Li *et al.* published a series of studies on the *in vivo* multimodal imaging with UCNPs in small animals.[26,30,121,122] They used hydrophilic azelaic acid functionalized $NaGdF_4$ UCNPs as a dual-modal *in vivo* imaging probe for combined UCL and MR imaging in mice. In their another work, $^{18}F$-labeled $NaGdF_4$, Yb/Er UCNPs were firstly fabricated and used for triple-modality PET, UCL, and MR imaging[26,30](**Figure 4**). Several other related approaches

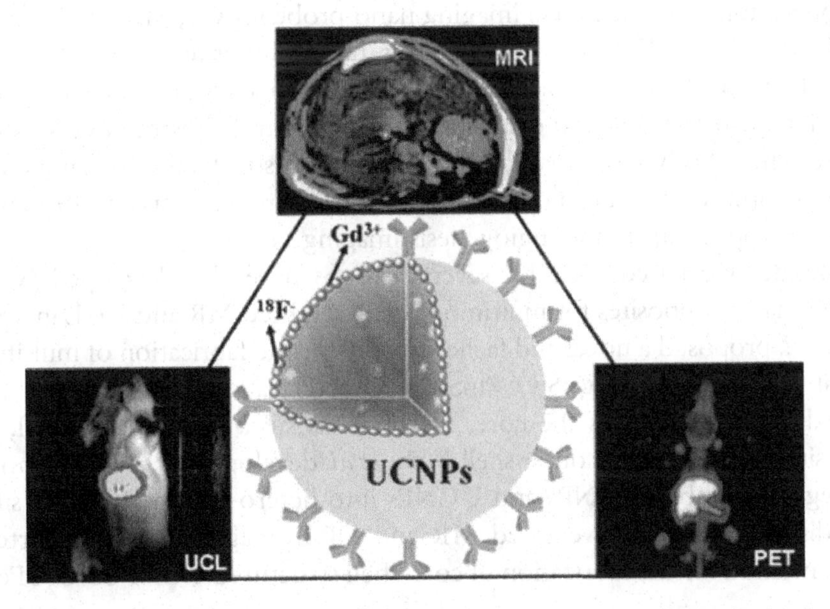


**Figure 4.** A scheme showing the use of $^{18}F$-labeled Gd-containing UCNPs for *in vivo* multimodal UCL, PET and MR imaging.[30]

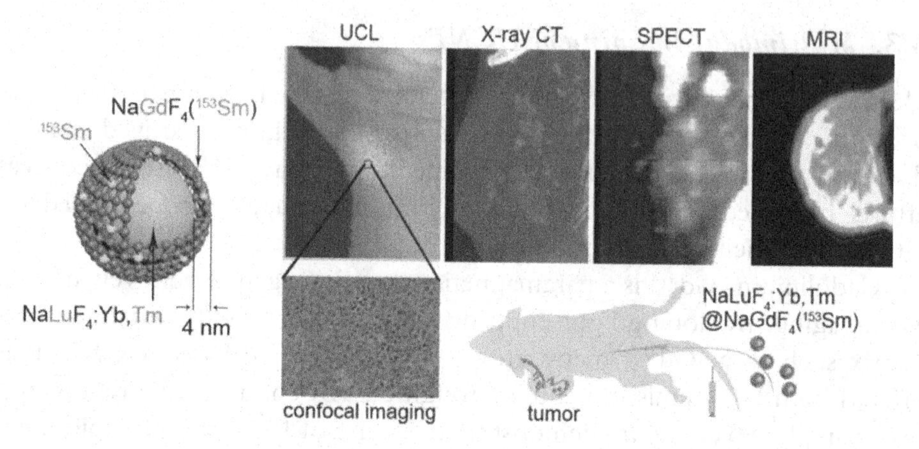


**Figure 5.** Core-shell lanthanide up-conversion nanophosphors as four-modal probes for tumor angiogenesis imaging.[124]

to develop UCNP-based multimodal imaging probes have also been reported by the Li's group,[123] Cui's group[29] and Shi's group.[27] Shi and coworkers reported the combination of Gd-doped UCNPs with Au nanoparticles based on a simple electrostatic adsorption mechanism, resulting in a sub-50 nm sized multifunctional nanostructure useful for UCL, MR and CT tri-modal imaging modalities.[27] Recently, Li's group demonstrated that the synthesis of core-shell nanocomposite $NaLuF_4$:Yb, Tm@$NaGdF_4$-($^{153}Sm^{3+}$) constituted a hexagonal $NaLuF_4$:Yb, Tm as the core and a 4 nm $Sm^{3+}$doped $NaGdF_4$ as the shell and its application as a four-modal imaging nano-probe for CT, MRI, SPECT, and UCL imaging[124] (**Figure 5**). Using this nanocomposite $NaLuF_4$:Yb, Tm@ $NaGdF_4$-($^{153}Sm^{3+}$) as a multimodal probe, they realized four-modal imaging in a small animal and achieved detailed information on the distribution from the cell to whole-body level, distribution in different soft tissues, dynamic long-term quantification data, and 3D information. Moreover, this nanocomposite has been applied in tumor angiogenesis imaging.

Besides Gd-based UCNPs, several groups have also developed IONP-UCNP nanocomposites for multimodal T2-weighted MR and UCL imaging. Lin *et al.* proposed a novel and facile strategy for the fabrication of multifunctional core-shell $Fe_3O_4$@n$SiO_2$@m$SiO_2$@UCNP nanostructures using silica coated $Fe_3O_4$ spheres as the core, mesoporous silica as the inner-shell, and deposited UCNPs as the outer shell.[28] Shi *et al.* developed a "neck-formation" strategy to combine IONPs and UCNPs into hetero-nanoparticles by silica-shielding for both T2-weighted MR and UCL imaging.[103] Wu reported a facile method for the fabrication of core-shell structured UCNP@$SiO_2$@$Fe_3O_4$

nanocomposite particles with good superparamagnetic and luminescent properties for cell imaging.[125] Li *et al.* developed a step-wise synthetic method to synthesize core-shell $Fe_3O_4$@UCNP nanostructure with multifunctional properties for multimodal imaging.[123] Recently, Liu's group also developed a novel class of MFNPs based on UCNP@IONP@Au with combined optical and magnetic properties useful in multi-modality *in vitro* and *in vivo* imaging.[104]

## 3.4. *UCNPs for cell labeling and in vivo tracking*

Owing to the auto-fluorescence free nature during UCL imaging, UCNPs have also been used in cell labeling and *in vivo* tracking with great sensitivity. In an earlier work, Zhang *et al.* reported the use of silica coated UCNP to label myoblast cells for *in vivo* dynamic tracking in mice.[126] In 2010, we used UCNPs as a novel class of cell labeling agents for *in vivo* tracking and multicolor imaging of cancer cells also in mice.[78] In a recent study, Li *et al.* fabricated sub-10 nm hexagonal $NaLuF_4$:$Gd^{3+}$, $Yb^{3+}$, and Tm3+ with strong UCL emission to label cancer cells. Excellent *in vivo* tracking sensitivities with detection limits down to 50 and 1000 UCNP-labeled cells after subcutaneous and intravenous injection, respectively, were achieved in this work[46] (**Figures 6(a) and 6(b)**).

In the recent years, mesenchymal stem cells (MSCs) have been extensively explored for potential applications in tissue engineering, immunotherapy, and gene therapy because they can differentiate into a variety of cell types including fat, bone, and cartilage under suitable conditions.[127–130] Owing to the above-mentioned unique optical properties, a number of groups including us have used UCNPs to label stem cells for *in vitro* and *in vivo* tracking.[108,131] In our recent study, we used UCNPs as an exogenous contrast agent to track mouse MSCs (mMSCs) *in vivo*. Oligo-arginine was conjugated to PEG coated UCNPs to enhance the nanoparticles uptake by mMSCs[131] (**Figure 6(c)**). It was found that the proliferation and differentiation of mMSCs were not notably affected by UCNP-labeling. Thanks to the absence of auto-fluorescence during UCL imaging, we were able to detect as few as ~10 cells injected into a mouse, achieving an ultra-high sensitivity which is many orders of magnitude higher than currently used exogenous stem cell labeling nanoagents such as QDs and IONPs (**Figure 6(d)**). *In vivo* translocation of UCNP-labeled mMSCs after intravenous injection could be well monitored by UCL imaging. In our latest study, we further used multifunctional UCNP@IONP@Au composite MFNPs previously developed for labeling and *in vivo* multimodal tracking of mMSCs.[132] More interestingly, utilizing the

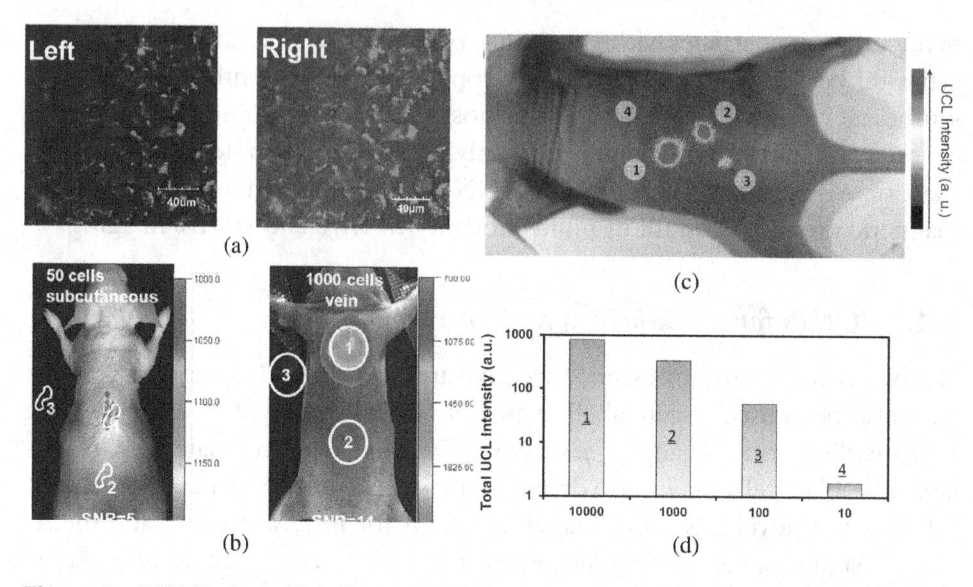


**Figure 6.** UCNPs for cell labeling and *in vivo* tracking **(a)** Confocal UCL image (Left) and its overlay with a bright field image (Right) of cells stained with 200 μg/ml $NaLuF_4$ UCNPs for 3 h at 37°C.[46] **(b)** *In vivo* UCL imaging of athymic nude mice after subcutaneous injection of 50 KB cells (Left) and tail-vein injection of 1000 KB cells (Right). The KB cells were pre-incubated with 200 μg/ml $NaLuF_4$ UCNPs for 3h at 37 °C before injection. **(c)** & **(d)** *In vivo* detection of UCNP-labeled mMSCs.[131] **(c)** A UCL image of a mouse subcutaneously injected with various numbers of mMSCs ($10\sim10^4$) labeled with UCNPs. **(d)** Quantification of UCL signals in **(c)**.

strong magnetism of those MFNPs, we uncovered that MFNP-labeled mMSCs were able to specifically migrate towards a wound nearby a magnet, offering enhanced tissue repairing efficacy. Thus, UCNPs and their nanocomposites show great potential in highly-sensitive stem cell tracking and manipulation, promising for future imaging-guided cell therapy applications.

## 4. Summary and Outlook

This section discusses the recent developments of the UCNPs focusing on their applications in bio-imaging. Despite the tremendous amount of exciting results reported in the past few years in this field, there are still many challenges ahead towards applications of UCNP-based imaging in the clinic. First of all, the quantum yield of UCL emission of UCNP is usually less than 1%, and becomes even lower as their sizes reduce, significantly limiting the use of these nano-probes in optical imaging. Secondly, the potential long-term toxicity of $Ln^{3+}$ doped UCNPs is another significant concern. Much more systematic investigations are still demanded before these nanomaterials can be finally used in the clinic. Thirdly, how the surface modifications and size affect the *in vivo* behaviors of UCNPs also need more investigation. Lastly, though multimodal imaging based on UCNPs and their nanocomposites have been demonstrated in several studies, better design of novel multifunctional agents based on UCNPs for simultaneous medical diagnosis and combined cancer therapies are still needed in the future to realize real-time monitoring of treatment progress as well as achieving synergistic therapeutic effects.

## References

1. Joubert M-F. *Opt Mater* 1999;11:181–203.
2. Auzel F. *Chem Rev* 2004;104:139–173.
3. Sandrock T, Scheife H, Heumann E, Hube G. *Opt Lett* 1997;22:808–810.
4. Downing E, Hesselink L, Ralston J, Macfarlane R. *Science* 1996;273:1185–1189.
5. Shionoya S, Yen WM. *Phosphor Handbook*. CRC Press, Boca Raton, Florida, 1999.
6. Rijke FVD, Zijlmans H, Li S, Vail T, Raap AK, Niedbala RS, Tanke HJ. *Nat Biotech* 2001;19:273–276.
7. Wang L, Yan R, Huo Z, Wang L, Zeng J, Bao J, Wang X, Peng Q, Li Y. *Angew Chem Int Ed* 2005;44:6054–6057.
8. Wang M, Hou W, Mi C-C, Wang W-X, Xu Z-R, Teng H-H, Mao C-B, Xu S-K. *Anal Chem* 2009;81:8783–8789.
9. Liu J, Liu Y, Liu Q, Li C, Sun L, Li F. *J Am Chem Soc* 2011;133:15276–15279.
10. Ju Q, Tu D, Liu Y, Li R, Zhu H, Chen J, Chen Z, Huang M, Chen X. *J Am Chem Soc* 2012;134:1323–1330.
11. Liu Q, Peng J, Sun L, Li F. *ACS Nano* 2011;5:8040–8048.
12. Liu C, Wang Z, Jia H, Li Z. *Chem Comm* 2011;47:4661–4663.
13. Zhang F, Shi Q, Zhang Y, Shi Y, Ding K, Zhao D, Stucky GD. *Adv Mater* 2011;23: 3775–3779.
14. Wang Y, Shen P, Li C, Wang Y, Liu Z. *Anal Chem* 2012;84:1466–1473.
15. Yi GS, Chow GM. *Adv Funct Mater* 2006;16:2324–2329.
16. Wang F, Liu X. *J Am Chem Soc* 2008;130:5642–5643.
17. Wang F, Han Y, Lim CS, Lu Y, Wang J, Xu J, Chen H, Zhang C, Hong M, Liu X. *Nature* 2010;463.
18. Mai H, Zhang Y, Si R, Yan Z, Sun L, You L, Yan C. *J Am Chem Soc* 2006;128:6426–6436.
19. Zhou J, Liu Z, Li F. *Chem Soc Rev* 2012;41:1323–1349.
20. Chatterjee DK, Gnanasammandhan MK, Zhang Y. *Small* 2010;6:2781–2795.
21. Wang F, Banerjee D, Liu Y, Chen X, Liu X. *Analyst* 2010;135:1839–1854.
22. Wu S, Han G, Milliron DJ, Aloni S, Altoe V, Talapin DV, Cohen BE, Schuck PJ. *Proc Natl Acad Sci* 2009;106:10917–10921.
23. Wang F, Liu X. *Chem Soc Rev* 2009;38:976–989.
24. Nyk M, Kumar R, Ohulchanskyy TY, Bergey EJ, P. Prasad PN. *Nano Lett* 2008;8: 3834–3838.
25. Xiong L, Chen Z, Tian Q, Cao T, Xu C, Li F. *Anal Chem* 2009;81:8687–8694.
26. Zhou J, Yu M, Sun Y, Zhang X, Zhu X, Wu Z, Wu D, Li F. *Biomaterials* 2011;32:1148–1156.

27. Xing H, Bu W, Zhang S, Zheng X, Li M, Chen F, He Q, Zhou L, Peng W, Hua Y, Shi J. *Biomaterials* 2012;33:1079–1089.
28. Gai S, Yang P, Li C, Wang W, Dai Y, Niu N, Lin J. *Adv Funct Mater* 2010;20:1166–1172.
29. He M, Huang P, Zhang C, Hu H, Bao C, Gao G, He R, Cui D. *Adv Funct Mater* 2011;21:4470–4477.
30. Liu Q, Sun Y, Li C, Zhou J, Li C, Yang T, Zhang X, Yi T, Wu D, Li F. *ACS Nano* 2011;5:3146–3157.
31. Wang C, Cheng L, Liu Z. *Biomaterials* 2011;32:1110–1120.
32. Qian HS, Guo HC, Ho PC-L, Mahendran R, Zhang Y. *Small* 2009;5:2285–2290.
33. Wang C, Tao H, Liang C, Zhuang L. *Biomaterials* 2011;32:6145–6154.
34. Cheng L, Yang K, Li Y, Chen J, Wang C, Shao M, Lee S-T, Liu Z. *Angew Chem Int Ed* 2011;50:7385–7390.
35. Boyer J-C, Vetrone F, Cuccia LA, Capobianco JA. *J Am Chem Soc* 2006;128:7444–7445.
36. Liang X, Wang X, Zhuang J, Peng Q, Li Y. *Adv Funct Mater* 2007;17:2757–2765.
37. Wang L, Zhang Y, Zhu Y. *Nano Res* 2010;3:317–325.
38. Zhang F, Wan Y, Yu T, Zhang F, Shi Y, Xie S, Li Y, Xu L, Tu B, Zhao D. *Angew Chem Int Ed* 2007;46:7969–7979.
39. Zhang F, Zhao D. *ACS Nano* 2009;3:159–164.
40. Liu C, Wang H, Li X, Chen D. *J Mater Chem* 2009;19:3546–3553.
41. Park YI, Kim JH, Lee KT, Jeon K-S, Na HB, Yu JH, Kim HM, Lee N, Choi SH, Baik SI, Kim H, Park SP, Park B-J, Kim YW, Lee SH, Yoon S-Y, Song IC, Moon WK, Suh YD, Hyeon T. *Adv Mater* 2009;21:4467–4471.
42. Kumar R., Nyk M, Ohulchanskyy TY, Flask CA, Prasad PN. *Adv Funct Mater* 2009;19:853–859.
43. Chen F, Bu W, Zhang S, Liu X, Liu J, Xing H, Xiao Q, Zhou L, Peng W, Wang L, Shi J. *Adv Funct Mater* 2011;21:4285–4294.
44. Chen G, Ohulchanskyy TY, Liu S, Law W-C, Wu F, Swihart MT, Ågren H, Prasad PN. *ACS Nano* 2012;6:2969–2977.
45. Wei Y, Lu F, Zhang X, Chen D. *Chem Mater* 2006;18:5733–5737.
46. Liu Q, Sun Y, Yang T, Feng W, Li C, Li F. *J Am Chem Soc* 2011;133:17122–17125.
47. Xia A, Chen M, Gao Y, Wu D, Feng W, Li F. *Biomaterials* 2012;33:5394–5405.
48. Yang T, Sun Y, Liu Q, Feng W, Yang P, Li F. *Biomaterials* 2012;33:3733–3742.
49. Zhou J, Zhu X, Chen M, Sun Y, Li F. *Biomaterials* 2012;33:6201–6210.
50. Scha¨fe H, Ptacek P, Zerzouf O, Haase M. *Adv Funct Mater* 2008;18:2913–2918.
51. Ehlert O, Thomann R, Darbandi M, Nann T. *ACS Nano* 2007;2:120–124.
52. Zhang Y-W, Sun X, Si R, You L-P, Yan C-H. *J Am Chem Soc* 2005;127:3260–3261.
53. Dong N, Pedroni M, Piccinelli F, Conti G, Sbarbati A, Hernandez JR, Maestro LM, Cruz MCIDL, Sanz-Rodriguez F, Juarranz A, Chen F, Vetrone F, Capobianco JA, Sole JAG, Bettinelli M, Jaque D, Speghini A. *ACS Nano* 2011;5:8665–8671.
54. Wang J, Wang F, Wang C, Liu Z, Liu X. *Angew Chem Int Ed* 2011;50:10369–10372.
55. Yanes AC, Santana-Alonso A, Méndez-Ramos J, del-Castillo J, Rodríguez VD. *Adv Funct Mater* 2011;21:3136–3142.
56. Mai H, Zhang Y, Sun L, Yan C. *J Phys Chem C* 2007;111:13721–13729.
57. Mai H-X, Zhang Y-W, Sun L-D, Yan C-H. *J Phys Chem C* 2007;111:13730–13739.
58. Wang X, Zhuang J, Peng Q, Li Y. *Nature* 2005;437:121–124.
59. Wang L, Li P, Zhuang J, Bai F, Feng J, Yan X, Li Y. *Angew Chem Int Ed* 2008;47:1054–1057.

60. Wang M, Liu J-L, Zhang Y-X, Hou W, Wu X-L, Xu S-K. *Mater Lett* 2009;63:325–327.
61. Wang M, Mi C-C, Liu J-L, Wu X-L, Zhang Y-X, Hou W, Li F, Xu S-K. *J Allloys and Compounds* 2009;485:L24–L27.
62. Sun Y, Chen Y, Tian L, Yu Y, Kong X, Zhao J, Zhang H. *Nanotechnology* 2007;18:275609.
63. Tian G, Gu Z, Zhou L, Yin W, Liu X, Yan L, Jin S, Ren W, Xing G, Li S, Zhao Y. *Adv Mater* 2012;24:1226–1231.
64. Yi G, Lu H, Zhao S, Ge Y, Yang W, Chen D, Guo L-H. *Nano Letters* 2004;4:2191–2196.
65. Patra A, Friend CS, Kapoor R, Prasad PN. *J Phys Chem B* 2002;106:1909–1912.
66. Patra A, Friend CS, Kapoor R, Prasad PN. *Chem Mater* 2003;15:3650–3655.
67. Vetrone F, Boyer J-C, Capobianco JA, Speghini A, Bettinelli M. *J Appl Phys* 2004;96: 661–667.
68. Luo X-X, Cao W-H. *J Alloys Compd* 2008;460:529–534.
69. Xu L, Yu Y, Li X, Somesfalean G, Zhang Y, Gao H, Zhang Z. *Opt Mater* 2008;30: 1284–1288.
70. Qin X, Yokomori T, Ju Y. *Appl Phys Lett* 2007;90:073104.
71. Xiong L, Yang T, Yang Y, Xu C, Li F. *Biomaterials* 2010;31:7078–7085.
72. Naccache R, Vetrone F, Mahalingam V, Cuccia LA, Capobianco JA. *Chem Mater* 2009;21:717–723.
73. Chen G, Ohulchanskyy TY, Law WC, Ågren H, Prasad PN. *Nanoscale* 2011;3:2003–2008.
74. Budijono SJ, Shan J, Yao N, Miura Y, Hoye T, Austin RH, Ju Y, Prud'homme RK. *Chem Mater* 2010;22:311–318.
75. Hilderbrand SA, Shao F, Salthouse C, Mahmood U, Weissleder R. *Chem Comm* 2009;28: 4188–4190.
76. Yi G-S, Chow G-M. *Chem Mater* 2007;19:341–343.
77. Rantanen T, Järvenpää M-L, Vuojola J, Kuningas K, Soukka T. *Angew Chem Int Ed* 2008;47:3811–3813.
78. Cheng L, Yang K, Zhang S, Shao M, Lee S. *Nano Res* 2010;3:722–732.
79. Zhang Q, Song K, Zhao J, Kong X, Sun Y, Liu X, Zhang Y, Zeng Q, Zhang H. *J Colloid Interface Sci* 2009;336:171–175.
80. Meiser F, Cortez C, Caruso F. *Angew Chem Int Ed* 2004;43:5954–5957.
81. Dong B, Xu S, Sun J, Bi S, Li D, Bai X, Wang Y, Wang L, Song H. *J Mater Chem* 2011;21:6193–6200.
82. Wang Z-L, Hao J, Chan HLW, Law G-L, Wong W-T, Wong K-L, Murphy MB, Su T, Zhang ZH, Zeng SQ. *Nanoscale* 2011;3:2175–2181.
83. Chen Q, Wang X, Chen F, Zhang Q, Dong B, Yang H, Liu G, Zhu Y. *J Mater Chem* 2011;21:7661–7667.
84. Cao T, Yang T, Gao Y, Yang Y, Hu H, Li F. *Inorg Chem Comm* 2010;13:392–394.
85. Shen J, Sun L-D, Zhang Y-W, Yan C-H. *Chem Commun* 2010;46:5731–5733.
86. Chen Z, Chen H, Hu H, Yu M, Li F, Zhang Q, Zhou Z, Yi T, Huang C. *J Am Chem Soc* 2008;130:3023–3029.
87. Hu H, Yu M, Li F, Chen Z, Gao X, Xiong L, Huang C. *Chem Mater* 2008;20:7003–7009.
88. Zhou H-P, Xu C-H, Sun W, Yan C-H. *Adv Funct Mater* 2009;19:3892.
89. Cheng L, Yang K, Chen Q, Liu Z. *ACS Nano* 2012;6:5605–5613.
90. Li L-L, Zhang R, Yin L, Zheng K, Qin W, P. R. Selvin PR, Lu Y. *Angew Chem Int Ed* 2012;51:6121–6125.
91. Hu H, Xiong L, Zhou J, Li F, Cao T, Huang C. *Chem Eur J* 2009;15:3577–3584.

92. Li Z, Zhang Y, Jiang S. *Adv Mater* 2008;20:4765–4769.
93. Chatterjee DK, Zhang Y. *Nannomedicine* 2008;3:73–82.
94. Yan C, Dadvand A, Rosei F, Perepichka DF. *J Am Chem Soc* 2010;132:8868–8869.
95. Jeong S, Won N, Lee J, Bang J, Yoo J, Kim SG, Chang JA, Kim J, Kim S. *Chem Comm* 2011;47:8022–8024.
96. Zhang H, Li Y, Ivanov IA, Qu Y, Huang Y, Duan X. *Angew Chem Int Ed* 2010;49: 2865–2868.
97. Zhang F, Braun GB, Shi Y, Zhang Y, Sun X, Reich NO, Zhao D, Stucky G. *J Am Chem Soc* 2010;132:2850–2851.
98. Veiseha O, Gunna JW, Zhang M. *Advanced Drug Delivery Reviews* 2010;62:284–304.
99. Mikhaylov G, Mikac U, Magaeva AA, Itin VI, Naiden EP, Psakhye I, Babes L, Reinhecke T, Peters C, Zeiser R, Bogyo M, Turk V, Psakhye SG, Turk B, Vasiljeva O. *Nat Nanotechnol* 2011;6:594–602.
100. Mi C, Zhang J, Gao H, Wu X, Wang M, Wu Y, Di Y, Xu Z, Xu S. *Nanoscale* 2010;2: 1141–1148.
101. Xia A, Gao Y, Zhou J, Li C, Yang T, Wu D, Wu L, Li F. *Biomaterials* 2011;32: 7200–7208.
102. Yu X, Shan Y, Li G, Chen K. *J Mater Chem* 2011;21:8104–8109.
103. Chen F, Zhang S, Bu W, Liu X, Chen Y, He Q, Zhu M, Zhang L, Zhou L, Peng PW, Shi J. *Chem Eur J* 2010;16:11254–11260.
104. Cheng L, Yang K, Li Y, Zeng X, Shao M, Lee S-T, Liu Z. *Biomaterials* 2012;33: 2215–2222.
105. Zijlmans HJMAA, Bonnet J, Burton J, Kardos K, Vail T, Niedbala RS, Tanke HJ. *Anal Biochem* 1999;267:30–36.
106. Yu M, Li F, Chen Z, Hu H, Zhan C, Yang H, Huang C. *Anal Chem* 2009;81: 930–935.
107. Boyer J-C, Manseau M-P, Murray JI, Veggel CJMV. *Langmuir* 2010;26:1157–1164.
108. Jalil RA, Zhang Y. *Biomaterials* 2008;29:4122–4128.
109. Lim SF, Riehn R, Ryu WS, Khanarian N, Tung C-K, Tank D, Austin RH. *Nano Lett* 2006;6:169–174.
110. Kobayashi H, Kosaka N, Ogawa M, Morgan NY, Smith PD, Murray CB, Ye X, Collins J, Kumar GA, Bell H, Choyke PL. *J Mater Chem* 2009;19:6481–6484.
111. Niu W, Wu S, Zhang S. *J Mater Chem* 2010;20:9113–9117.
112. Yu X, Li M, Xie M, Chen L, Li Y, Wang Q. *Nano Res* 2010;3:51–60.
113. Cheng L, Yang K, Shao M, Lee S-T, Liu Z. *J Phys Chem C* 2011;115:2686–2692.
114. Wang M, Mi C-C, Wang W-X, Liu C-H, Wu Y-F, Xu Z-R, Mao C-B, Xu S-K. *ACS Nano* 2009;3:1580–1586.
115. Xiong L-Q, Chen Z-G, Yu M-X, Li F-Y, Liu C, Huang C-H. *Biomaterials* 2009;30:5592–5600.
116. Yu X-F, Sun Z, Li M, Xiang Y, Wang Q-Q, Tang F, Wu Y, Cao Z, Li W. *Biomaterials* 2010;31:8724–8731
117. Selvan ST, Patra PK, Ang CY, Ying JY. *Angew Chem Int Ed* 2007;46:2448–2452.
118. Kim J, Lee JE, Lee SH, Yu JH, Lee JH, Park TG, Hyeon T. *Adv Mater* 2008;20:478–483.
119. Liu Y, Ai K, Liu J, Yuan Q, He Y, Lu L. *Angew Chem Int Ed* 2012;51:1437–1442.
120. Xu H, Cheng L, Wang C, Ma X, Li Y, Liu Z. *Biomaterials* 2011;32:9364–9373.

121. Zhou J, Sun Y, Du X, Xiong L, Hua H, Li F. *Biomaterials* 2010;31:3287–3295.
122. Sun Y, Yu M, Liang S, Zhang Y, Li C, Mou T, Yang W, Zhang X, Li B, Huang C, Li F. *Biomaterials* 2011;32:2999–3007.
123. Zhu X, Zhou J, Chen M, Shi M, Feng W, Li F. *Biomaterials* 2012;33:4618–4627.
124. Sun Y, Zhu X, Peng J, Li F. *ACS Nano* 2013;7:11290–11300.
125. Hu D, Chen M, Gao Y, Li F, Wu L. *J Mater Chem* 2011;21:11276–11282.
126. Idris NM, Li Z, Ye L, Sim EKW, Mahendran R, Ho PC-L, Zhang Y. *Biomaterials* 2009;30:5104–5113.
127. Wang H, Cao F, Cao ADeY, Contag C, Gambhir SS, Wu JC, Chen X. *Stem Cells* 2009;27:1548–1558.
128. Prockop DJ, Gregory CA, Spees JL. *Proc Natl Acad Sci* 2003;100:11917–11923.
129. Chamberlain G, Fox J, Ashton B, Middleton J. *Stem Cells* 2007;25:2739–2749.
130. Kim T, Momin E, Choi J, Yuan K, Zaidi H, Kim J, Park M, McMahon N Lee MT, Quinones-Hinojosa A, Bulte JWM, Hyeon T, Gila AA. *J Am Chem Soc* 2011;133:2955–2961.
131. Wang C, Cheng L, Xu H, Liu Z. *Biomaterials* 2012;33:4872–4881.
132. Cheng L, Wang C, Ma X, Wang Q, Cheng Y, Wang H, Li Y, Liu Z. *Adv Funct Mater* 2012;DOI:10.1002/adfm.201201733.

# Chapter 2

# Redox-Sensitive Polymeric Nanoparticles for Intracellular Drug Delivery

*Huiyun Wen**,§, *Yongyong Li*† *and Xin Zhao*‡

**School of Chemical Engineering, Northwest University, Xi'an, 710069, P. R. China*
†*The Institute for Biomedical Engineering and Nano Science, Tongji University, Shanghai, 200092, P. R. China*
‡*Harvard Medical School, MA, Cambridge, 02139, USA*
§*huiyunwen@126.com*

Polymeric nanoparticles (NPs) have recently emerged as promising candidates for targeted and controlled drug delivery. Theoretically, ideal NPs are capable of possessing high drug loading levels, the ability to deliver drugs to specific pathological sites within the body, and the capacity to rapidly release their drug load at the site of action. However, intracellular barriers such as cellular internalization and endosomal escape have previously hindered drug release in the cytosol or other targeted areas, thereby limiting efficient therapy. Interestingly, intracellular compartments such as the cytosol and nucleus are environments that possess an intrinsically high redox potential. To this end, it is possible to incorporate redox-responsive signals into the polymer architectures of NPs *via* disulfide linkers or other crosslinking agents. These modifications can thus lead to a more controlled and/or triggered intracellular release of therapeutic agents. Currently, such redox polymeric NPs have been designed and have been based on various nanostructures, including micelles, vesicles, polymersomes, and dendrimers. In this chapter, we highlight several recent and exciting advances in the world of redox-sensitive polymeric NPs that have been designed for intracellular drug delivery, with a focus on their design, drug release performance, and therapeutic benefits.

## 1. Introduction

Cancer diagnostics and therapeutics have received increasing attention over the years and a wide range of therapeutic and diagnostic protocols are being proposed and developed as a result. Conventional chemotherapy, one of the frontline strategies for the clinical treatment of cancer, relies on the systemic administration of therapeutic agents.[1] However, most therapeutic agents used in conventional chemotherapy are small molecular weight (MW) compounds that are typically less than 500 Da. Their small size results in several inherent disadvantages,[2] including a shorter half-life and a higher overall clearance rate in the blood stream. Additionally, these drugs are not specifically targeted to the tumor tissue and are indiscriminately distributed throughout the body. This lack of specificity results in a severe risk of systemic toxicity and side effects.[3–5] As a consequence, the therapeutic efficacy of conventional chemotherapeutic agents is unsatisfactory due to the relatively low dose of drug that is actually delivered to the pathological region. Alternatively, the administration of excessively large doses of drugs to compensate for this lack of specificity can lead to increased toxicity and induction of multidrug resistance (MDR) in patients.[6] Thus, the development of new, highly specific drug delivery systems and novel strategies of improving drug efficacy are critical to cancer chemotherapeutics.[7]

Polymeric nanoparticles (NP) based delivery vehicles show great promise for cancer therapeutics due to their tunable structure and capacity for modification through versatile engineering methods.[8,9] Particulate delivery systems, including micelles,[10] vesicles,[11] polymersomes,[12] nanogels[13] and non-covalently modified polymeric NPs, are those in which drugs are physically incorporated in NPs. Additionally, therapeutic agents can be covalently linked to polymers to form drug–polymer conjugates,[14] protein–polymer conjugates,[15,16] or DNA–polymer conjugates.[17] The polymers and their derivatives are significant in the formulation of NPs and can be classified into two categories: synthetic and natural.[10,18–20] Commonly used synthetic polymers include polyethylene glycol (PEG),[21–24] poly $\varepsilon$-caprolactone (PCL), polylactide (PLA), polyamino acids and N-(2-hydroxypropyl) methacrylamide copolymers, while natural polymers used in drug delivery systems include albumin, dextran, gelatin, alginate, collagen and chitosan. Furthermore, recent advances in polymer chemistry have facilitated the precise control of polymer architecture and composition in the synthesis of polymeric NPs. Controlled polymerization techniques include reversible addition-fragmentation chain-transfer polymerization (RAFT),[25–27] nitroxide-mediated polymerization (NMP),[28] atom transfer radical polymerization (ATRP),[29] ring-opening metathesis polymerization (ROMP),[30] N-carboxyanhydride (NCA) ring-opening polymerization[31] and "click" chemistry.[32] These methods allow for the synthesis of polymers with controllable architectures, controllable MWs, and low polydispersities. Moreover, these recent developments in polymer chemistry have led to the potential for new opportunities in polymeric, NP based drug delivery systems.

The platform material (core), the surface polymers, and the encapsulated payload (drug, protein, or active agents) are the three main design components necessary for polymeric NPs.[32,33] In order to ensure that drugs are discharged into extra-vascular spaces and accumulate in tumor tissue, a passive targeting strategy is utilized *via* the Enhanced Permeability and Retention (EPR) effect.[34–36] If further fine-tuning of drug delivery is needed, polymeric NPs can also be conjugated or grafted with affinity ligands to increase the selectivity for tumor cells. A wide variety of active agents can also be incorporated with therapeutic drugs into these NPs for simultaneous diagnosis and treatment of a tumor. These can vary from macromolecules to imaging agents such as proteins, nucleic acids, organic dyes and quantum dots.[37,38] Overall, polymeric NPs offer the following advantages compared to conventional small molecule compounds: (1) prolonged circulation time in the blood stream, (2) enhancement of tumor specificity and therapeutic efficiency *via* engineered active or passive targeting mechanisms, (3) a possibility of overcoming MDR.[37,39,40]

Potential targeted drug delivery systems made by polymeric NPs need to overcome multiple biobarriers, including both extracellular and intracellular.[3] Indicators that a drug has overcome extracellular barriers include a longer circulation time in the blood, preferential accumulation at tumor site, and higher tumor specificity.[37] However, intracellular barriers, including cellular internalization, endosomal escape, and drug release into the cytosol or other intracellular locations are also significant barriers that can hinder targeted drug release.[41] Recently, stimuli responsive polymeric NPs that release drugs in response to intracellular tumor signals have been used as a promising platform to overcome the aforementioned barriers.[42] Stimuli-responsive moieties can be intimately tied to the NP polymer architecture to trigger its assembly and/or disassembly for controlled drug delivery. This ability to switch drug delivery "on and off" is rather complex as it mandates a particular bond cleavage, conformational, or solubility change to a desired microenvironment stimulus.[43,44] For instance, stimuli-responsive NPs may be sensitive to specific internal stimuli, such as a lowered pH or a higher glutathione (GSH) potential.[18,45] This ability is particularly relevant for intracellular tumor microenvironments, where

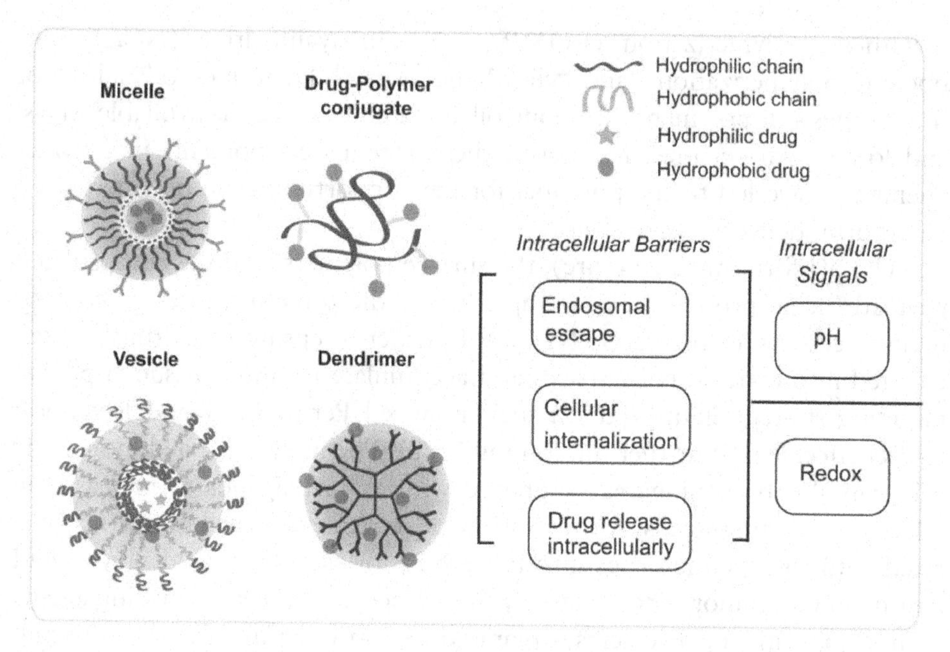


**Figure 1.** Schematic illustration of representative polymeric NPs triggered drug release by intracellular biological stimuli across several intracellular barriers.

changes associated with its relatively low pH and high GSH potential can be utilized. Intracellular pH sensitivity can either trigger the release of the transported drug into late endosomes or lysosomes, or promote the escape of these nanocarriers from the lysosomes to the cytoplasm. Additionally, the high GSH concentration in tumor cells can also trigger intracellular drug release (**Figure 1**).

As a result, there has been much recent interest in the design of novel, stimuli sensitive polymeric NPs capable of releasing anticancer drugs in response to a tumor intracellular signal — particularly those in high redox condition. Subsequent sections of this chapter will focus on triggers dependent on the redox potential found as a result of the high GSH levels in intracellular tumor microenvironments. We will also highlight the NP assembly and disassembly mechanisms, as well as key trends in redox-responsive polymer design and recent advances in redox-responsive polymeric NPs for intelligent, anticancer drug delivery. The goal of such systems is to mimic intelligent biological systems and, ultimately, to moderate the complex system responses to desired levels. The versatility and untapped potential of redox-sensitive polymeric materials makes them one of the most exciting fields of emerging cancer nanotechnology.

## 2. Cellular Redox Environment

When compared to normal tissue, tumor tissue is found to possess a variety of unique microstructural features and physicochemical properties, including vascular abnormalities, low acidity, abnormal temperature gradients, and over-expressed proteins and enzymes.[46] These markedly different intracellular microenvironments can be used by NP technology to moderate different drug release profiles. To this end, important release factors include the acidic pH found inside both endosomes and lysosomes (pH 4.5–6.5)[47,48] and the high GSH-caused reductive microenvironments in the cytoplasm and in the endosomes/lysosomes.[49] Incorporating elements like relevant pH, temperature, and redox potential stimuli allows for the possibility of focusing on desired therapeutic organs. Integrating polymeric NPs can promote cellular uptake, while also allowing for precise triggering and optimization of drug release at the target disease site. Finally, it also enables the regulation of the intracellular fates of delivered drugs.

In addition to well-documented pH gradients that facilitate drug release following endocytosis into acidic sub-cellular compartments, new strategies are being attempted to explore redox-sensitive drug release mechanisms. As a result, the existence of redox gradients between intracellular (reducing) and extracellular (oxidizing) environments has been widely studied.[50,51] Recently, a vast number of redox-mediated polymeric NPs have emerged.

To understand the process of redox-mediated drug release behavior, a comprehensive understanding of the cellular redox environment is necessary. Schafer,[52] for example, explicitly defined this environment as: "...a linked set of redox couples as found in a biological fluid, organelle, cell, or tissue is the summation of the products of the reduction potential and reducing capacity of the linked redox couples present." *In vivo*, the cellular redox environment is regulated by GSH tripeptides. The intracellular tumor compartments (e.g., cytosol, mitochondria, and cell nucleus) contain 100 to 1000 times higher GSH levels (approximately 2–10 mM) than in human plasma (typically 1–2 $\mu$M) and extracellular blood (approximately 2–20 $\mu$M).[53] Additionally, the cytosolic GSH level in some tumor cells has been found to be at least four times higher than in normal cells.[53] This higher GSH level creates a reductive microenvironment inside cells. Importantly, this sharp contrast in GSH concentration between tumoral and normal cells may serve as an ideal stimulus for intracellular destabilization of polymeric NPs to achieve rapid drug release (**Figure 2**).

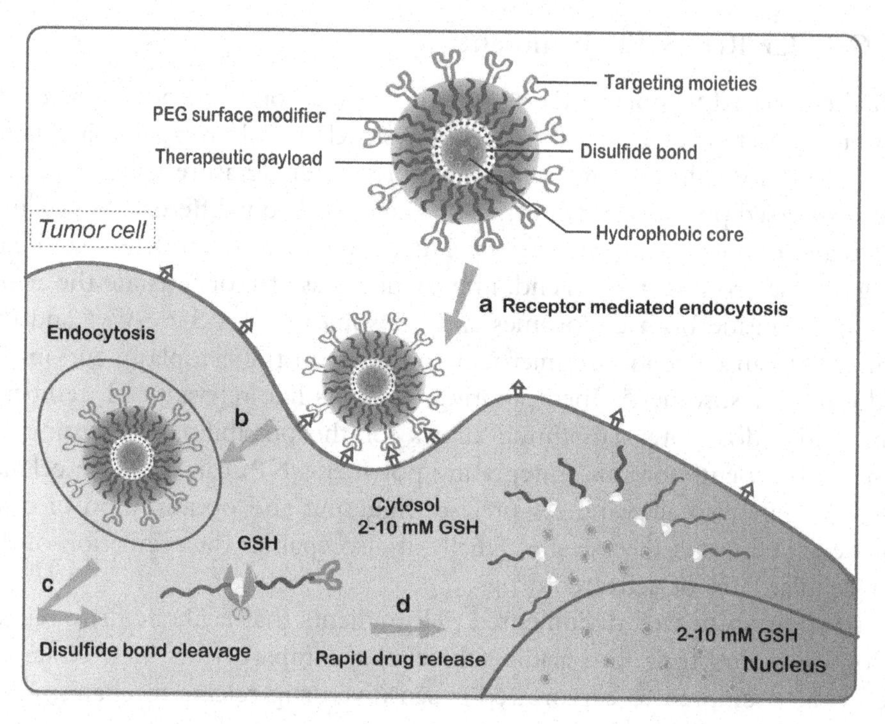


**Figure 2.** Schematic illustration of GSH-responsive NPs for intracellular drug release. It includes cellular internalization, endocytosis, redox triggered disulfide cleavage and rapid drug release.

## 3. Redox Trigger Employing Drug Delivery Systems

Redox-sensitive polymeric NPs have received increasing interest over the past years. Many redox-sensitive polymeric NPs incorporate similar design principles as those used in pH-responsive systems containing acid labile moieties. A disulfide bond (–S–S–) is the most widely studied, reduction-responsive linkage and serves as a particularly attractive target, as it is rapidly reduced to thiols in the reductive environment found in intracellular compartments.[54] The disulfide bond is relatively stable in blood circulation, but can be cleaved in a reductive environment, thereby allowing the rapid release of the loaded anticancer drug.[55,56] Because of this, disulfide bonds can play a significant role in drug delivery by utilizing the redox potential differences between the extracellular and the intracellular environments.

Disulfide linkage can be obtained through direct polymerization of disulfide-containing monomers, by the oxidative coupling of thiol-functionalized precursors (thiomers), and through the use of disulfide-containing cross-linkers.[57,58] The commonly used disulfide bond containing agents, such as cystamine,[59–61] dithiodipropionic acid,[62] bis (2-methacryloyloxyethyl) disulfide,[63] dimethyl 3,3-dithiobispropionimidate (DTBP),[64,65] 3,3'-dithiobis (sulfosuccinimidyl propionate) (DTSSP),[66] dithiobis (succinimidylpropionate) (DSP),[67] 2-(pyridyldithio)-ethylamine (PDA),[68] cystamine bisacrylamide (CBA),[69] and N-Succinimidyl 3-(2-pyridyldithio) propionate (SPDP)[70,71] have been used to connect or crosslink the polymeric NPs. **Table 1** summarizes some representative polymeric NPs that bear disulfide bonds or other crosslinking agents with inherent redox sensitivity.

### 3.1. *Redox-sensitive micelles*

Redox-responsive polymeric NPs, such as micelles, often contain a disulfide linkage between the hydrophilic and hydrophobic segments, thereby creating "shell-sheddable" micelles.[89] Upon exposure to high levels of GSH, the micelles will destabilize through cleavage of their disulfide bond, thus facilitating release of the loaded drug. The "shell-sheddable" micelles thus enhance therapeutic efficacy by releasing a drug once the shell is detached in a reductive environment.

For example, Zhong *et al.*[73] reported on biodegradable micelles that had sheddable PEG shells. These shells were based on disulfide-linked PEG–SS–PCL co-polymer and allowed for the triggered intracellular release of Doxorubicin (DOX). The prepared PEG–SS–PCL micelles showed a faster DOX release profile within 12 hours under 10 mM dithiothreitol (DTT) than that of control micelles containing PEG–PCL. Both a more rapid intracellular release of DOX and a higher anticancer efficacy were achieved in a reductive environment using this system.

Similarly, our group has also conducted a series of studies utilizing shell-sheddable micelles. We synthesized and evaluated the biological efficacy of micelles that rapidly disassemble and release an encapsulated drug under tumor-relevant GSH levels (**Figure 3**).[76] The unique design included a redox-sensitive PEG shell grafted onto a poly ε-benzyloxycarbonyl-L-lysine (PzLL) core *via* a disulfide linkage (mPEG–SS–PzLL). Changes in size indicated that the micelles were disassembling within 4 hours in response to 10 mM GSH following detachment of the PEG shell. Redox stimulus-induced micellar rearrangements accelerated *in vitro* release of DOX from micelles three to five times faster than in the absence of GSH. Furthermore, confocal microscopy demonstrated that DOX-loaded micelles that contained a disulfide-linked PEG shell were efficiently internalized into Michigan Cancer Foundation-7 human breast cancer cells (MCF-7). Then, they effectively released DOX into both the cytoplasm and nucleus in the presence of a GSH-reducing environment. Importantly, inhibition of cell proliferation was directly correlated with increased GSH concentrations due to the rapid release of DOX.

Table 1. Overview of redox-responsive polymeric NPs.

| Polymeric NPs | Polymers | Drug | Size | Disulfide bond or Crosslink agent | Ref. |
|---|---|---|---|---|---|
| Micelle | PEG–PLys–PPha | MTX | 48 nm | DTSSP | 72 |
| | PCL–SS–PCL | DOX | 240 nm | PCL–SH + PEG–SS–Py | 73 |
| | PCL–SS–PEEP | DOX | 70 nm | PCL–SH + PEEP–SS–Py | 74 |
| | PEG–$L_2$–PCL | DOX | 30 nm | DTT | 75 |
| | PEG–SS–PzLL | DOX | 300 nm | DTDP | 76 |
| | mPEG–SS–PLeu | DOX | 160 nm | Cystamine | 77 |
| Vesicle | PEG–SS–PPS | Biomolecular drugs | 120 nm | PEG–monothiolate + PPS–monothiolate | 78 |
| | PzLL–SS–PEG–SS–PzLL | DOXHCl | 380 nm | Cystamine | 79 |
| | SA2 amphiphilic oligopeptides | — | 75 nm | Cysteines | 80 |
| Prodrug | CPT–SS–PEG–SS–CPT | CPT | 226 nm | Cystamine | 81 |
| | mPEG–PPS–SS–TG | Tioguanine (TG) | 16–35 nm | PEG–thiol+tioguanine disulfide | 82 |
| | PAMAM–NAC conjugate | NAC | — | SPDP | 83 |
| Dendrimer | G3-OH, G4-OH, and G5-OH-PAMAM-gold NPs | CPP, 6-MP, DOX, CPT | 3 nm | Thiol-containing, thiolated drugs +Au–S bond | 84 |
| | Hyperbranched polyphosphate (HPHDP) | DOX | 38–78 nm | 2-[(2-hydroxyethyl)-disulfanyl]ethoxy-2-oxo-1,3,2-dioxaphospholane | 85 |
| | HPHSEP-star-PEP | DOX | 70–104 nm | 2-ethoxy-2-oxo-1,3,2-dioxaphospholane (EP) | 86 |
| Other NPs | Nanorings: $HPAA_{12}$ | DNA | 150 nm | CBA | 87 |
| | MSNs–SS–mPEG | Fluorescein dye | 200 nm | MSNs–SH+ mPEG–SS–Py | 88 |

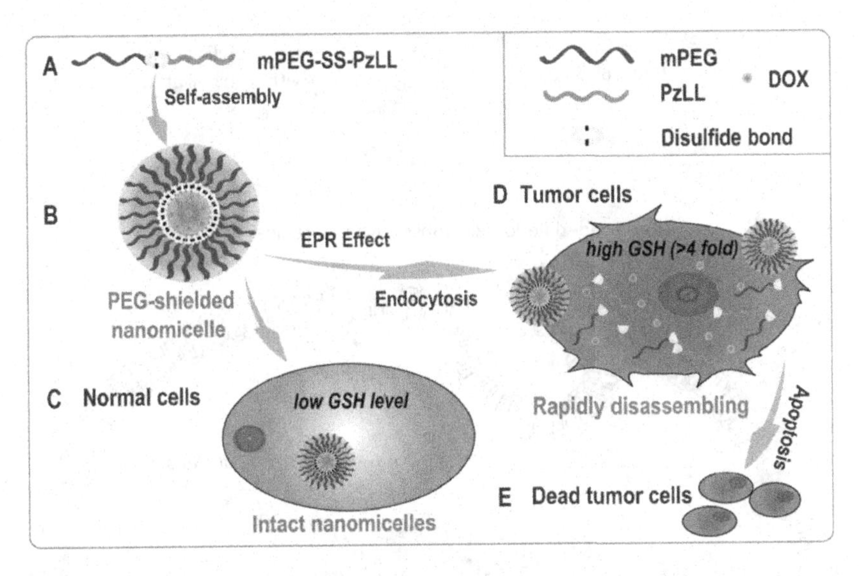


**Figure 3.** Predicted antitumor activity of redox-sensitive DOX-loaded mPEG–SS–PzLL micelles.[76] [Reproduced with permission from Ref. 76. Copyright The Royal Society of Chemistry.]

Another approach for the design of disulfide-linked micelles is to introduce crosslinking agents bearing thiol groups into block co-polymer micelles. For instance, Lee *et al.*[72] synthesized cross-linked polymer micelles that self-assembled from triblock co-polymer of PEG-b-PLys-b-PPha (PEG-b-poly (L-lysine)-b-poly (L-phenylalanine)). The PLys middle shells were cross-linked by adding a disulfide-containing cross-linker DTSSP. The cross-links in the PLys shells increased micellar stability and prevented drug leakage during blood circulation. Additionally, the disulfide-cross-linked micelles allowed the facilitated release of entrapped methotrexate (MTX) into the cytoplasm in response to a higher intracellular GSH level.

The redox-responsive micelles were also developed to overcome a second problem common to conventional anticancer therapies: MDR. PCL–SS–PEEP micelles self-assembled from PCL and polyethyl ethylene phosphate (PEEP) *via* a single disulfide bond and were shown to overcome MDR by enhanced drug accumulation and cellular retention in breast cancer cells (**Figure 4**).[90] Enhanced DOX release profile in MCF-7/ADR cells (where ADR is adriamycin-resistant) was achieved through the successful shell detachment of the micelles. This detachment was triggered by the high levels of GSH found in the intracellular environment. These results indicate that redox-sensitive micelles exhibit promising advantages in overcoming MDR in cancer cells as well as improving overall therapeutic efficiency of anticancer drugs.

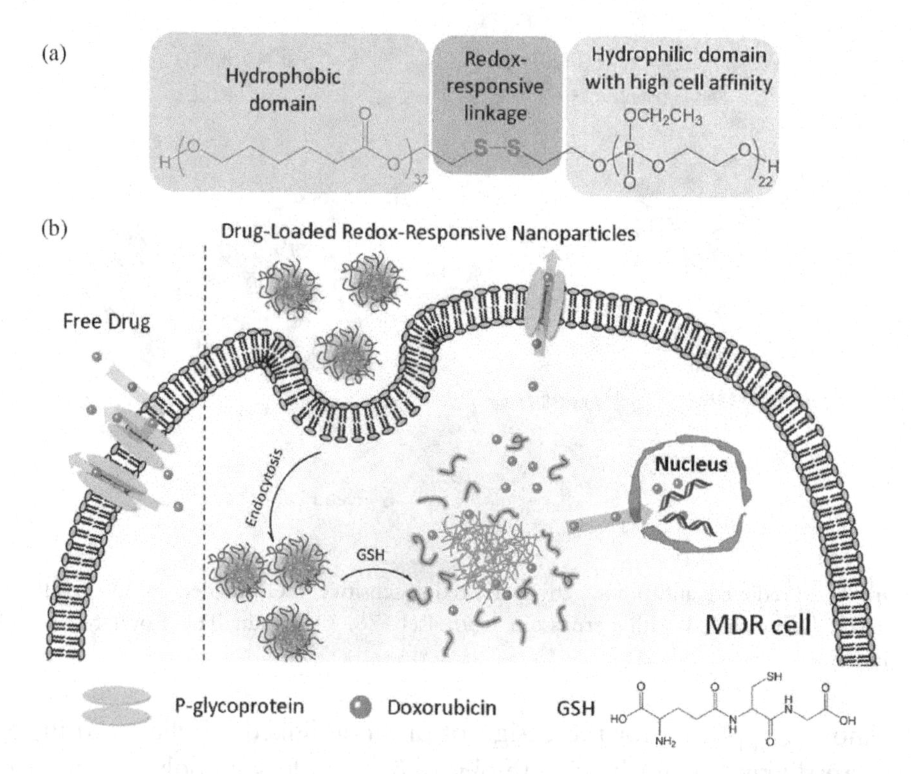


**Figure 4.** Chemical structure of disulfide-bridged PCL–SS–PEEP block co-polymer **(a)** and schematic illustration of redox-responsive nanoparticles for overcoming MDR of cancer cells **(b)**.[90] [Reprinted with permission from Ref. 90. Copyright 2011 American Chemical Society.]

## 3.2. *Redox-sensitive vesicles*

Polymer vesicles, also called polymersomes, are spherical shell structures formed by amphiphilic block co-polymers.[91] In the past few years, such vesicles have been widely investigated as drug carriers because of their unique capability to entrap aqueous therapeutic agents.[92] Redox-sensitive materials bearing disulfide bonds have also been introduced to form redox-sensitive vesicles.

As an illustration, Kim *et al.*[93] designed and synthesized a reduction-sensitive and biocompatible vesicle, SSCB [6] VC, that self-assembled from an amphiphilic cucurbit[6]uril (CB[6]) derivative that contained disulfide bonds between hexaethylene glycol (HEG) segment and a CB[6] core. CB[6] is the host molecule and a member of the cucurbit[$n$] uril ($n$ = 5–10; CB[$n$]) family.[94] The most interesting feature was that the hydrophobic cavity of CB[6] offered a specific site for inclusion of hydrocarbon molecule through a host–guest interaction. Moreover, the polar carbonyl groups on the CB[6] core showed a high binding affinity towards polyamines that occurred through charge-dipole and hydrogen-bonding interactions. Thus, the surface of the vesicle could be decorated with folate- and fluorescein isothiocyana FITC-spermidine conjugates as targeting ligands (T) and imaging probes (I), respectively. It was also found that the vesicle structure was stable when circulating in the blood. However, upon folate-receptor mediated cell internalization, the vesicle structure was able to readily collapse and release DOX due to the highly reductive cytoplasmic environment. Thus, the cytotoxic capabilities of the vesicle increased dramatically.

In addition to the vesicle work of Kim *et al.*, reduction-sensitive vesicles formed from block co-polymeric macroamphiphiles have also been reported. In this instance, such macroamphiphiles have been formed that could protect biomolecules in the extracellular environment, but also release their drug into the early endosome rather than the lysosome. The redox-sensitive block co-polymers PEG–SS–PPS were synthesized by intervening disulfide linkage between the PEG chain and hydrophobic polypropylene sulfide (PPS) chain.[78] Vesicles formed from this block co-polymer and bearing this disulfide bond were rapidly disrupted due to the high redox potential difference between the extracellular and intracellular environments. Moreover, cellular uptake, disruption, and biomolecular drug release were observed within 10 minutes of administration. These structural characteristics plus the biomolecule's time course are strong evidence that the drug was released in an early endosome of the endolysosomal processing cycle.

Recently, Li *et al.*[79] developed reduction-cleavable vesicles that self-assembled from the triblock co-polymers PEG and PzLL. These vesicles contained disulfide bonds as novel hydrophilic drug delivery carriers for overcoming the MDR of cancer cells (**Figure 5**). Structural characteristics of the vesicles were confirmed with both TEM and confocal laser scanning microscopy (CLSM), whereby they were observed to have an obvious hollow structure surrounded by a thin outer layer. The average diameter of the vesicles was 380 nm with a polydispersity index PDI of 0.25. The critical aggregation concentration of obtained vesicles was 14 mg/L and was measured by loading hydrophobic florescence probe pyrene in the membranes of the vesicles. In addition, Doxirubicin hydrochloride (DOX·HCl) was entrapped in the vesicles as an exemplar hydrophilic drug. When the vesicles were exposed to 10 mM GSH, they were found to have a three-time higher drug release. Furthermore, the inclusion of DOX·HCl into these vesicles effectively reduced HeLa cell viability and enhanced nuclear accumulation of DOX·HCl in HeLa cells treated with 10 mM GSH. These effects were induced by the disassembly of the vesicle's structure under the highly reductive intracellular conditions. Furthermore,

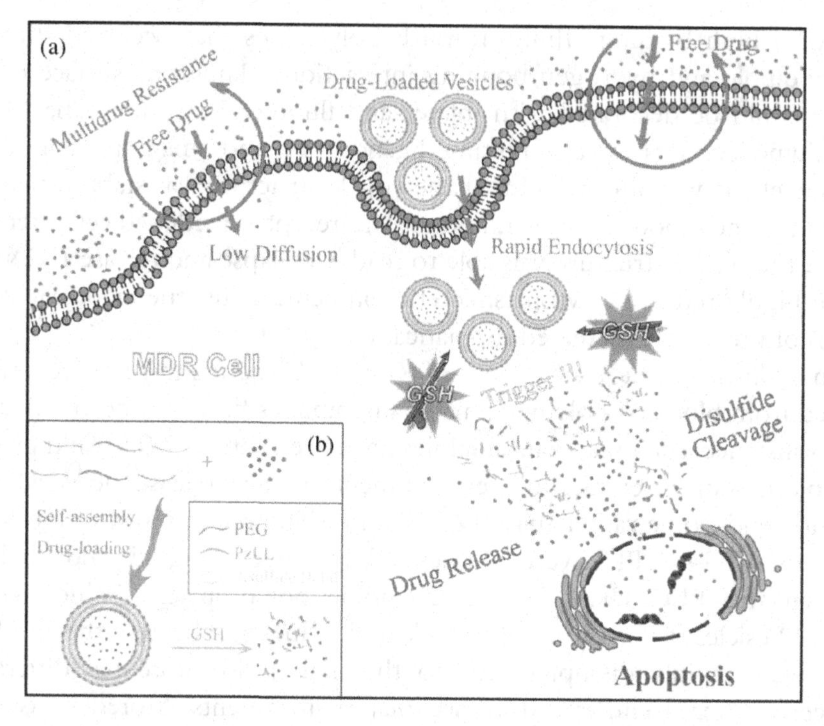


**Figure 5.** (**a**) Schematic illustration of redox-sensitive vesicles overcoming MDR of cancer cell by intracellular rapid drug release; (**b**) Self-assembly behavior of PzLL–SS–PEG–SS–PzLL triblock co-polymers and their drug release behavior triggered by GSH.[79] [Reprinted with permission from Ref. 79. Copyright 2013 American Chemical Society.]

when hydrophilic gemcitabire hydrochloride GC·HCl was included instead, it potentiated the inhibitory effect of GC·HCl-resistant M. D. Anderson-metastatic breast-231 (MDA-MB-231) breast cancer cells. These results indicate that the prepared redox-sensitive vesicles may provide a means to overcome MDR in cancer cells.

## 3.3. *Redox-sensitive prodrugs*

Prodrugs fabricated by conjugating therapeutic drugs onto the backbone of block co-polymers *via* cleavable linkages (e.g., disulfide, hydrazone) can release parent drugs under specific conditions.[46] Kataoka *et al.*[95] pioneered this particular prodrug design by conjugating DOX onto co-polymer chains *via* an acidic cleavable hydrazone bond. This bond allows for the release of DOX when the molecule is inside acidic environments such as endosomes or lysosomes. Very similar to the fabrication of pH-sensitive prodrugs, redox-sensitive prodrugs exhibit GSH-triggered cleavage of disulfide bonds within intracellularly reductive environments such as the cytosol or nucleus (2–10 mM GSH). However, these linkages are only affected by highly reductive conditions, as they have been found to be relatively stable when circulating in the blood.

Building on this earlier work, our group has covalently attached camptothecin (CPT) onto PEG polymers with a disulfide linkage.[81] The resulting CPT–SS–PEG–SS–CPT co-polymer then self-assembled into a micelle bearing a CPT core and was found to be relatively stable when circulating in the blood. However, highly reductive conditions caused cleavage of the disulfide bond, thereby initiating micellar rearrangement and accelerating the release of the entrapped drug. Over a time course of 240 hours, only 0.1% of CPT was released in Phosphate-buffered saline (PBS) solution. This is contrasted with a rapid release rate of 4% per hour during the first 10 hours when CPT–SS–PEG–SS–CPT prodrugs were exposed to reductive conditions of 10 mM DTT. Additional measurements revealed that the prodrugs effectively decreased HepG2 cancer cell viability, thus demonstrating an increased intracellular accumulation and pharmacological efficacy of CPT when it was released from GSH-redox-sensitive prodrug micelles. Consequently, the novel prodrug design encapsulated in the CPT–SS–PEG–SS–CPT molecule is predicted as being capable of both delivering anti-cancer drugs to desired tumor sites and releasing its cargo in a redox-dependent manner.

Previously, various iterations of these disulfide-containing prodrugs have been developed, which contain a disulfide linkage connecting folate to the anticancer drug, thus allowing the active drug to be released upon disulfide cleavage under intracellular GSH concentrations.[97–99] For instance, Santra *et al.*[96] had reported a folate-DOX prodrug with a disulfide linker (DSP) between fluorescent DOX and folic acid (Fol) for targeted delivery to folate receptor-expressing cancer cells (**Figure 6**). Interestingly, they found that when DOX was covalently linked to Fol, both the fluorescence and cytotoxicity of DOX were quenched (OFF). More specifically, they observed that DOX fluorescence was quenched five-fold after binding to Fol. This result indicated that Fol was not only a targeting ligand, but also a quencher for DOX. Santra *et al.* then synthesized an active probe, DOX–S–S–Fol, which had a cleavable disulfide bond, and a non-cleavable, control probe, DOX–C–C–Fol, which had a stable linker (disuccinimidyl suberate, or DSS) (**Figure 6(a)**). Upon Fol-mediated cell internalization, the DOX–SS–Fol prodrug became fluorescent (ON) and cytotoxic due to the GSH-triggered-cleavage of the disulfide bond (**Figure 6(b)**). However, the DOX–C–C–Fol control probe remained quenched (OFF) and had no cytotoxicity (**Figure 6(c)**). These results further confirmed that the released DOX from the DOX–SS–Fol had a greater cytotoxicity when compared to free DOX.

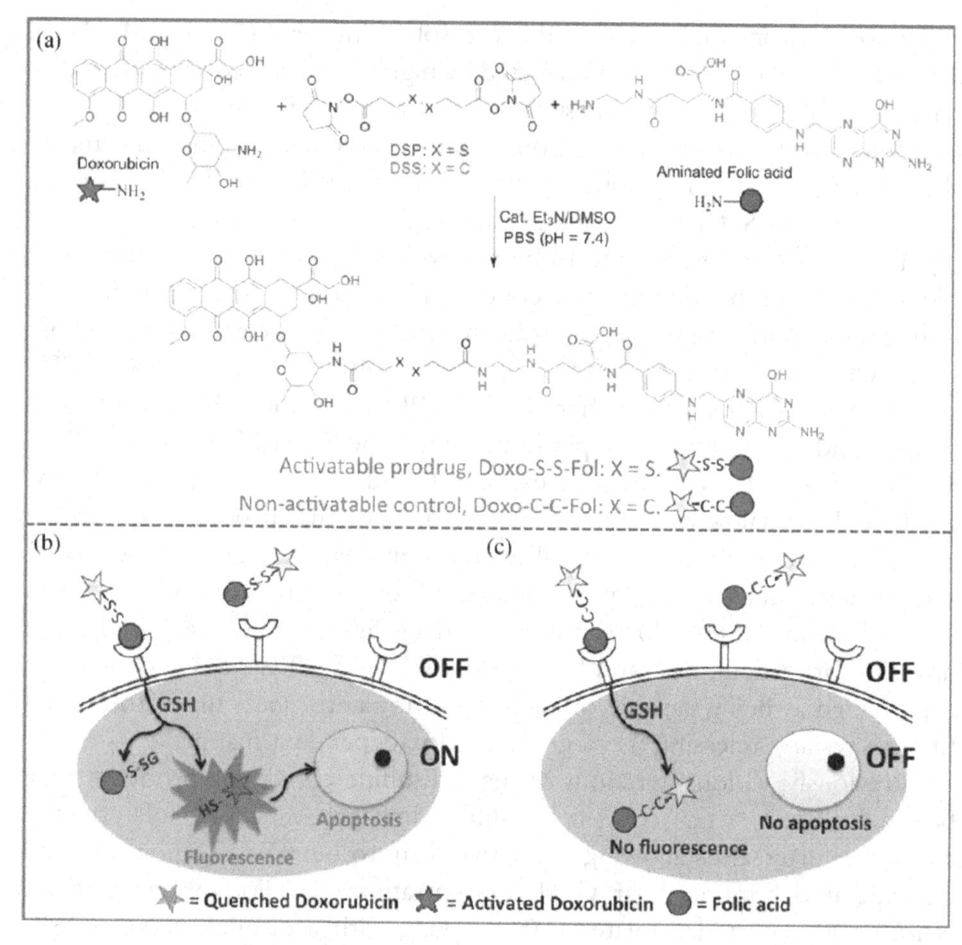


**Figure 6.** GSH-responsive prodrugs for targed cancer imaging and therapy. **(a)** Syntheses of DOX–S–S–Fol prodrug and DOX–C–C–Fol prodrug; **(b)** Upon cell internalization, DOX released from the DOX–S–S–Fol prodrug and fluorescence appeared (ON) due to the cleavage of disulfide bond triggered by intracellular GSH; **(c)** Upon cell internalization, DOX–C–C–Fol prodrug was stable and kept quenched fluorescence (OFF) due to the absence of a cleaving mechanism.[96] [Reprinted with permission from Ref. 96. Copyright 2011 American Chemical Society.]

## 3.4. *Redox-sensitive dendrimers*

Dendrimers are currently one of the most promising and versatile drug carriers. Structurally, they contain repeatedly branching, polymeric macromolecules that extend from a core site. This pattern results in a near perfect three-dimensional geometric pattern.[100] Polyamidoamine (PAMAM) is one of the most widely investigated dendrimers because it is capable of binding numerous targeting, imaging, or therapeutic agents within its tertiary amine and amide linkages.

Recently, Wang[84] synthesized a GSH-sensitive PAMAM dendrimer-encapsulated gold NPs (DEGNPs) and utilized them as carriers of thiol-containing or thiolated anticancer drugs (e.g., captopril (CPP), 6-mercaptopurine (6-MP), DOX and cisplatin (CPT)). The resulting gold NPs were synthesized within hydroxyl-terminated generation 3, 4, and 5 PAMAM dendrimers and had an average size of 3 nm. DEGNPs were able to entrap thiol-containing or thiolated drugs through the high affinity present in an A–S bond. The entrapped drugs showed an "OFF–ON" release behavior in the presence of thiol-reductive agents. That DEGNPs were able to release their entrapped drugs as a result of intracellular GSH concentrations was further confirmed *via* flow cytometry and CLSM studies. These results are promising and indicate that dendrimer-based, GSH-sensitive drug delivery system provides a new method for the fabrication of interior-functionalized, dendrimer–drug conjugates.

Similarly, Yan *et al.*[85] synthesized a series of hyperbranched homopolymers (HPHDP) with alternating hydrophobic disulfide and hydrophilic polyphosphate units. This hyperbranched, homopolymer was a novel precursor for molecular self-assembly, since most of the previously reported precursors were either amphiphilic linear systems or dendritic block co-polymers.[101,102] Additionally, the HPHDP with different MW successfully self-assembled into micelles with a multi-core/shell structure. When measured, their sizes ranged from 30–78 nm with a very narrow size distribution. The resulting HPHDP micelles showed redox-sensitive assembly and disassembly behavior in highly reductive environments. Furthermore, the HPHDP micelles that had entrapped hydrophobic drugs could be transported into the nuclei of cancer cells rapidly and efficiently. The intracellular GSH concentration would then trigger the release of the transported drug.

Additionally, Yan *et al.* obtained redox-sensitive micelles that self-assembled from amphiphilic HPHSEP-star-$PEP_X$ (hyperbranched multiarm co-polyphosphates) with a tunable size ranging from 70 to 100 nm. This was achieved by modulating the MW of the PEP multiarm.[86] The HPHSEP-star-$PEP_X$ micelles structure can then be destroyed when exposed to reductive conditions, followed by its drug release. Meanwhile, DOX-loaded HPHSEP-star-$PEP_X$ micelles exhibited higher inhibition of cellular proliferation in HeLa cells that had been pre-treated with GSM-reduced ethyl ester GSH-OET, suggesting the successful delivery of DOX into the nucleus.

## 4. Dual Stimuli Engineered Redox Sensitivity

It is well known that tumor cells are associated with different basal pH and temperature profiles when compared to normal cells.[103] Healthy blood is kept at a constant pH of 7.4. However, most tumor tissues are slightly acidic, with a pH falling in the range of 6.5–7.0. Moreover, intracellular compartments are also more acidic than healthy tissues, such as the early endosome (pH 6.0–6.5), late endosome (pH 5.0–6.0) and lysosome (pH 4.5–5.0).[104] Finally, the temperature in tumor tissues is higher than in normal tissues ($T = 37°C$).

As highlighted above, the merits of polymeric NPs engineered with redox triggers have inspired the development of drug delivery systems that are responsive to two or more stimuli. Engineering dual- or multi-stimuli responsiveness is advantageous as it allows for more sophisticated and intricate drug delivery. In short, it provides for more efficient loading of the desired drug as well as a more precisely controlled release of the therapeutic cargo at tumor sites.[42] The most widely investigated dual or multiple stimuli combinations are as follows: pH/redox, T/redox, and pH/temperature/redox.

## 4.1. *pH/redox-sensitive polymeric NPs*

Dual sensitive, core-cross-linked polypeptide micelles can be combined with reductive and pH-sensitive triggers. When self-assembled from PEG-P (LL-CCA/LA) lipoic acid (LA) and cis-1, 2-cyclohexanedicarboxylic acid (CCA) and decorated with PEG-b-PLys block co-polymers, they are capable of achieving the intracellular release of DOX (**Figure** 7).[105] The micelles were cross-linked in the presence of 10 mol% DTT, delivering fast micelle dissociation and DOX release behavior in a 10 mM GSH environment. In addition, the acidic pH of the endosomal compartment facilitated the cleavage of the acid-labile amide bond between PLL and CCA, thus promoting intracellular drug release. The prepared pH/redox dual sensitive micelles were found to be stable with a limited drug release during blood circulation, but rapid drug release in response to cytoplasmic GSH and endosomal pH conditions. Considering these factors, the pH/redox-sensitive polypeptide micelles are highly recommended as an advanced platform for enhanced cancer therapies.

Cyclic benzylidene acetal groups (CBAs) functionalized polymeric NPs[107–109] have been considered as an effective strategy for intracellular drug release due to the acid-labile acetal groups that can be hydrolyzed under acidic conditions. Recently, Zhao[106] developed a graft co-polymer NPs with both a pH- and reduction-induced disassembling property for enhanced intracellular curcumin release. The graft co-polymer PTTMA-g-SS-PEG was synthesized *via* RAFT co-polymerization and a coupling reaction based on acid-labile CBA-functionalized poly(2,4,6-trimethoxybenzylidene-1,1,1-tris(hydroxymethyl) ethane methacrylate (PTTMA) and disulfide linked PEG

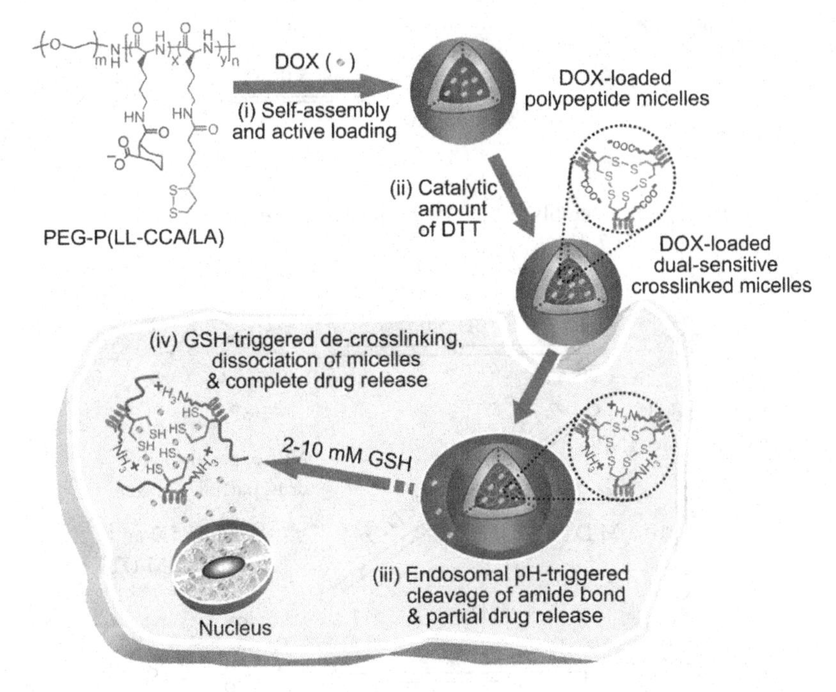


**Figure 7.** Illustration of reduction and pH dual-sensitive core-crosslinked PEG-P(LL-CCA/LA) micelles for loading and triggered intracellular release of DOX.[105] [Reprinted with permission from Ref. 105. Copyright 2013 Elsevier.]

chains. The obtained PTTMA-g-SS-PEG NPs were stable at pH 7.4 with less than 15.0% of curcumin (CUR) released at this pH, but disassembled in at a pH of 5.0 and/or 10 mM DTT conditions. At a pH of 5.0, the hydrolysis of CBAs in the hydrophobic PTTMA core induced the NPs expansion and partial drug release. Furthermore, the presence of 10 mM DTT allowed disulfide bond cleavage and the PEG shell to subsequently detach from the NPs. This action also allowed for partial drug release. Finally, the NPs exhibited a structural disassembly that rapidly released 94.3% of the entrapped drug under highly reductive (10 mM DTT) and highly acidic (pH 5.0) environment conditions. These conditions mimic the tumor microenvironment in endosome/lysosome and cytoplasm, further demonstrating their usefulness as a therapeutic platform (**Figure 8**).

A biodegradable polypeptide used as the co-polymer backbone and based on a highly packed interlayer-cross-linked micelle (HP-ICM) and engineered with dual reduction and pH sensitivity was reported as a method for intracellular drug release.[110] The HP-ICM was self-assembled from a triblock co-polymer of mPEG-PAsp-(MEA)-PAsp (DIP) (PAsp (MEA): 2-mercaptoethylamine (MEA)-grafted poly(l-aspartic acid), and PAsp(DIP): 2-(di-isopropylamino)ethylamine

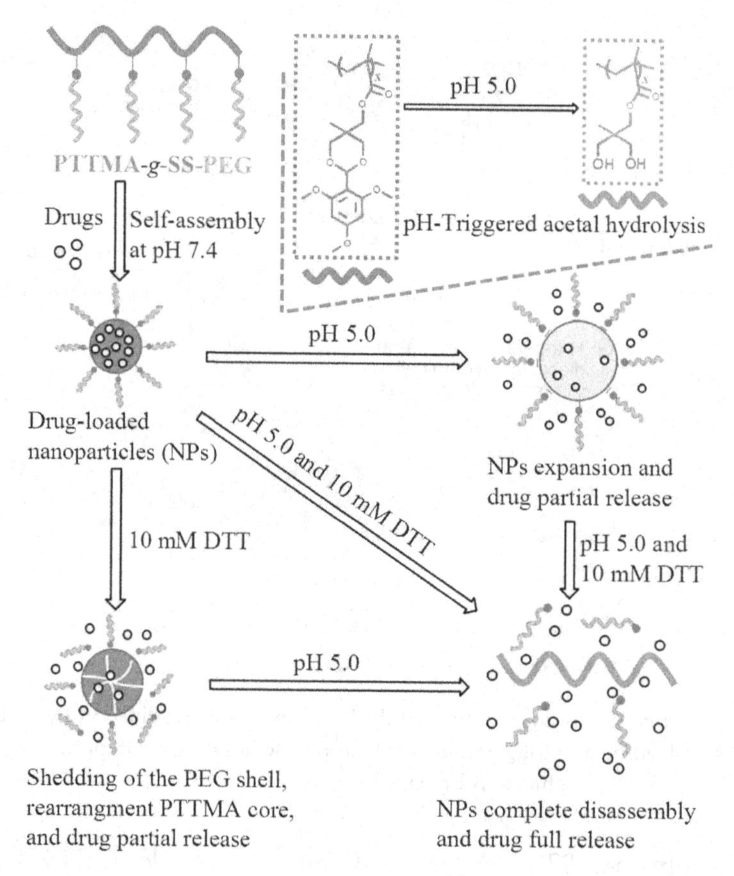


**Figure 8.** Schematic illustration of disassembly of PTTMA-g-SS-PEG NPs for entrapped drug release induced by reduction and pH dual stimuli.[106] [Reprinted with permission from Ref. 106. Copyright 2013 American Chemical Society.]

(DIP)-grafted poly(l-aspartic acid)). The novel HP-ICM was composed of three functionalities: a PEG surface (to prolong the blood circulation), a highly packed, pH-sensitive partially hydrated core (to load drugs), and a disulfide-cross-linked interlayer (to stabilize the core at neutral pH). The DOX loaded HP-ICM was stable when circulating in the blood, whereas disassembly of the HP-ICM in both acidic and high GSH environment induced rapid DOX release into the lysosome and DOX accumulation in the nucleus.

## 4.2. *T/redox-sensitive polymeric NPs*

It has been reported that a significant hyperthermic environment exists in tumor tissues, at a range of 40–43°C.[111] Temperature sensitive polymers represent an important class of stimuli-responsive materials for future drug delivery. Poly

(N-isopropylacrylamide) (PNIPAAm)-containing co-polymers are the most widely investigated temperature-sensitive polymers for drug delivery and have an lower critical solution temperature LCST of 32°C. At this temperature, it exhibits a reversible volume phase transition caused by the coil-to-globule transition.[112–114]

Zhong *et al.*[115] developed dual temperature and redox-sensitive cross-linked vesicles that were self-assembled from the triblock co-polymers PEO–PAA–PNIPAM and prepared by RAFT polymerization. The PEO–PAA–PNIPAM co-polymers were water-soluble at room temperature, and quickly self-assembled into 220 nm vesicles at 37°C (above their LCST). Cystamine was then used to cross-link the interfaces of the vesicles *via* carbodiimide chemistry. The dual-sensitive cross-linked vesicles were stable in a diluted state, in an organic solvent, and at high salt conditions as well as in water with varying temperatures. However, they rapidly disassembled under reductive conditions that were analogous to those found in the intracellular tumor environment. Furthermore, they synthesized a second temperature and reduction sensitive PEG–PAA–PNIPAM of a different MW by increasing the temperature above its LCST and subsequently crosslinking with cystamine.[116] Various exemplar proteins (e.g., BSA, lysozyme, cytochrome C, and ovalbumin) were then loaded into the polymersomes, all with a high loading efficiency. The PEG–PAA–PNIPAM vesicles successfully avoided leakage of the loaded protein (e.g., BSA, lysozyme and cytochrome C) over a time span of 11 hours in PBS solution at 37°C. Conversely, fast protein release was observed when the vesicles were in the reductive environment found intracellularly (this due to DTT-triggered de-crosslinking). Finally, the dual-sensitive polymersomes could successfully and efficiently deliver the previously loaded FITC-cytochrome C into the cytosol of MCF-7 cells.

## 5. Multiple Stimuli Engineered Redox Sensitivity

Recently, Huang *et al.*[117] reported thermo-, pH-, and reduction-sensitive polymeric micelles that were based on a block co-polymer of p (PEGMEMA-co-Boc-Cyst-MMAm)-block-PEG and used for paclitaxel (PTX) delivery. The micelles were prepared by a quick heating process and were stable in blood circulation. However, under the weak acidic conditions found in tumor tissues, the micelles' surface became positively charged due to ionization of the imidazole group, thus facilitating cellular uptake through adsorptive endocytosis. The micelles disassembled in the lower endosomal pH due to the complete ionization of the imidazole groups, followed by rapid drug release into the cytoplasm. Simultaneously, the cleavage of disulfide bonds in the micelles also promoted drug release due to the high intracellular GSH

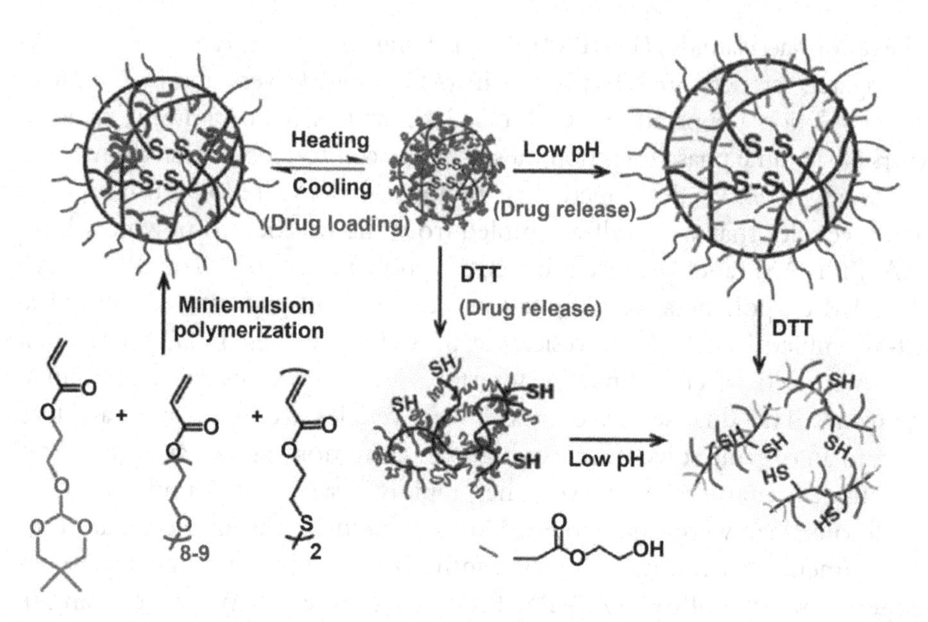


**Figure 9.** Schematic illustration of thermo/pH/redox triply responsive nanogels for drug loading and triggered drug release.[118] [Reprinted with permission from Ref. 118. Copyright 2011 Elsevier.]

levels. In conclusion, the thermo-, pH- and reduction-sensitive polymeric micelles are an attractive platform to achieve the fast intracellular release of anticancer drugs and enhanced therapeutic efficacy.

Li[118] prepared triply (thermo-, pH-, and redox-) responsive nanogels by the miniemulsion co-polymerization of monomethyl oligo (ethylene glycol) acrylate (OEGA) and acid-labile ortho ester-containing acrylic monomer (DMDEA) with a disulfide-containing cross-linker, bis (2-acryloyloxyethyl) disulfide (BADS) (**Figure 9**). OEG chains endowed the nanogels with a thermo-responsive property and the dehydration of OEG chains induced shrinkage of the nanogels in water. The nanogels were around 50 nm below their volume phase transition temperatures, and shrunk greatly to 17–35 nm when the temperature was increased to 37°C. The hydrophobic drug was loaded by heating above the phase transition temperature of the nanogels. Ortho ester groups that were hydrolyzed in mild acidic media endowed the nanogels with a pH-responsive property. Coupled with the nanogels swelling behavior, this allowed for drug release into either the late endosome or lysosome. Moreover, the disulfide-containing linkage in the nanogels endowed them with a redox-responsive property allowing for cleavage in a reductive environment. Reduction-cleavage of the disulfide linkage in the nanogels also facilitated intracellular drug release. The *in vitro* release profiles of PTX, Nile

Red and DOX indicated that nanogels avoided drug leakage at a pH of 7.4 (equivalent to blood pH), while drug release accelerated in a pH of 5.0 or 6.0 (equivalent to intracellular acidic environments). Moreover, both PTX and Nile Red release was accelerated in 20 mM DTT, but this effect was not as sensitive as that seen under acidic conditions.

## 6. Conclusion and Perspectives

The past several years have seen rapid progress in the fabrication of polymeric NPs engineered with single, dual, and multistimuli detectors for drug delivery. The diversity in chemical structure and composition of the responsive polymers that can be assembled into different polymeric NPs, and the various stimuli that exist at the tumor site, allow for greater flexibility in the design of stimuli-sensitive NPs. Additionally, the incorporation of affinity ligands and imaging agents into the stimuli responsive polymeric NPs will allow them to serve as multifunctional vehicles and greatly expand their application for diagnostics and therapeutics. In this chapter, the design of polymeric NPs sensitive to intracellular tumor conditions (e.g., reductive environments) may represent an attractive strategy for intracellular drug delivery.

However, while a vast variety of redox-sensitive polymeric NPs have been reported, only a small number of systems have been tested in preclinical, *in vivo* models. Unfortunately, this means that a very limited number of polymeric NPs with redox functionality have reached the clinical stage. Before this stage is even met with a robust array of NPs, a more thorough understanding of significant parameters such as polymer biodegradability, biocompatibility, drug release mechanism, NPs' blood circulation times, therapeutic response, and biological barriers is needed.

As we have reviewed in this chapter, the immense progress made in polymer chemistry and the merits of intracellular drug delivery has led to the design of redox-sensitive polymeric NPs. We expect that our review of these redox-sensitive NPs will further aid researchers in developing polymeric NPs that are capable of reaching the clinical setting.

## Acknowledgment

We gratefully acknowledge financial support from the National Natural Science Foundation of China (NSFC 21306152) and the Natural Science Basic Research Plan in Shaanxi Province of China (Program No. 2014JQ2067.)

## References


1. Langer R. Drug delivery and targeting. *Nature* 1998;392:5–10.
2. Ferrari M. Cancer nanotechnology: opportunities and challenges. *Nat Rev Cancer* 2005;5:161–171.
3. Li L, Sun J, He Z. Deep penetration of nanoparticulate drug delivery systems into tumors: challenges and solutions. *Curr Med Chem* 2013;20:2881–2891.
4. Pietroiusti A, Campagnolo L, Fadeel B. Interactions of engineered nanoparticles with organs protected by internal biological barriers. *Small* 2013;9:1557–1572.
5. Moros M, Mitchell SG, Grazu V, de la Fuente JM. The fate of nanocarriers as nanomedicines *in vivo*: important considerations and biological barriers to overcome. *Curr Med Chem* 2013;20:2759–2778.
6. Dong X, Mumper RJ. Nanomedicinal strategies to treat multidrug-resistant tumors: current progress. *Nanomedicine* (*Lond.*) 2010;5:597–615.
7. Peer D, Karp JM, Hong S, Farokhzad OC, Margalit R, Langer R. Nanocarriers as an emerging platform for cancer therapy. *Nat Nanotechnol* 2007;2:751–760.
8. Park JH, Lee S, Kim JH, Park K, Kim K, Kwon IC. Polymeric nanomedicine for cancer therapy. *Prog Polym Sci* 2008;33:113–137.
9. Rao JP, Geckeler KE. Polymer nanoparticles: preparation techniques and size-control parameters. *Prog Polym Sci* 2011;36:887–913.
10. Kataoka K, Harada A, Nagasaki Y. Block co-polymer micelles for drug delivery: design, characterization and biological significance. *Adv Drug Deliv Rev* 2012;64:37–48.
11. Discher DE, Eisenberg A. Polymer vesicles. *Science* 2002;297:967–973.
12. Liao JF, Wang C, Wang YJ, Luo F, Qian ZY. Recent advances in formation, properties, and applications of polymersomes. *Curr Pharm Des* 2012;18:3432–3441.
13. Chacko RT, Ventura J, Zhuang JM, Thayumanavan S. Polymer nanogels: a versatile nanoscopic drug delivery platform. *Adv Drug Deliv Rev* 2012;64:836–851.
14. Pasut G, Veronese FM. Polymer–drug conjugation, recent achievements and general strategies. *Prog Polym Sci* 2007;32:933–961.
15. Gonzalez-Toro DC, Thayumanavan S. Advances in polymer and polymeric nanostructures for protein conjugation. *Eur Polym J* 2013;49:2906–2918.
16. Gombotz WR, Pettit DK. Biodegradable polymers for protein and peptide drug-delivery. *Bioconjug Chem* 1995;6:332–351.
17. Peng L, Wu CS, You MX, Han D, Chen Y, Fu T, Ye M, Tan WH. Engineering and applications of DNA-grafted polymer materials. *Chem Sci* 2013;4:1928–1938.
18. Joglekar M, Trewyn BG. Polymer-based stimuli-responsive nanosystems for biomedical applications. *Biotechnol J* 2013;8:931–945.
19. Torchilin VP. Structure and design of polymeric surfactant-based drug delivery systems. *J Control Release* 2001;73:137–172.
20. Stuart MAC, Huck WTS, Genzer J, Muller M, Ober C, Stamm M, Sukhorukov GB, Szleifer I, Tsukruk VV, Urban M, Winnik F, Zauscher S, Luzinov I, Minko S. Emerging applications of stimuli-responsive polymer materials. *Nat Mater* 2010;9:101–113.
21. Thakur S, Tekade RK, Kesharwani P, Jain NK. The effect of polyethylene glycol spacer chain length on the tumor-targeting potential of folate-modified PPI dendrimers. *J Nano Res* 2013;15:1625–1640.

22. Pasut G, Veronese FM. State of the art in PEGylation: the great versatility achieved after forty years of research. *J Control Release* 2012;161:461–472.
23. Bhadra D, Bhadra S, Jain P, Jain NK. Pegnology: a review of PEG-ylated systems. Pharmazie 2002;57:5–29.
24. Veronese FM, Pasut G. PEGylation, successful approach to drug delivery. *Drug Discov Today* 2005;10:1451–1458.
25. Kim H, Kang YJ, Kang S, Kim KT. Monosaccharide-responsive release of insulin from polymersomes of polyboroxole block co-polymers at neutral pH. *J Am Chem Soc* 2012;134:4030–4033.
26. Lokitz BS, York AW, Stempka JE, Treat ND, Li YT, Jarrett WL, McCormick CL. Aqueous RAFT synthesis of micelle-forming amphiphilic block co-polymers containing N-Acryloylvaline. Dual mode, temperature/pH responsiveness, and “locking” of micelle structure through interpolyelectrolyte complexation. *Macromolecules* 2007;40:6473–6480.
27. Li YT, Lokitz BS, McCormick CL. RAFT synthesis of a thermally responsive ABC tri-block co-polymer incorporating N-acryloxysuccinimide for facile *in situ* formation of shell cross-linked micelles in aqueous media. *Macromolecules* 2006;39:81–89.
28. Delplace V, Tardy A, Harrisson S, Mura S, Gigmes D, Guillaneuf Y, Nicolas J. Degradable and comb-like PEG-based co-polymers by nitroxide-mediated radical ring-opening polymerization. *Biomacromolecules* 2013;14:3769–3779.
29. Wang Y, Alb AM, He JB, Grayson SM. Neutral linear amphiphilic homopolymers prepared by atom transfer radical polymerization. *Polym Chem* 2014;5:622–629.
30. Gueugnon F, Denis I, Pouliquen D, Collette F, Delatouche R, Heroguez V, Gregoire M, Bertrand P, Blanquart C. Nanoparticles produced by ring-opening metathesis polymerization using norbornenyl-poly(ethylene oxide) as a ligand-free generic platform for highly selective *in vivo* tumor targeting. *Biomacromolecules* 2013;14:2396–2402.
31. Huang J, Heise A. Stimuli responsive synthetic polypeptides derived from N-carboxyanhydride (NCA) polymerisation. *Chem Soc Rev* 2013;42:7373–7390.
32. Luo Z, Tikekar RV, Nitin N. Click chemistry approach for imaging intracellular and intratissue distribution of curcumin and its nanoscale carrier. *Bioconjug Chem* 2014;25:32–42.
33. Jabr-Milane L, van Vlerken L, Devalapally H, Shenoy D, Komareddy S, Bhavsar M, Amiji M. Multi-functional nanocarriers for targeted delivery of drugs and genes. *J Control Release* 2008;130:121–128.
34. Maeda H. The enhanced permeability and retention (EPR) effect in tumor vasculature: the key role of tumor-selective macromolecular drug targeting. *Adv Enzyme Regul* 2001;41:189–207.
35. Maeda H, Wu J, Sawa T, Matsumura Y, Hori K. Tumor vascular permeability and the EPR effect in macromolecular therapeutics: a review. *J Control Release* 2000;65:271–284.
36. Maeda H, Bharate GY, Daruwalla J. Polymeric drugs for efficient tumor-targeted drug delivery based on EPR-effect. *Eur J Pharm Biopharm* 2009;71:409–419.
37. Elsabahy M, Wooley KL. Design of polymeric nanoparticles for biomedical delivery applications. *Chem Soc Rev* 2012;41:2545–2561.
38. Kamaly N, Xiao ZY, Valencia PM, Radovic-Moreno AF, Farokhzad OC. Targeted polymeric therapeutic nanoparticles: design, development and clinical translation. *Chem Soc Rev* 2012;41:2971–3010.
39. Sanvicens N, Marco MP. Multifunctional nanoparticles — properties and prospects for their use in human medicine. *Trends Biotechnol* 2008;26:425–433.
40. Pridgen EM, Langer R, Farokhzad OC. Biodegradable, polymeric nanoparticle delivery systems for cancer therapy. *Nanomed* 2007;2:669–680.

41. Nie SM. Understanding and overcoming major barriers in cancer nanomedicine. *Nanomed* 2010;5:523–528.
42. Cheng R, Meng FH, Deng C, Klok HA, Zhong ZY. Dual and multi-stimuli responsive polymeric nanoparticles for programmed site-specific drug delivery. *Biomaterials* 2013;34:3647–3657.
43. Mura S, Nicolas J, Couvreur P. Stimuli-responsive nanocarriers for drug delivery. *Nat Mater* 2013;12:991–1003.
44. Ganta S, Devalapally H, Shahiwala A, Amiji M. A review of stimuli-responsive nanocarriers for drug and gene delivery. *J Control Release* 2008;126:187–204.
45. Alvarez-Lorenzo C, Concheiro A. Bioinspired drug delivery systems. *Curr Opin Biotechnol* 2013;24:1167–1173.
46. Ge ZS, Liu SY. Functional block co-polymer assemblies responsive to tumor and intracellular microenvironments for site-specific drug delivery and enhanced imaging performance. *Chem Soc Rev* 2013;42:7289–7325.
47. He X, Li J, An S, Jiang C. pH-sensitive drug-delivery systems for tumor targeting. *Ther Deliv* 2013;4:1499–1510.
48. Gil ES, Hudson SM. Stimuli-reponsive polymers and their bioconjugates. *Prog Polym Sci* 2004;29:1173–1222.
49. Cheng R, Feng F, Meng FH, Deng C, Feijen J, Zhong ZY. Glutathione-responsive nanovehicles as a promising platform for targeted intracellular drug and gene delivery. *J Control Release* 2011;152:2–12.
50. Sun HL, Meng FH, Cheng R, Deng C, Zhong ZY. Reduction-sensitive degradable micellar nanoparticles as smart and intuitive delivery systems for cancer chemotherapy. *Expert Opin Drug Deliv* 2013;10:1109–1122.
51. Wei H, Zhuo RX, Zhang XZ. Design and development of polymeric micelles with cleavable links for intracellular drug delivery. *Prog Polym Sci* 2013;38:503–535.
52. Schafer FQ, Buettner GR. Redox environment of the cell as viewed through the redox state of the glutathione disulfide/glutathione couple. *Free Radic Biol Med* 2001;30:1191–1212.
53. Meng FH, Hennink WE, Zhong Z. Reduction-sensitive polymers and bioconjugates for biomedical applications. *Biomaterials* 2009;30:2180–2198.
54. Saito G, Swanson JA, Lee KD. Drug delivery strategy utilizing conjugation via reversible disulfide linkages: role and site of cellular reducing activities. *Adv Drug Deliv Rev* 2003;55:199–215.
55. Li CG, Zhao S, Li JJ, Yin YJ. Polymeric biomaterials containing thiol/disulfide bonds. *Prog Chem* 2013;25:122–134.
56. Bauhuber S, Hozsa C, Breunig M, Gopferich A. Delivery of nucleic acids *via* disulfide-based carrier systems. *Adv Mater* 2009;21:3286–3306.
57. Oh JK, Tang CB, Gao HF, Tsarevsky NV, Matyjaszewski K. Inverse miniemulsion ATRP: a new method for synthesis and functionalization of well-defined water-soluble/cross-linked polymeric particles. *J Am Chem Soc* 2006;128:5578–5584.
58. Singh S, Topuz F, Hahn K, Albrecht K, Groll J. Embedding of active proteins and living cells in redox-sensitive hydrogels and nanogels through enzymatic cross linking. *Angew Chem Int Ed* 2013;52:3000–3003.
59. Li YT, Smith AE, Lokitz BS, McCormick CL. *In situ* formation of gold-"decorated" vesicles from a RAFT-synthesized, thermally responsive block co-polymer. *Macromolecules* 2007;40:8524–8526.

60. Cai XJ, Dong CY, Dong HQ, Wang GM, Pauletti GM, Pan XJ, Wen HY, Mehl I, Li YY, Shi DL. Effective gene delivery using stimulus-responsive catiomer designed with redox-sensitive disulfide and acid-labile imine linkers. *Biomacromolecules* 2012;13:1024–1034.
61. Wen HY, Dong CY, Dong HQ, Shen AJ, Xia WJ, Cai XJ, Song YY, Li XQ, Li YY, Shi DL. Engineered redox-responsive PEG detachment mechanism in PEGylated nano-graphene oxide for intracellular drug delivery. *Small* 2012;8:760–769.
62. Breunig M, Lungwitz U, Liebl R, Goepferich A. Breaking up the correlation between efficacy and toxicity for nonviral gene delivery. *Proceedings of the National Academy of Sciences of the United States of America* 2007;104:14454–14459.
63. Tsarevsky NV, Matyjaszewski K. Combining atom transfer radical polymerization and disulfide/thiol redox chemistry: a route to well-defined (bio)degradable polymeric materials. *Macromolecules* 2005;38:3087–3092.
64. Xu XW, Smith AE, McCormick CL. Facile 'one-pot' preparation of reversible, disulfide-containing shell cross-linked micelles from a RAFT-synthesized, pH-responsive triblock co-polymer in water at room temperature. *Aust J Chem* 2009;62:1520–1527.
65. Wang YX, Chen P, Shen JC. The development and characterization of a glutathione-sensitive cross-linked polyethylenimine gene vector. *Biomaterials* 2006;27:5292–5298.
66. Koo AN, Min KH, Lee HJ, Lee SU, Kim K, Kwon IC, Cho SH, Jeong SY, Lee SC. Tumor accumulation and antitumor efficacy of docetaxel-loaded core-shell-corona micelles with shell-specific redox-responsive cross-links. *Biomaterials* 2012;33:1489–1499.
67. Gosselin MA, Guo WJ, Lee RJ. Efficient gene transfer using reversibly cross-linked low molecular weight polyethylenimine. *Bioconjug Chem* 2001;12:989–994.
68. Zugates GT, Anderson DG, Little SR, Lawhorn IEB, Langer R. Synthesis of poly(beta-amino esters)s with thiol-reactive side chains for DNA delivery. *J Am Chem Soc* 2006;128:12726–12734.
69. Lin C, Zhong ZY, Lok MC, Jiang XL, Hennink WE, Feijen J, Engbersen JFJ. Novel bioreducible poly(amido amine)s for highly efficient gene delivery. *Bioconjug Chem* 2007;18:138–145.
70. Kakizawa Y, Harada A, Kataoka K. Environment-sensitive stabilization of core-shell structured polyion complex micelle by reversible cross linking of the core through disulfide bond. *J Am Chem Soc* 1999;121:11247–11248.
71. Vachutinsky Y, Oba M, Miyata K, Hiki S, Kano MR, Nishiyama N, Koyama H, Miyazono K, Kataoka K. Antiangiogenic gene therapy of experimental pancreatic tumor by sFlt-1 plasmid DNA carried by RGD-modified crosslinked polyplex micelles. *J Control Release* 2011;149:51–57.
72. Koo AN, Lee HJ, Kim SE, Chang JH, Park C, Kim C, Park JH, Lee SC. Disulfide-cross-linked PEG-poly(amino acid)s co-polymer micelles for glutathione-mediated intracellular drug delivery. *Chem Commun (Camb)* 2008:6570–6572.
73. Sun HL, Guo BN, Cheng R, Meng FH, Liu HY, Zhong ZY. Biodegradable micelles with sheddable poly (ethylene glycol) shells for triggered intracellular release of doxorubicin. *Biomaterials* 2009;30:6358–6366.
74. Tang LY, Wang YC, Li Y, Du JZ, Wang J. Shell-detachable micelles based on disulfide-linked block co-polymer as potential carrier for intracellular drug delivery. *Bioconjug Chem* 2009;20:1095–1099.
75. Xu YM, Meng FH, Cheng R, Zhong ZY. Reduction-sensitive reversibly crosslinked biodegradable micelles for triggered release of doxorubicin. *Macromol Biosci* 2009;9: 1254–1261.

76. Wen HY, Dong HQ, Xie WJ, Li YY, Wang K, Pauletti GM, Shi DL. Rapidly disassembling nanomicelles with disulfide-linked PEG shells for glutathione-mediated intracellular drug delivery. *Chem Commun* 2011;47:3550–3552.
77. Ren TB, Xia WJ, Dong HQ, Li YY. Sheddable micelles based on disulfide-linked hybrid PEG-polypeptide co-polymer for intracellular drug delivery. *Polymer* 2011;52:3580–3586.
78. Cerritelli S, Velluto D, Hubbell JA. PEG–SS–PPS: reduction-sensitive disulfide block copolymer vesicles for intracellular drug delivery. *Biomacromolecules* 2007;8:1966–1972.
79. Ren TB, Wu W, Jia MH, Dong HQ, Li YY, Ou ZL. Reduction-cleavable polymeric vesicles with efficient glutathione-mediated drug release behavior for reversing drug resistance. *Acs Appl Mater Interfaces* 2013;5:10721–10730.
80. van Hell AJ, Crommelin DJA, Hennink WE, Mastrobattista E. Stabilization of peptide vesicles by introducing inter-peptide disulfide bonds. *Pharm Res* 2009;26:2186–2193.
81. Li XQ, Wen HY, Dong HQ, Xue WM, Pauletti GM, Cai XJ, Xia WJ, Shi DL, Li YY. Self-assembling nanomicelles of a novel camptothecin prodrug engineered with a redox-responsive release mechanism. *Chem Commun* 2011;47:8647–8649.
82. van der Vlies AJ, Hasegawa U, Hubbell JA. Reduction-sensitive tioguanine prodrug micelles. *Mol Pharm* 2012;9:2812–2818.
83. Kurtoglu YE, Navath RS, Wang B, Kannan S, Romero R, Kannan RM. Poly (amidoamine) dendrimer–drug conjugates with disulfide linkages for intracellular drug delivery. *Biomaterials* 2009;30:2112–2121.
84. Wang X, Cai X, Hu J, Shao N, Wang F, Zhang Q, Xiao J, Cheng Y. Glutathione-triggered “off-on” release of anticancer drugs from dendrimer-encapsulated gold nanoparticles. *J Am Chem Soc* 2013;135:9805–9810.
85. Liu JY, Huang W, Pang Y, Huang P, Zhu XY, Zhou YF, Yan DY. Molecular self-assembly of a homopolymer: an alternative to fabricate drug-delivery platforms for cancer therapy. *Angew Chem Int Ed* 2011;50:9162–9166.
86. Liu JY, Pang Y, Huang W, Zhu ZY, Zhu XY, Zhou YF, Yan DY. Redox-responsive polyphosphate nanosized assemblies: a smart drug delivery platform for cancer therapy. *Biomacromolecules* 2011;12:2407–2415.
87. You YZ, Yu ZQ, Cui MM, Hong CY. Preparation of photoluminescent nanorings with controllable bioreducibility and stimuli-responsiveness. *Angew Chem Int Ed* 2010;49: 1099–1102.
88. Cui YN, Dong HQ, Cai XJ, Wang DP, Li YY. Mesoporous silica nanoparticles capped with disulfide-linked PEG gatekeepers for glutathione-mediated controlled release. *Acs Appl Mater Interfaces* 2012;4:3177–3183.
89. Romberg B, Hennink WE, Storm G. Sheddable coatings for long-circulating nanoparticles. *Pharm Res* 2008;25:55–71.
90. Wang YC, Wang F, Sun TM, Wang J. Redox-responsive nanoparticles from the single disulfide bond-bridged block co-polymer as drug carriers for overcoming multidrug resistance in cancer cells. *Bioconjug Chem* 2011;22:1939–1945.
91. Li MH, Keller P. Stimuli-responsive polymer vesicles. *Soft Matter* 2009;5:927–937.
92. Discher BM, Won YY, Ege DS, Lee JCM, Bates FS, Discher DE, Hammer DA. Polymersomes: tough vesicles made from diblock co-polymers. *Science* 1999;284:1143–1146.
93. Park KM, Lee DW, Sarkar B, Jung H, Kim J, Ko YH, Lee KE, Jeon H, Kim K. Reduction-sensitive, robust vesicles with a non-covalently modifiable surface as a multifunctional drug-delivery platform. *Small* 2010;6:1430–1441.

94. Kim K, Selvapalam N, Ko YH, Park KM, Kim D, Kim J. Functionalized cucurbiturils and their applications. *Chem Soc Rev* 2007;36:267–279.
95. Bae Y, Fukushima S, Harada A, Kataoka K. Design of environment-sensitive supramolecular assemblies for intracellular drug delivery: polymeric micelles that are responsive to intracellular pH change. *Angew Chem Int Ed Engl* 2003;42:4640–4643.
96. Santra S, Kaittanis C, Santiesteban OJ, Perez JM. Cell-specific, activatable, and theranostic prodrug for dual-targeted cancer imaging and therapy. *J Am Chem Soc* 2011;133: 16680–16688.
97. Henne WA, Doorneweerd DD, Hilgenbrink AR, Kularatne SA, Low PS. Synthesis and activity of a folate peptide camptothecin prodrug. *Bioorg Med Chem Lett* 2006;16: 5350–5355.
98. Low PS, Henne WA, Doorneweerd DD. Discovery and development of folic-acid-based receptor targeting for imaging and therapy of cancer and inflammatory diseases. *Acc Chem Res* 2008;41:120–129.
99. Leamon CP, Reddy JA, Vlahov IR, Vetzel M, Parker N, Nicoson JS, Xu LC, Westrick E. Synthesis and biological evaluation of EC72: a new folate-targeted chemotherapeutic. *Bioconjug Chem* 2005;16:803–811.
100. Bharali DJ, Khalil M, Gurbuz M, Simone TM, Mousa SA. Nanoparticles and cancer therapy: a concise review with emphasis on dendrimers. *Int J Nanomedicine* 2009;4:1–7.
101. Wang XS, Guerin G, Wang H, Wang YS, Manners I, Winnik MA. Cylindrical block co-polymer micelles and co-micelles of controlled length and architecture. *Science* 2007;317:644–647.
102. Zhou YF, Yan DY. Supramolecular self-assembly of amphiphilic hyperbranched polymers at all scales and dimensions: progress, characteristics and perspectives. *Chem Commun* 2009:1172–1188.
103. Yin Q, Shen JN, Zhang ZW, Yu HJ, Li YP. Reversal of multidrug resistance by stimuli-responsive drug delivery systems for therapy of tumor. *Adv Drug Deliv Rev* 2013;65: 1699–1715.
104. Gao GH, Li Y, Lee DS. Environmental pH-sensitive polymeric micelles for cancer diagnosis and targeted therapy. *J Control Release* 2013;169:180–184.
105. Wu LL, Zou Y, Deng C, Cheng R, Meng FH, Zhong ZY. Intracellular release of doxorubicin from core-crosslinked polypeptide micelles triggered by both pH and reduction conditions. *Biomaterials* 2013;34:5262–5272.
106. Zhao JQ, Liu JJ, Xu SX, Zhou JH, Han SC, Deng LD, Zhang JH, Liu JF, Meng AM, Dong AJ. Graft co-polymer nanoparticles with pH and reduction dual-induced disassemblable property for enhanced intracellular curcumin release. *Acs Appl Mater Interfaces* 2013;5:13216–13226.
107. Gillies ER, Jonsson TB, Frechet JMJ. Stimuli-responsive supramolecular assemblies of linear-dendritic co-polymers. *J Am Chem Soc* 2004;126:11936–11943.
108. Gillies ER, Frechet JMJ. A new approach towards acid sensitive co-polymer micelles for drug delivery. *Chem Commun* 2003:1640–1641.
109. Griset AP, Walpole J, Liu R, Gaffey A, Colson YL, Grinstaff MW. Expansile nanoparticles: synthesis, characterization, and *in vivo* efficacy of an acid-responsive polymeric drug delivery system. *J Am Chem Soc* 2009;131:2469–2471.
110. Dai J, Lin SD, Cheng D, Zou SY, Shuai XT. Interlayer-crosslinked micelle with partially hydrated core showing reduction and pH dual sensitivity for pinpointed intracellular drug release. *Angew Chem Int Ed* 2011;50:9404–9408.

111. Issels RD. Hyperthermia adds to chemotherapy. *Eur J Cancer* 2008;44:2546–2554.
112. Zhang XZ, Wu DQ, Chu CC. Synthesis, characterization and controlled drug release of thermosensitive IPN–PNIPAAm hydrogels. *Biomaterials* 2004;25:3793–3805.
113. Wei H, Cheng SX, Zhang XZ, Zhuo RX. Thermo-sensitive polymeric micelles based on poly (N-isopropylacrylamide) as drug carriers. *Prog Polym Sci* 2009;34:893–910.
114. Rzaev ZMO, Dincer S, Piskin E. Functional co-polymers of N-isopropylacrylamide for bioengineering applications. *Prog Polym Sci* 2007;32:534–595.
115. Xu HF, Meng FH, Zhong ZY. Reversibly crosslinked temperature-responsive nano-sized polymersomes: synthesis and triggered drug release. *J Mater Chem* 2009;19:4183–4190.
116. Cheng R, Meng FH, Ma SB, Xu HF, Liu HY, Jing XB, Zhong ZY. Reduction and temperature dual-responsive crosslinked polymersomes for targeted intracellular protein delivery. *J Mater Chem* 2011;21:19013–19020.
117. Huang XG, Jiang XL, Yang QZ, Chu YF, Zhang GY, Yang Y, Zhuo RX. Triple-stimuli (pH/thermo/reduction) sensitive co-polymers for intracellular drug delivery. *J Mater Chem B* 2013;1:1860–1868.
118. Qiao ZY, Zhang R, Du FS, Liang DH, Li ZC. Multi-responsive nanogels containing motifs of ortho ester, oligo (ethylene glycol) and disulfide linkage as carriers of hydrophobic anti-cancer drugs. *J Control Release* 2011;152:57–66.

Chapter 3

# Design of Biocompatible PEIs as Artificial Gene Delivery Vectors

*Ting Shi, Chao Lin and Peng Zhao*

*East Hospital, The Institute for Biomedical Engineering and Nano Science, Tongji University School of Medicine, Shanghai 200092, P. R. China*

Polyethylenimine (PEI) is one of the most widely used polymeric gene delivery vectors due to its strong gene binding ability and unique buffering capacity for endosomal escape. However, high transfection ability of PEI is associated with low cytotoxicity. In the past two decades, synthetic methods have been proposed to construct PEI derivatives with improved biocompatibility for safe and potent gene delivery. One method is an integration of PEI with biodegradable polymers such as polyethylene glycol. Another method is the preparation of degradable PEIs, which have chemically hydrolytic linkers. In this chapter, we outline these biocompatible PEIs and their design conception as well as discuss the effects of their physiochemical and gene delivery properties on transfection efficacy and toxicity. It is anticipated that rapid development of biocompatible PEIs will have a great impact on PEI-based gene therapy in the future.

## 1. Introduction

Gene therapy has emerged as a novel method to treat a broad spectrum of human diseases caused by gene defects.[1] For successful gene therapy, gene delivery vectors play an important role in escorting exogenous genes into targeted somatic cells. Gene delivery vectors are normally classified into two types: viral and non-viral vectors. Viral vectors have originated from natural viruses with eliminated pathogenicity. Although a lot of viral vectors such as retrovirus, lentivirus, and adenovirus are extremely efficient for gene delivery and studied in pre-clinical trials, several inherent drawbacks including random integration into host genome, immunogenicity, limited gene-carrying capacity, and high cost seriously hamper further clinical application of viral vectors.[2] Moreover, an accident in clinical gene therapy trials, the death of volunteer (Jesse Gelsinger) accelerates advancement of safe and efficient gene delivery vectors.[3]

In this context, non-viral gene vectors such as cationic lipids and polymers have received more and more attention in the past two decades.[4,5] Among different non-viral vectors, polyethylenimine (PEI) is one of the most efficient polymeric gene delivery vectors partially due to its strong gene binding ability and unique buffer capacity for endosomal escape ("proton sponge" effect).[6] Further studies indicate that the transfection efficiency and cytotoxicity of PEI are influenced by different attributes such as molecular structure and topological structure. For example, high molecular-weight PEI (HMW-PEI) possesses higher transfection efficiency, but higher cytotoxicity as compared to low molecular-weight PEI (LMW-PEI). Besides, linear HMW-PEIs are normally less toxic as compared to branched HMW-PEIs. The mechanism underlying the cytotoxicity of HMW-PEIs is elucidated in terms of two-stage cytotoxicity.[7] The first stage involves destabilization of cellular membranes by free PEI, which is dissociated from complexes, leading to irreversible necrosis-related cytotoxicity. The second stage includes the interaction of PEI-based polyplexes with negatively-charged mitochondrial membranes, causing cellular apoptosis. The cytotoxicity occurring in this stage can be minimized if cationic polymers are degraded intracellularly as a result of lowered charge density of LMW polycations.

Over past two decades, much effort has been made in the development of biocompatible PEIs as safe and potent gene delivery vectors. Two major methods have been proposed. One method involves the integration of LMW-PEIs with biocompatible or biodegradable polymers such as polyethylene glycol (PEG) and polylactides. Another approach is the design of biodegradable HMW-PEIs which have chemically hydrolytic linkers. However, to the best of our knowledge, these PEI derivatives are not summarized yet. In this chapter, we thus thoroughly review design and preparation of biocompatible or degradable PEIs and discuss the effects of their physiochemical and gene delivery properties on transfection efficacy and cytotoxicity as well as those challenging works which should be addressed in the future. It can be anticipated that further development of biodegradable PEIs will have great impact on gene therapy in the near future.

## 2. Biocompatible PEIs as Gene Delivery Vectors

A number of synthetic polymers have been found to have low cytotoxicity and good biocompatibility. As such, it is hypothesized that an incorporation of these polymers with LMW-PEIs may afford PEI derivatives which have an improved biocompatibility. Based on this concept, biocompatible PEIs have been designed and investigated for non-viral gene delivery. These PEIs are synthesized by either coupling of LMW-PEIs with biocompatible polymers by carbodiimide chemistry. Herein, we outline biocompatible PEIs and discuss their gene delivery properties, transfection ability and cytotoxicity

## 2.1. *Pluronics-based PEIs*

Pluronic triblock copolymers are a class of amphiphilic copolymers comprising of one hydrophilic ethylene oxide (EO) block and two hydrophobic propylene oxide (PO) blocks which are arranged in an EOx–POy–EOx style. Due to their good biocompatibility, pluoronics were used as structural elements to integrate with polycations, affording polycation-based gene delivery systems.[8,9] Kabanov and other researchers have found that pluronics have unique ability to incorporate with cell membranes as a result of the presence of hydrophobic PO blocks,[10–12] thereby inducing enhanced cell interaction and DNA transport. This finding thus inspires people to design pluronics-based PEIs for enhanced gene uptake as well as low cytotoxicity. As a typical example, Fan *et al.* synthesized a group of PEI derivatives containing different types of pluronics, i.e., P123, P105, F68, L61, being denoted as pluronic-g-PEI (**Figure 1**).[13] They showed that the polymers had the ability to bind DNA to form complexes with the polymer/DNA mass ratio of 0.8/1. Besides, the pluronics-based PEIs degraded slowly at pH 7.4 and 37°C within 60 hours and the degradation rate depended on the hydrophilicity of pluronics. It was also found that pluronic-g-PEI displayed higher cell viability

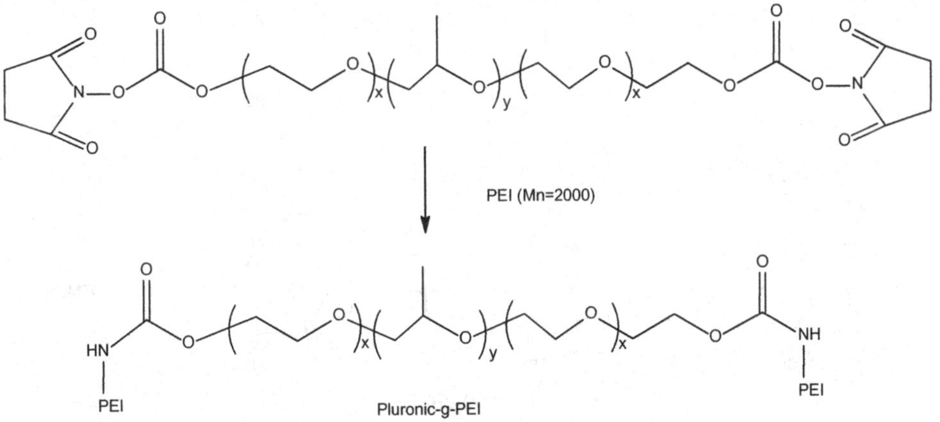

**Figure 1.** Synthesis of pluronic-based PEIs with LMW-PEI ($M_n$ = 2000).

at varying concentrations from 4 to 32 $\mu$g/ml against HeLa cell lines as compared to 25 kDa-PEI. The highest level of gene expression was obtained for P105-based PEI in HeLa cell lines and comparable with 25 kDa-PEI, being one of the most efficient polymeric gene delivery vectors.

## 2.2. *Chitosan-based PEIs*

Chitosan is one of the most widely used natural cationic polysaccharides. It has good biocompatibility and fairly low cytotoxicity. However, chitosan is poorly water-soluble at physiological pH and thus limits its bio-application. Since PEI is well soluble in water and thus chemical modification of chitosan with PEI can favorably improve solubility of chitosan and also lead to new biocompatible PEIs.

Liu *et al.* prepared chitosan-based PEIs (denoted as OTMCS–PEI) comprising amphiphilic N-octyl-N-quaternary chitosan and 2 kDa LMW-PEI (**Figure 2**).[14] OTMCS–PEI completely degraded in 0.1 M Phosphate buffered

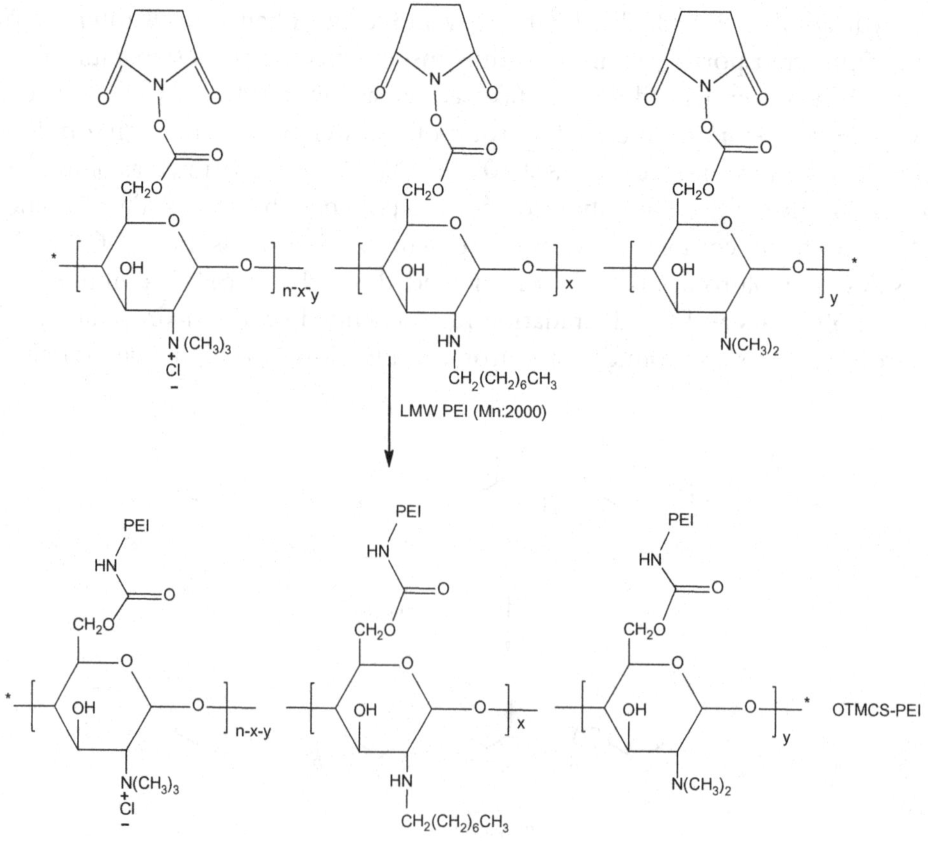

**Figure 2.** Synthesis of PEI-grafted chitosan.

saline (PBS) after 60 hours and the degradation rate was in line with a zero-order model. According to this model, the half-life of OTMCS–PEI was about 30 hours. OTMCS–PEI could condense DNA at the weight ratio of 0.6, with particle sizes of around 150–200 nm and zeta potentials from +10 to +30 mV. Cytotoxicity assay revealed that more than 60% of the cells maintain survival at polymer concentration from 4 to 48 $\mu$g/mL as compared to 25 kDa-PEI. OTMCS–PEI induced 320-fold higher transfection efficiency than 2 kDa-PEI and 10-fold higher than 25 kDa-PEI against HeLa cell lines. Moreover, the transfection efficiency of OTMCS–PEI was 2–3 times higher than 25 kDa-PEI in liver.

## 2.3. *PEG-based PEIs*

PEG is one of the most used biocompatible polymers. As such, chemical modification of PEI with PEG may generate biocompatible PEIs for safe gene delivery.

Namgung *et al.* synthesized novel star-shaped PEIs by chemically coupling 2.5 kDa-PEI to multiarm 10 kDa-PEG methylacrylar (mAPEG) at two composition ratios, yielding MAPEG–$LPEI_6$ and MAPEG–$LPEI_3$, respectively (**Figure 3**).[15] The two polymers could effectively condense DNA into polyplex (<200 nm) with about +3 mV of zeta potential at mass ratio 2/1 and 1/1 for MAPEG–$LPEI_3$ and MAPEG–$LPEI_6$, respectively. These MAPEG–LPEI polymers could protect DNA from enzymatic degradation within 24 hours. *In vitro* gene transfection showed that MAPEG–$LPEI_6$ caused 10-fold higher transfection efficiency against HeLa cells as compared to 25 kDa-LPEI. In other cell lines, NIH 3T3 and PC-3, MAPEG–$LPEI_6$ showed comparable transfection efficiency to 25 kDa-LPEI (human prostate cells), implying that transfection ability of MAPEG–LPEIs is cell-dependent. However, MAPEG–$LPEI_6$ was more efficient for gene delivery than MAPEG–$LPEI_3$ against the cells. Polyplexes of MAPEG–$LPEI_3$ displayed better cyto-compatibility (about 80%) in HeLa cells at the mass ratios in the range of 5/1–30/1 as compared to those of MAPEG–$LPEI_6$ polyplexes and 25 kDa-LPEI.

## 2.4. *Dextran-based PEIs*

Dextran is a hydrophilic biocompatible polymer. It has a lot of hydroxygen side groups amenable to different chemical modification. This provides possibility to chemically modify dextran with PEI, yielding dextran-based PEIs for non-viral gene delivery. Chu *et al.* modified dextran with epichlorohydrin (ECH) and the resulting oxy-modified dextran is further grafted with 0.8 kDa

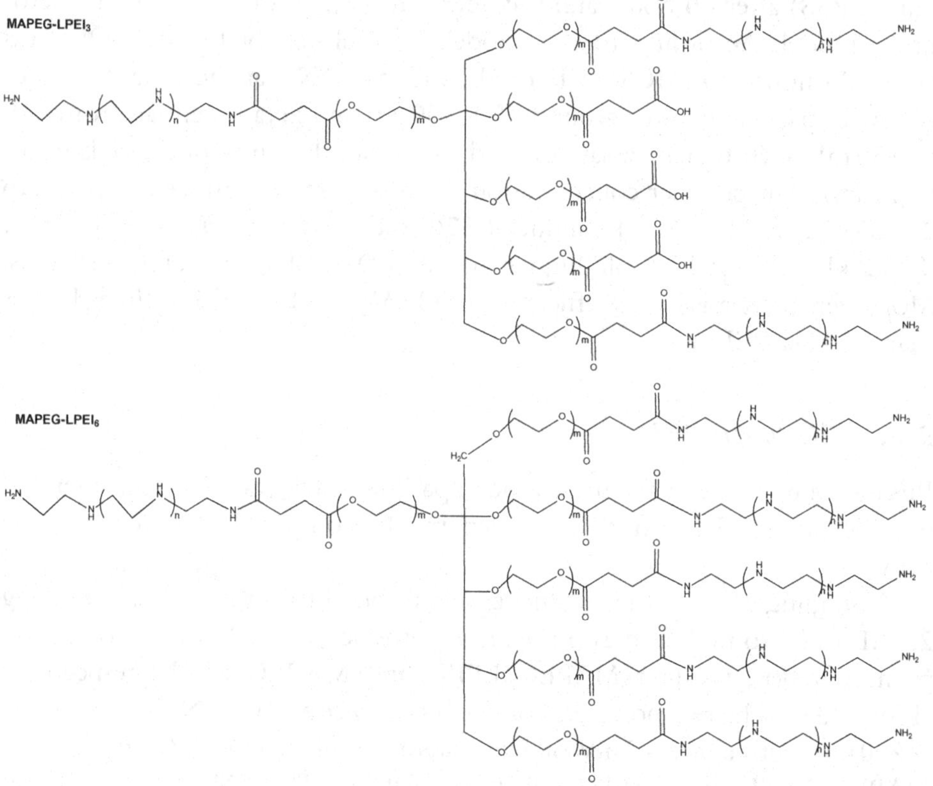

MAPEG-LPEI$_3$

MAPEG-LPEI$_6$

**Figure 3.** Chemical structures of six-arm PEG-modified PEIs.

or 1.8 kDa-PEI (**Figure 4**).[16] These dextran-grafted-PEIs (dextran-g-PEIs) showed efficient DNA binding ability at and above the mass ratio of 3/1. Besides, dextran-g-PEI1800 was capable of strongly condensing DNA at and above the mass ratio of 1/1. But, a higher mass ratio (e.g., 3/1) was needed for DNA binding with dextran-g-PEI800. The hydrodynamic size of dextran-g-PEI1800-based complexes was in the range of 100–200 nm at the weight ratio of 3/1, whereas the size of dextran-g-PEI800-based polyplexes was below 200 nm at and above the mass ratio of 15/1. Moreover, the complexes showed moderate surface charges ranging from +10 to +20 mV. These favorable gene delivery properties offered these dextran-based PEIs with minimized red blood cell aggregation and hemolytic activity as well as low cytotoxicity towards HeLa, 293T (human embryon kidney) and HepG2 (human liver hepatocellular) cell lines. *In vitro* transfection experiments manifested that dextran-g-PEI1800 was efficient to deliver green fluorescent protein (GFP)-encoding DNA into these cells, yielding 20–60% GFP-positive cells in the

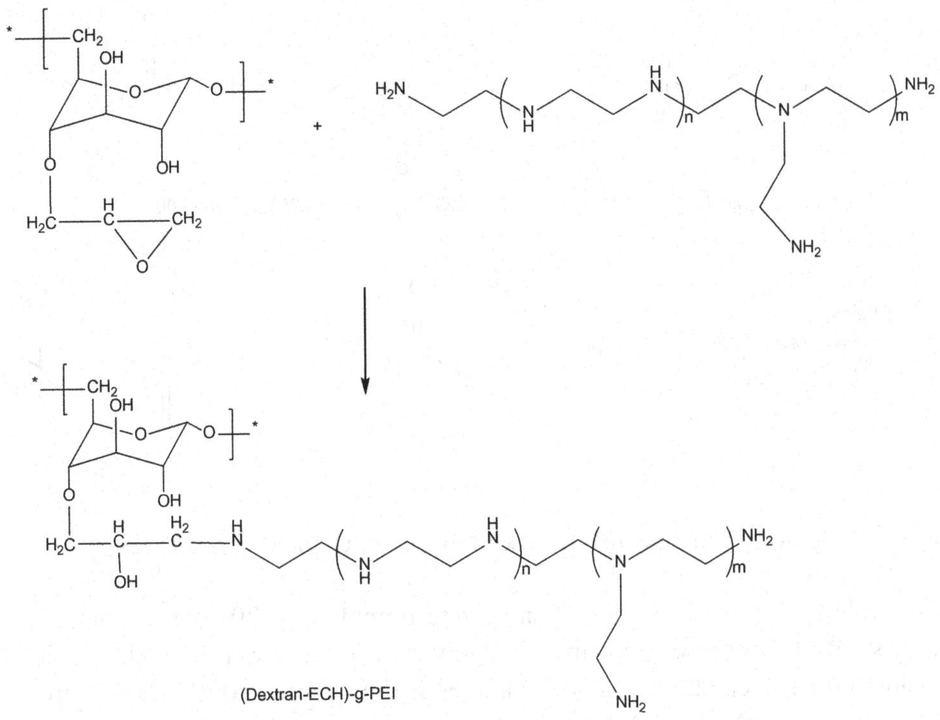

(Dextran-ECH)-g-PEI

**Figure 4.** Synthesis of dextran-based PEIs.

serum. The complexes of dextran-g-1800 kDa-PEIs also exerted detectable transfection efficacy in the tumor xenografted in a Balb/c nude mouse model.

## 3. Hydrolytically Degradable PEIs as Gene Delivery Vectors

### 3.1. *Ester-based PEIs*

The incorporation of ester linker into LMW-PEIs is a routine method to prepare ester-based PEIs. These hydrolytically degradable PEIs are normally obtained by Michael-type addition of diacrylate to LMW-PEIs.[17] By this approach, a number of ester-based PEIs have been developed as non-viral gene delivery vectors.

Park *et al.* synthesized a group of hydrolytic PEG-alt-PEIs by Michael-type addition of PEG diacrylate (Mn: 258, 575 and 700) to linear 423 Da-PEI (**Figure 5**).[18,19] These copolymers degraded rapidly at 37°C in 0.1 M PBS buffer at pH 7.4 and the degradation completed after 72 hours. PEI-alt-PEG (Mn: 575) exhibited more rapid degradation with a half-life of 8 hours as compared to 25 hours for PEI-alt-PEG (Mn: 700). The copolymer/DNA complexes were formed at the N/P (nitroge to phosphate mol. ratio) ratio of 12/1 with

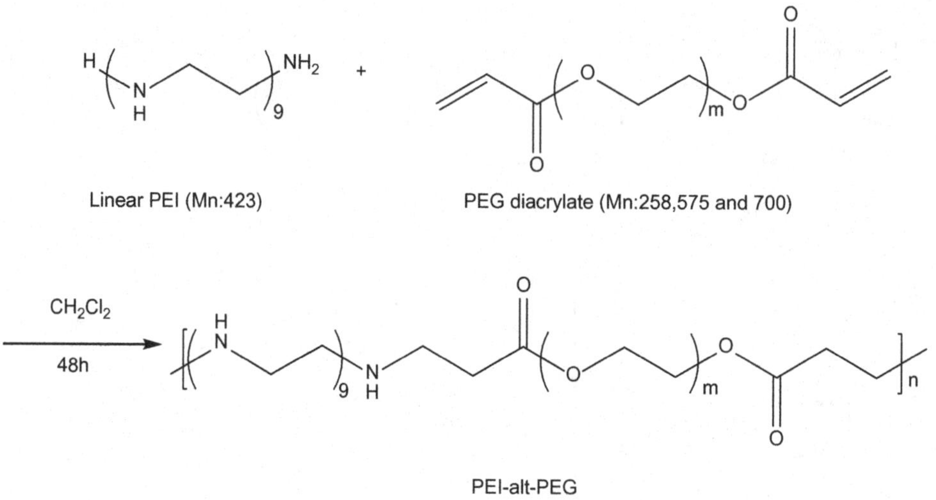

Linear PEI (Mn:423)

PEG diacrylate (Mn:258,575 and 700)

$CH_2Cl_2$

48h

PEI-alt-PEG

**Figure 5.** Synthesis of ester-based PEI with PEG diacrylate crosslinker.

nanoscale sizes (~150 nm) and positive zeta potential (~+20 mV). Cytotoxicity assay showed that these copolymers had low cytotoxicity (over 80% cell viability) against HeLa, HepG2 and MG63 cell lines at the polymer concentrations from 1 to 30 μg/ml. Against HeLa cells, the copolymers displayed inferior transfection ability to 25 kDa-PEI. However, PEI–alt–PEG (Mn: 258) induced higher efficiency in HepG2 than 25 kDa-PEI when the N/P ratio of 27/1 was applied. Interestingly, in MG63 (human osteoblast-like) cells, the transfection efficiency of PEI–alt–PEG (Mn: 258) was much higher than that of 25 kDa-PEI at the N/P ratio of 27. The results of this study suggest that the transfection efficiency of the copolymers depended on cell type, PEG molecular weight and N/P ratio.

Instead of PEG diacrylate, 1,3-butanediol diacrylate was used by Forrest *et al.* to generate hydrolytic PEIs with branched 800 Da-PEI *via* Michael-type addition (**Figure 6**).[20] The 1,3-butanediol diacrylate can react with primary or secondary amines of PEI800 at different mole ratios, resulting in the formation of branched PEIs with varying molecular weights in the range of 14–30 kDa. These polymers efficiently condensed DNA at the mass ratio of 3/1 into nano-complexes with the particle sizes of about 40–60 nm. These polymers were rapidly degradable *via* hydrolysis of ester at physiological conditions with half-life of 4 hours. *In vitro* transfection experiment showed that the ester-crosslinked PEI caused 2–16-fold higher level of gene expression than 25 kDa-PEI against breast carcinoma cells, MDA-MB-231, C2C12 (mouse myoblast cell) and myoblasts, meanwhile maintaining high cell viability (80–100%). These results indicate that 1,3-butanediol diacrylate-crosslinked PEIs have a high potential for safe and efficient gene delivery.

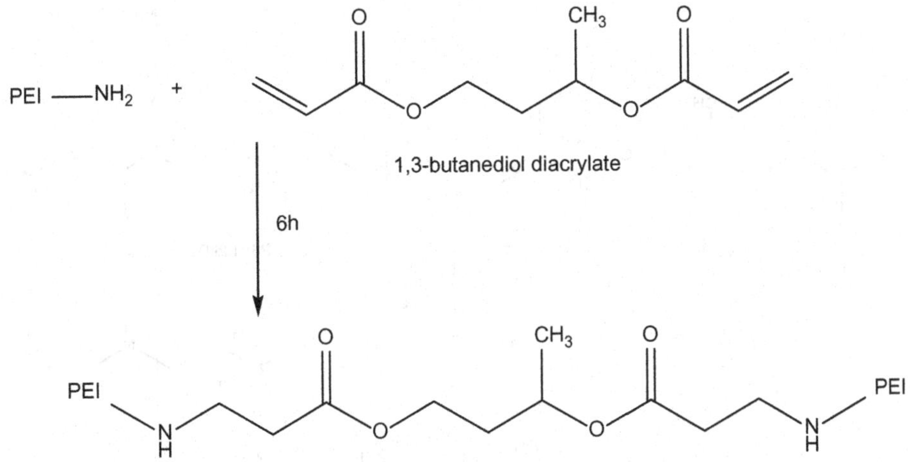

**Figure 6.** Synthesis of ester-based PEIs with 1,3-butanediol diacrylate crosslinker.

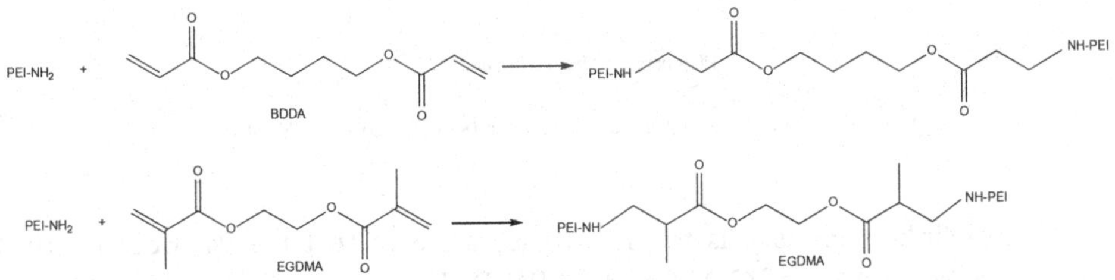

**Figure 7.** Synthesis of ester-based PEIs with BDDA and EGDMA.

Another work from Dong *et al.* also showed the preparation of hydrolytic branched PEIs by Michael-type addition of 1,4-butanediol diacrylate (BDDA) or ethyleneglycol dimethacrylate (EGDMA) to branched 2 kDa-PEI (**Figure** 7).[21] Compared to 25 kDa branched PEI, BDDA-crosslinked PEIs could form smaller particles with DNA (70–150 nm) and induce efficient expression of reporter gene as compared to 25 kDa-PEI, that is, 19-fold in B16F10 cells, 17-fold in 293T cells, 2.3-fold in NIH 3T3 (mouse fibroblast) cells. Importantly, they were essentially non-toxic at their optimal conditions for *in vitro* gene transfection. *In vivo* gene transfection test in a mouse model further showed that the muscle site transfected with BDDA-crosslinked PEI maintained normal structure and detectable gene expression.

Arote *et al.* crosslinked 1.2 kDa-PEI with glycerol dimethacrylate(GDM) as a crosslinker *via* Michael-type addition reaction, yielding PEI-based poly (amino ester)s (denoted as PAEs) (**Figure 8**).[22] These PAEs condensed DNA into nanosized particles at physiological pH with the size of below 150 nm and zeta potential in the range of +30 to +55 mV. Besides, the half-life of

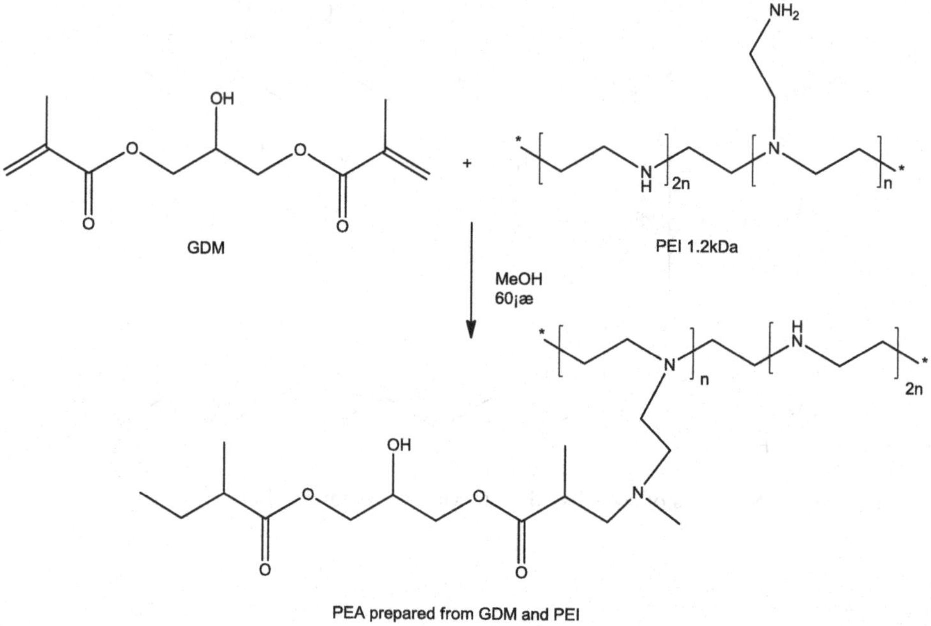

**Figure 8.** Synthesis of ester-based PEIs with GDM crosslinker.

GDM-crosslinked PAEs was found in the range of 9–10 days. Because the PAEs were composed of GDM and LMW-PEI, they displayed relatively low cytotoxicity against HeLa, HepG2 and 293T cell lines, with about 70% cell viability at a concentration of 40 $\mu$g/ml. The PAEs induced higher transfection efficiency compared to 25 kDa-PEI *in vitro*. Especially, GDM-crosslinked PEIs at a GDM/PEI mole ratio of 1/4 showed the best transfection ability in HepG2 cells, that is, ~30 times higher level of gene expression compared to that of 25 kDa-PEI at an N/P ratio of 30.

It appeared that chemical structure of crosslinkers in ester-containing PEIs may affect resulting transfection efficiency. For example, Ahn *et al.* showed that, although PEG–PEI copolymers were degradable by ester hydrolysis and revealed 80% cell viability, their transfection efficiencies were much lower as compared to 25 kDa-PEI.[23] Moreover, Park *et al.* showed that the MW of PEG has an effect on transfection efficiency. The efficiency decreased markedly with increasing PEG MWs from 258 to 700 kDa.[18] Besides, Yu *et al.* applied PEG dimethacrylate (PEGDMA), a more hydrophobic crosslinker agent than PEG diacrylate, for the preparation of PEG-cr-PEIs containing LMW-PEIs (0.6k, 1.2k, and 1.8k) (**Figure 9**).[24,25] PEG-cr-PEI copolymers degraded rapidly at pH 7.4 with half-life of 6.8–10 hours which was shorter than that at pH 5.6, suggesting non-acid-catalysis degradation of

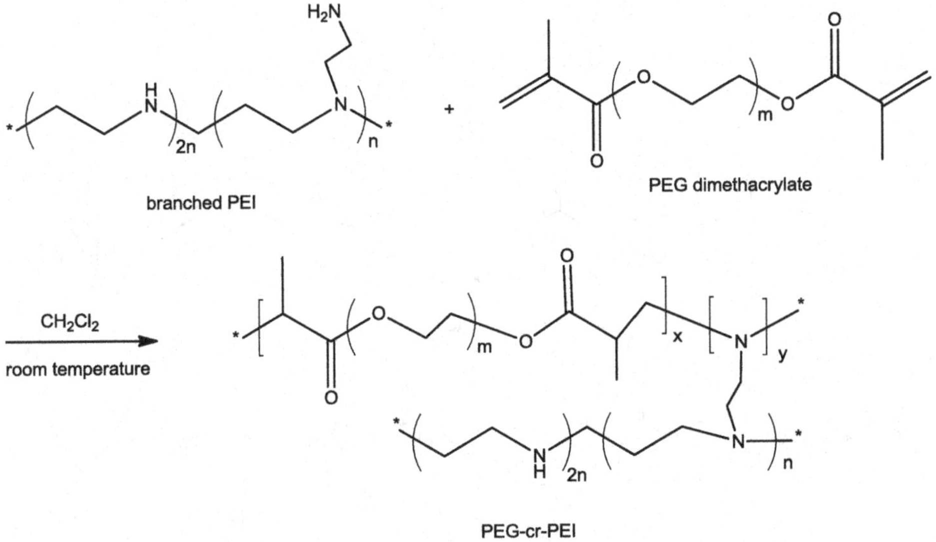

branched PEI

PEG dimethacrylate

$CH_2Cl_2$

room temperature

PEG-cr-PEI

**Figure 9.** Synthesis of PEG dimethyacrylate-crosslinked PEIs.

the copolymers. The PEG-cr-PEIs showed good ability to bind DNA, yielding nanoscale PEG-cr-PEI/DNA complexes with zeta potentials above +20 mV and nanoscale size below 250 nm at and above N/P ratios of 10/1. The cytotoxicity of PEG-cr-PEI/DNA complexes was lower than that of 25 kDa-PEI-based complexes in 293T, HeLa and HepG2 cells at N/P ratios from 5/1 to 40/1. *In vitro* transfection assay indicated that the most efficient transfection of PEG-cr-PEI was comparable to 25 kDa-PEI against HeLa or HepG2 cells but lower in 293T cells.

Polycaprolactone (PCL) is a biocompatible, hydrophobic polymer. Arote *et al.* prepared PCL–PEIs by Michael-type addition of PCL diacrylate to different LMW-PEI (Mn = 0.6, 1.2, 1.8 and 2.5 k) (**Figure 10**).[26] These PCL-crosslinked PEIs could effectively condense DNA at N/P ratio of 10 into nanoscale complexes with the particle sizes below 200 nm and zeta potential ranging from +20 to ~+48 mV. The half-life of these PEIs was 4.5–5 days at physiological pH. Importantly, they were essentially non-toxic against 293 T, HepG2 and HeLa cells. For example, at an N/P ratio of 30/1, the polyplexes of polyester amines (PEAs) induced ~80% cell viability against HepG2 and HeLa cell lines. By contrast, cell viability decreased to 20 to ~40% for 25 kDa-PEI. These PCL-crosslinked–PEIs revealed efficient transfection activity in the cell lines. The most efficient gene transfection was observed for the PCL-PEI-1.2 k, which induced 15–25 fold higher efficiency than 25 kDa-PEI. Also, the complexes of PCL–PEI were efficient for transfection in a mouse model after aerosol administration.

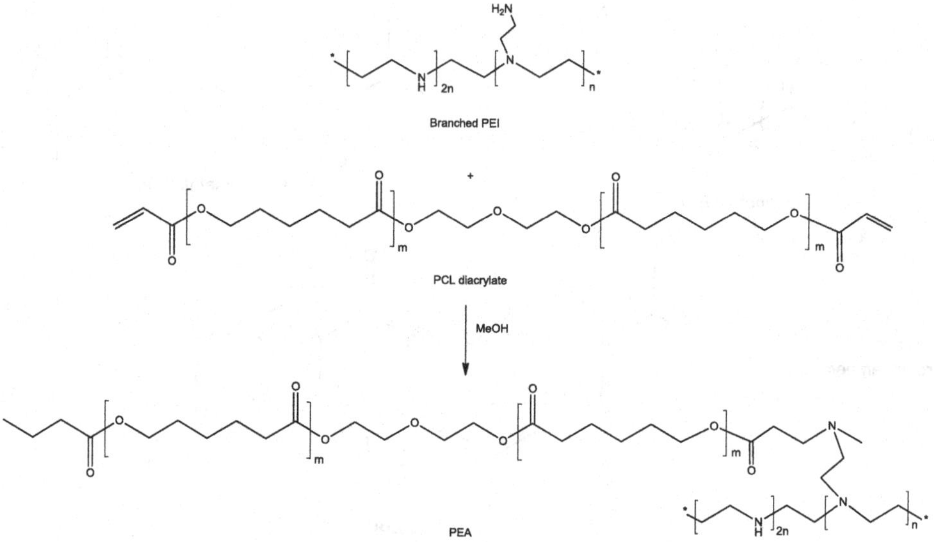

Figure 10. Synthesis of ester-based PEIs with PCL diacrylate crosslinker.

In addition to PEG diacrylate and PCL diacrylate, Poloxamers diacrylates were also used as crosslinking agents for preparation of hydrolytic PEIs. Kabanov *et al.* reported Poloxamer-crosslinked–PEI by Michael-type addition of Poloxamer diacrylate to LMW-PEIs.[27] Kim *et al.* synthesized hydrolytic PEIs by Michael-type addition reaction of Poloxamer diacrylate and 1.8 kDa-PEI.[28] (**Figure 11**). These PEAs were hydrolytically degradable in 0.1 M PBS buffer within 70 hours. They condensed DNA into nanoscale complexes with average particle sizes below 150 nm at N/P ratios from 10/1 to 50/1. These PEIs showed lower cytotoxicity in A549, 293T, and HepG2 cell lines, compared with 25 kDa-PEI at polymer concentrations ranging from 2 to 40 μg/mL. It was found that these branched PEIs led to higher cell viability as compared to 25 kDa-PEI, although the cytotoxicity was cell-dependent. The transfection efficiency of the PEAs was dependent on cell lines and the content of poloxamer. Poloxamer-crosslinked-PEIs showed higher transfection efficiencies in three cell lines as compared to 25 kDa-PEI and minor serum-dependency in A549 (human lung adenoma) cells when the content of the poloxamer in the PEIs was modulated to 30%.

Polyhydroxyalkanoates (PHAs) are class of degradable non-toxic polymers and have been widely applied in biomedical devices such as sutures and surgical mesh.[29,30] However, the lack of cationic residues and hydrophobic nature of PHAs limit their applications in gene delivery.[31] Zhou *et al.* prepared a group of mP3/4HB-g-bPEI copolymers by Michael-type addition reaction of P3/4HB acrylate (polyhydroxyalkanoates) to 25 kDa-PEI (**Figure 12**).[32] They found that

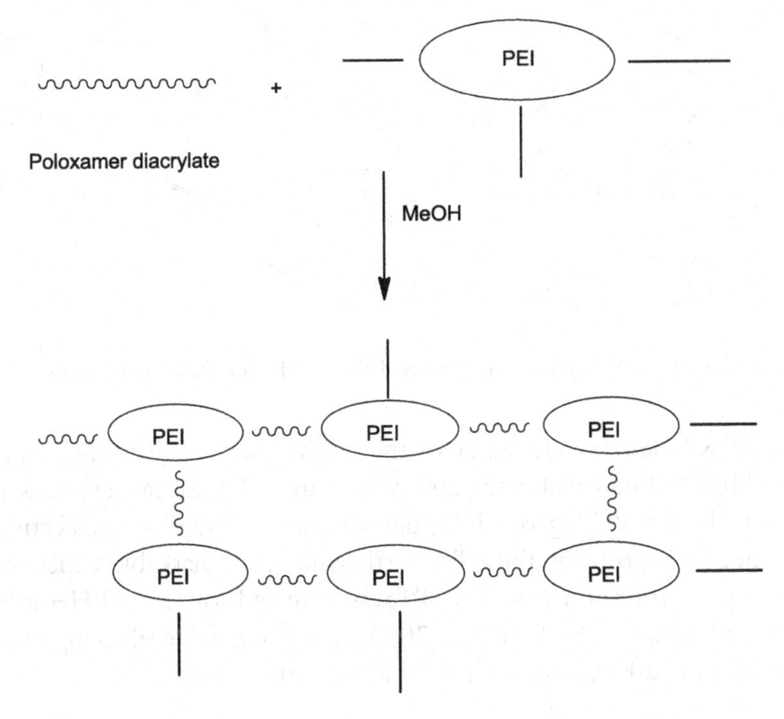


**Figure 11.** Synthesis of ester-based PEIs with poloxamer diacrylate crosslinker.

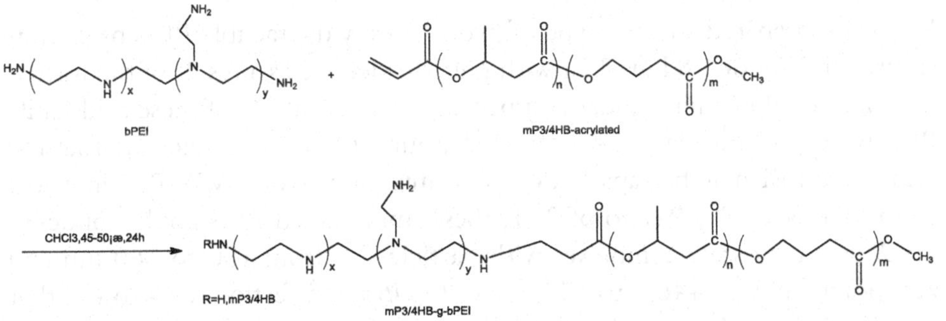

bPEI + mP3/4HB-acrylated

CHCl3,45-50¡æ,24h

R=H,mP3/4HB

mP3/4HB-g-bPEI

**Figure 12.** Synthesis of ester-based PEIs with polyhydroxyalkanoates.

mP3/4HB-g-bPEIs could serve as non-viral vectors for siRNA delivery because siRNA could be strongly condensed by mP3/4HB-g-bPEIs, released from their complexes and efficiently protected from nuclease degradation. The particle sizes and zeta potentials of the complexes of mPHA-g-bPEI/siRNA was determined to be less than 200 nm and +33 to +43 mV, respectively. Moreover, mPHA-g-bPEI copolymers displayed a low cytotoxicity as compared to unmodified 25 kDa-PEI and caused efficient cellular uptake of Cy3–siRNA (small interfere RNA) in A549 cells by flow cytometry and confocal microscopy

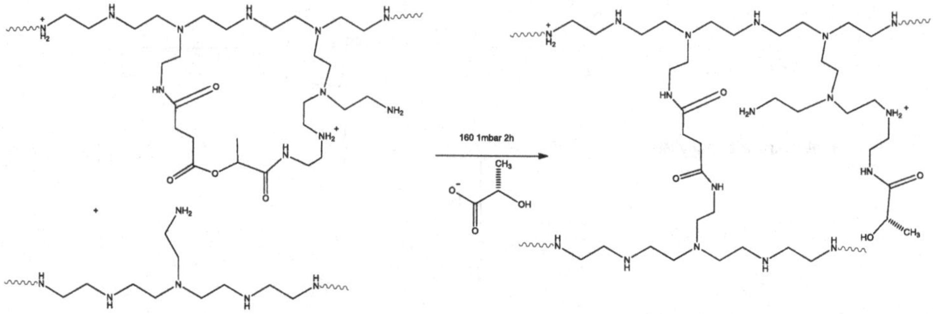

**Figure 13.** Synthesis of imine-based PEIs with glutadialdehyde linker.

analysis. SiRNA delivery efficiency of the copolymers was evaluated in A549-Luc and MCF-7-Luc (luciferase-expressing human breast cancer) cells It was shown that these mPHA-g-bPEI copolymers revealed higher transfection efficiency than 25 kDa-bPEI in the cells. Furthermore, a remarkable knockdown of luciferase gene expression (up to 85%) was detected for the mPHA-g-bPEI/siRNA complexes at an N/P ratio of 20/1 with comparable silencing efficiency to Lipofectamine 2000, a commercial transfection agent.

### 3.2. *Imine-based PEIs*

Kim *et al.* prepared another type of hydrolytically degradable PEIs by chemical crosslinking of 1.8 kDa-PEI with glutadialdehyde as a result of the formation of acid-labile imine linker (**Figure 13**).[33] The half-life of these acid-labile PEIs were 1.1 hours at pH 4.5 and 118 hours at pH 7.4, suggesting that the acid-labile PEI may be rapidly degraded into non-toxic LMW-PEI in acidic endosome. With N/P ratio of 3/1, these imine-based PEIs could condense DNA into nanoscale complexes with particle size from 130 to 160 nm and zeta potential from +46.1 to +50.9 mV. *In vitro* transfection test showed that the acid-labile PEIs showed much higher transfection efficiency than 1.8 kDa-PEI and comparable transfection efficiency to 25 kDa-PEI against 293T cells. Cytotoxicity assay revealed that the acid-labile PEI showed ~90% cell viability at the N/P ratio of 40/1, whereas 25 kDa-PEI was cytotoxic with only 60% cells maintaining viability. It was expected that the acid-labile PEI was less toxic than PEI, owing to the degradation of acid-labile linkage.

## 4. Conclusions

PEI with biocompatible or degradable groups are promising as non-viral vectors for gene delivery. From non-degradable PEI to its biocompatible counterparts, people more and more reach virus-mimicking, safe and potent polymeric gene delivery vectors. Further understanding on structure–activity relationships of these PEI derivatives should be focused, in order to achieve further clinical translation in the future.

## Acknowledgments

The authors, Ting Shi, Peng Zhao and Chao Lin, for contributing figure drawings, writing and discussion of this chapter. This review writing is supported by the grants from the Shanghai Municipal Natural Science Foundation (13ZR1443600) Fundamental Research Funds for the Central Universities.

## References

1. Verma IM, Somia N. Gene therapy — promises, problems and prospects. *Nature* 1997;389:239–242.
2. Pouton CW, Seymour LW. Key issues in non-viral gene delivery. *Adv Drug Deliv Rev* 2001;46:187–203.
3. Thomas CE, Ehrhardt A, Kay MA. Progress and problems with the use of viral vectors for gene therapy. *Nat Rev Genet* 2003;4:346–358.
4. Li S, Huang L. Non-viral gene therapy: promises and challenges. *Gene Ther* 2000;7: 31–34.
5. Pack DW, Hoffman AS, Pun S, Stayton PS. Design and development of polymers for gene delivery. *Nat Rev Drug Discov* 2005;4:589–593.
6. Boussif O, Lezoualch F, Zanta MA, Mergny MD, Scherman D, Demeneix B, Behr JP. A versatile vector for gene and oligonucleotide transfer into cells in culture and *in vivo* — polyethylenimine. *Proceedings of the National Academy of Sciences of the United States of America* 1995;92:7297–7301.
7. Moghimi SM, Symonds P, Murray JC, Hunter AC, Dekska G, Szcwczyk A. A two stage poly(ethylenimine)-mediated cytotoxicity: implications for gene transfer/therapy. *Molec Ther* 2005;11:990–995.
8. Liu T, Yu X, Kan B, Guo Q, Wang X, Shi S, Guo G, Luo F, Zhao X, Wei Y, Qian Z. Enhanced gene delivery using biodegradable poly (ester amine)s (PEAs) based on low-molecular-weight polyethylenimine and poly(epsilon-caprolactone)-pluronic-poly(epsilon-caprolactone). *J Biomed Nanotechnol* 2010;6:351–359.
9. Lee D-E, Hong Y-D, Choi K-H, Lee S-Y, Park P-H, Choi S-J. Preparation and evaluation of Tc-99m-labeled cyclic arginine-glycine-aspartate (RGD) peptide for integrin targeting. *Appl Radiat Isot* 2010;68:1896–1902.
10. Kabanov AV, Batrakova EV, Sriadlbhatla S, Yang ZH, Kelly DL, Alakov VY. Polymer genomics: shifting the gene and drug delivery paradigms. *J Control Release* 2005;101:259–271.
11. Ko YT, Kale A, Hartner WC, Papahadjopoulos-Sternberg B, Torchilin VP. Self-assembling micelle-like nanoparticles based on phospholipid-polyethyleneimine conjugates for systemic gene delivery. *J Control Release* 2009;133:132–138.
12. Sahay G, Batrakova EV, Kabanov AV. Different internalization pathways of polymeric micelles and unimers and their effects on vesicular transport. *Bioconjug Chem* 2008;19: 2023–2029.

13. Fan W, Wu X, Ding B, Gao J, Cai Z, Zhang W, Yin D, Wang X, Zhu Q, Liu J, Ding X, Gao S. Degradable gene delivery systems based on pluronics-modified low-molecular-weight polyethylenimine: preparation, characterization, intracellular trafficking, and cellular distribution. *Int J Nanomedicine* 2012;7:1127–1138.
14. Liu C, Zhu Q, Wu W, Xu X, Wang X, Gao S, Liu K. Degradable copolymer based on amphiphilic N-octyl-N-quaternary chitosan and low-molecular weight polyethylenimine for gene delivery. *Int J Nanomedicine* 2012;7:5339–5350.
15. Namgung R, Kim J, Singha K, Kim CH, Kim WJ. Synergistic effect of low cytotoxic linear polyethylenimine and multiarm polyethylene glycol: study of physicochemical properties and *in vitro* gene transfection. *Mol Pharm* 2009;6:1826–1835.
16. Chu M, Dong C, Zhu H, Cai X, Dong H, Ren T, Su J, Li Y. Biocompatible polyethylenimine-graft-dextran catiomer for highly efficient gene delivery assisted by a nuclear targeting ligand. *Polym Chem* 2013;4:2528–2539.
17. Lynn DM, Langer R. Degradable poly (beta-amino esters): synthesis, characterization, and self-assembly with plasmid DNA. *J Am Chem Soc* 2000;122:10761–10768.
18. Park MR, Han KO, Han IK, Cho MH, Nah JW, Choi YJ, Cho CS. Degradable polyethylenimine-alt-poly (ethylene glycol) copolymers as novel gene carriers. *J Control Release* 2005;105:367–380.
19. Park MR, Kim HW, Hwang CS, Han KO, Choi YJ, Song SC, Cho MH, Cho CS. Highly efficient gene transfer with degradable poly(ester amine) based on poly(ethylene glycol) diacrylate and polyethylenimine *in vitro* and *in vivo*. *J Gene Med* 2008;10:198–207.
20. Forrest ML, Koerber JT, Pack DW. A degradable polyethylenimine derivative with low toxicity for highly efficient gene delivery. *Bioconjug Chem* 2003;14:934–940.
21. Dong W, Li S, Jin G, Sun Q, Ma D, Hua Z. Efficient gene transfection into mammalian cells mediated by cross-linked polyethylenimine. *Int J Mol Sci* 2007;8:81–102.
22. Arote RB, Hwang S-K, Yoo M-K, Jere D, Jiang H-L, Kim Y-K, Choi Y-J, Nah J-W, Cho M-H, Cho CS. Biodegradable poly(ester amine) based on glycerol dimethacrylate and polyethylenimine as a gene carrier. *J Gene Med* 2008;10:1223–1235.
23. Ahn CH, Chae SY, Bae YH, Kim SW. Biodegradable poly (ethylenimine) for plasmid DNA delivery. *J Control Release* 2002;80:273–282.
24. Yu J-H, Quan J-S, Huang J, Nah J-W, Cho C-S. Degradable poly (amino ester) based on poly (ethylene glycol) dimethacrylate and polyethylenimine as a gene carrier: molecular weight of PEI affects transfection efficiency. *J Mater Sci Mater Med* 2009;20:2501–2510.
25. Yu J-H, Huang J, Nah J-W, Cho M-H, Cho C-S. Degradable poly (ester amine) based on poly (ethylene glycol) dimethacrylate and polyethylenimine as a gene carrier. *J Appl Polym Sci* 2007;115:1189–1198.
26. Arote R, Kim T-H, Kim Y-K, Hwang S-K, Jiang H-L, Song H-H, Nah J-W, Cho M-H, Cho C-S. A biodegradable poly (ester amine) based on polycaprolactone and polyethylenimine as a gene carrier. *Biomaterials* 2007;28:735–744.
27. Kabanov AV, Lemieux P, Vinogradov S, Alakhov V. Pluronic block copolymers: novel functional molecules for gene therapy. *Adv Drug Deliv Rev* 2002;54:223–233.
28. Kim TH, Cook SE, Arote RB, Cho M-H, Nah JW, Choi YJ, Cho CS. A degradable hyperbranched poly (ester amine) based on poloxamer diacrylate and polyethylenimine as a gene carrier. *Macromol Biosci* 2007;7:611–619.
29. Rai R, Keshavarz T, Roether JA, Boccaccini AR, Roy I. Medium chain length polyhydroxyalkanoates, promising new biomedical materials for the future. *Mater Sci Eng R* 2011;72:29–47.

30. Kircheis R, Wightman L, Schreiber A, Robitza B, Rossler V, Thomas M, Ge Q, Lu JJ, Chen J, Klibanov AM. Cross-linked small polyethylenimines: while still nontoxic, deliver DNA efficiently to mammalian cells *in vitro* and *in vivo*. *Pharm. Res* 2005;22:373–380.
31. Blow N. Delivering the future. *Nature* 2007;450:1117–1122.
32. Zhou L, Chen Z, Chi W, Yang X, Wang W, Zhang B. Mono-methoxy-poly (3-hydroxybutyrate-co-4-hydroxybutyrate)-graft-hyper-branched polyethylenimine copolymers for siRNA delivery. *Biomaterials* 2005;33:2334–2344.
33. Kim YH, Park JH, Lee M, Kim YH, Park TG, Kim SW. Polyethylenimine with acid-labile linkages as a biodegradable gene carrier. *J Control Release* 2005;103:209–219.

# Chapter 4

# Magnetic $Fe_3O_4$ Nanoparticles for Cancer Photothermal Therapy

*Maoquan Chu*

*School of Life Science and Technology, Tongji University, 1239 Siping Road, Shanghai 200092, P. R. China*
*Research Center for Translational Medicine at Shanghai East Hospital, Tongji University, 150 Jimo Road, Shanghai 200120, P. R. China*

The use of magnetic iron oxide nanoparticles for cancer hyperthermal therapy induced by alternating magnetic field (AMF) is widely accepted. However, AMF applied for clinical therapy requires high currents and voltages. In recent years, a novel strategy of magnetic iron oxide nanoparticle-mediated hyperthermia by near-infrared (NIR) light irradiation has been reported by several research groups. This is a safe, simple and highly efficient strategy for cancer therapy that may be translated into clinical practice. Herein, we discuss the current knowledge on the photothermal therapy (PTT) of the magnetic $Fe_3O_4$ nanoparticles *in vitro* and *in vivo* induced by NIR laser irradiation. The magnetic resonance imaging guided cancer PTT and the limitations of this cancer PTT were also reviewed.

## 1. Photothermal Therapy

Normal average body temperature in man is maintained at 37°C. The multiplication rates of viruses, bacteria and cancer cells in the human body usually decrease at higher temperatures. Therefore, when the body gets ill it will instinctively defend itself by increasing its temperature (a process commonly known as a fever) to fight disease-causing agents such as viruses. Furthermore, the level of expression of heat shock proteins will increase in tandem with body temperature; this enhances the body's disease fighting ability. Based on the phenomenon of natural fever, thermal therapy (also called hyperthermia) has been developed and applied in clinical practice. It has been demonstrated that cancer cells can be destroyed at 42.5°C,[1] whereas normal cells can survive at 44°C.[2] In addition, as compared with normal tissue, tumor tissue has a poor blood supply and slow blood flow, which results in poor heat dissipation in tumors. Hyperthermia is therefore often used for cancer treatment.

This treatment technique includes whole-body hyperthermia and local hyperthermia. Because an important challenge for hyperthermia is to minimize damage to healthy tissue,[3] local hyperthermia has attracted much attention in the past decades. Several modalities, including microwaves, ultrasound and laser irradiation, have been used for local intracorporal heat generation.[4] The use of laser with infrared or near-infrared (NIR) wavelengths for the treatment of disease including cancer is termed photothermal therapy (PTT).[5] NIR laser light can penetrate deeply into tissue, and is less harmful to normal tissue due to the lower energy of longer wavelength light. Therefore, NIR laser induced PTT has been widely utilized for cancer therapy. As compared with photodynamic therapy (PDT), PTT generates heat to kill target cancer cells; it does not require the generation of reactive oxygen species to damage the cells or tissues.

In recent years, NIR light-absorbing nanomaterials applied for PTT have attracted great interest in scientific research, because such nanomaterials when localized inside tumors by means of intratumoral injection or targeted delivery can specifically destroy the tumors; this is because the specific absorbance of the NIR light by the nanomaterials results in only minimal damage to the normal tissues surrounding the tumor.

## 2. NIR Light-Absorbing Nanomaterials for PTT

The red or NIR light-absorbing nanomaterials of great interest currently being studied for cancer PTT include: inorganic nanomaterials including gold nanostructures (e.g., gold nanoshells,[6,7] gold nanocages,[8–10] gold nanorods,[10–13] gold nanostars,[14,15] gold nanohexapods;[13] palladium nanoplates[16]; carbon-based nanomaterials (e.g., carbon nanotubes,[17–19] carbon nanohorns,[20,21] graphene,[22–24] activated carbon;[25] semiconductor nanomaterials (e.g., CdTe,[26] Cu2-xSe[27] and CuS[28]); magnetic $Fe_3O_4$ nanoparticles[29–32]; and polymer nanomaterials.[33] All of these nanomaterials can rapidly convert laser light energy into heat.

A possible mechanism for photothermal conversion is associated with the localized surface plasmon resonance (LSPR) response of the metal containing nanoparticles to NIR laser irradiation. When a metal containing nanoparticle is irradiated using light, the free electrons around the metal surface will exhibit a collective coherent oscillation. A dipole oscillation along the direction of the electric field of the light will be formed by this surface electron oscillation.[34] When the amplitude of the oscillation reaches a maximum at a specific frequency, the metal containing nanoparticles will strongly absorb the incident light; this phenomenon is called surface plasmon resonance (SPR).[35–40] SPR in nanostructures is known as LSPR.[41] According to the Drude free electron model[42]:

$$\omega_p = \sqrt{\frac{ne^2}{\mu\varepsilon_0}}$$

where $\omega_p$ is the plasma frequency, $\mu$ is the effective mass, $n$ is the charge carrier density, $e$ is the electron charge, and $\varepsilon_0$ is the permittivity of vacuum. Based on this equation, the SPR band is much stronger for noble metals (e.g., Au and Ag) than other metals, since the charge carrier density of typical noble metals is about $10^{23}$ electrons/cm$^3$.[43] Therefore, although the SPR effect has been previously demonstrated for both noble metals[44] and conducting metal oxides (the charge carrier density is about $10^{21}$ electrons/cm$^3$), the SPR effect of noble metals is strongest among other materials. This is one of the main reasons why gold nanostructures have been extensively investigated for cancer PTT in the past decade.[6–15]

However, biodegradability and toxicity has been the major hurdle that has prevented most of the nanomaterials from being used clinically. In order to search for biocompatible nanomaterials with low or no toxicity for use in cancer PTT, the photothermal conversion properties and PTT efficiency of carbon-based materials, polymer materials and magnetic nanoparticles and others, as mentioned above, have been investigated. Because temperatures in excess of 42.5°C may kill or weaken cancer cells, the photothermal conversion efficiencies of these gold-free nanomaterials are high enough to damage cancer cell structures.

## 3. Magnetic Iron Oxide Nanoparticles for Cancer PTT

### 3.1. *Biomedical applications of the magnetic iron oxide nanoparticles and the advantages of PTT using magnetic iron oxide nanoparticles*

As compared with the carbon-based materials and polymer materials mentioned above, magnetic iron oxide nanoparticles ($Fe_3O_4$ and $\gamma$-$Fe_2O_3$) have been safely used clinically,[45–48] and numerous types of these nanoparticles already have

Federal Drug Administration (FDA) approval for clinical use.[49] The use of magnetic iron oxide nanoparticles for cancer therapy is therefore attractive for clinical applications, for both doctors and patients. In the past decades, magnetic iron oxide nanoparticles have been widely applied in biomedical research, including drug delivery, magnetic resonance (MR) imaging, tissue repair, immunoassay, detoxification of biological fluids, cell separation and hyperthermia therapy.[50] The use of magnetic nanoparticles as heating mediators for cancer hyperthermal therapy is of particular interest, because anything that potentially improves the techniques used for cancer therapy is welcome, and this physical treatment has fewer side effects relative to chemotherapy or radiotherapy.[50] The principle of magnetic hyperthermia is that the magnetic nanoparticles induced by an alternating magnetic field (AMF) can generate heat by loss of hysteresis. The temperature of this local hyperthermia can be easily controlled by changing the AMF parameters or the amount of the nanoparticles in the tumors.

Although magnetic hyperthermia induced by AMF is currently used clinically,[45–48] this technique requires high currents and voltages since a larger air volume has to be 'filled up' using the AMF.[49] Moreover, the magnetic fields (termed H-fields) that can be tolerated are limited by local discomfort during the treatment of pelvic and thoracic tumors; this may result from boundary effects between tissues of different dielectric constants and conductivity, or skin reactions caused by increased current density.[50] Additionally, patients with metallic implants situated <30 cm from the treatment area have to be excluded during magnetic hyperthermia.[50]

In recent years, a novel strategy for heat generation using magnetic nanoparticles triggered by laser irradiation for application in cancer therapy has attracted great attention.[29–32] $Fe_3O_4$ nanoparticles exhibit black or close to black color, whereas $Fe_2O_3$ nanoparticles exhibit brown ($\gamma$-$Fe_2O_3$) or yellow ($\alpha$-$Fe_2O_3$) color. The optical absorption coefficient of $Fe_3O_4$ nanoparticles in the NIR band is higher than that of $Fe_2O_3$ nanoparticles. Consequently, $Fe_3O_4$ nanoparticles have frequently been used for cancer PTT.[29–32] PTT involving the magnetic nanoparticles induced by NIR light is more beneficial than hyperthermal therapy induced by AMF; this is because NIR light has an extensive safety profile in studies regarding disease diagnostics and therapy, and NIR light irradiation is relatively simple, convenient, and has a low cost.

### 3.2. *Photothermal conversion of the magnetic iron oxide nanoparticles*

The photothermal conversion efficiency of a material is related to its optical absorption coefficient. High photothermal conversion efficiency at the NIR

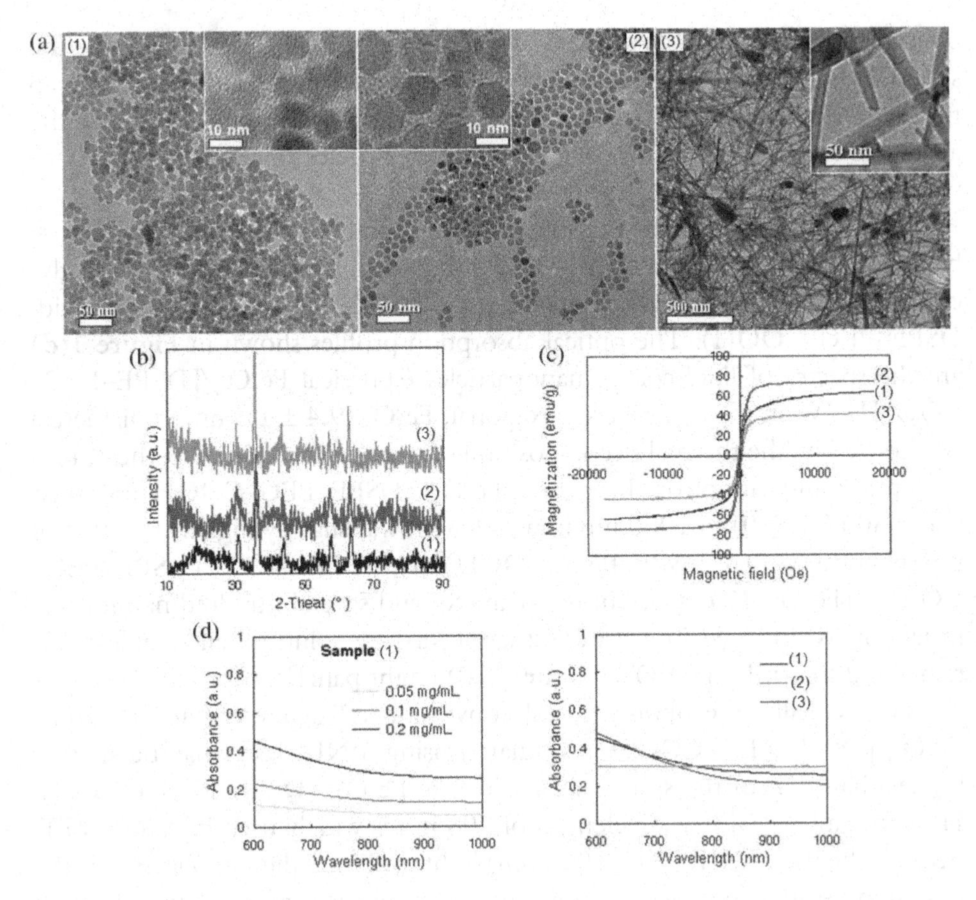


**Figure 1.** $Fe_3O_4$ nanoparticles. **(a)** high-resolution transmission electron microscope (HRTEM) images of spherical, hexagonal, and wire-like $Fe_3O_4$ nanoparticles. **(b)** and **(c)** are X-ray diffraction (XRD) patterns and magnetization of the nanoparticles shown in **(a)**, respectively. **(d)** Absorption spectra of the spherical $Fe_3O_4$ nanoparticles in different concentrations (left), and hexagonal and wire-like $Fe_3O_4$ nanoparticles containing 0.2 mg/mL of $Fe_3O_4$ (right). The samples 1 and 3 are synthesized in aqueous solutions, and the sample 2 is synthesized in 1-octadecene. [Reused with permission from [31], Elsevier.]

wavelength is caused by the high NIR optical absorption coefficient. The optical absorption coefficient is controlled by the material components, crystallinity, particle morphology, size and surface properties. As shown Figure 1, for spherical $Fe_3O_4$ nanoparticles (9.1±1.9 nm in diameter) with a face-centered cubic phase and superparamagnetic properties synthesized by co-precipitation of iron chloride salts with ammonia, their aqueous suspensions show a broad and continuous absorption spectrum from ultraviolet (UV) to the NIR band; the NIR absorption intensity increases with increasing concentration of $Fe_3O_4$ (**Figure 1(d)** (left panel)).[31] NIR absorption of the spherical $Fe_3O_4$ nanoparticles is ideal for *in vivo* cancer PTT due to the deep tissue penetration characteristics of NIR light. It is difficult to detect the absorption spectrum of the bare $Fe_3O_4$ nanoparticles because of their poor water solubility.

It should be noted that the spherical $Fe_3O_4$ nanoparticles described above have been coated with distearoyl-N-[3-carboxypropionoyl poly (ethylene glycol) succinyl] phosphatidylethanolamine (DSPE–$PEG_{2000}$ carboxylic acid, DSPE–PEG–COOH). The optical absorption profiles shown in **Figure 1(d)** are the spectra of lipid-coated nanoparticles (spherical $Fe_3O_4$/[DSPE–PEG–COOH]). When the cores were hexagonal, $Fe_3O_4$ (9.4 ± 1.3 nm in diameter) synthesized by the thermal decomposition of metal-oleate complex, the optical absorption intensity of the hexagonal $Fe_3O_4$/(DSPE–PEG–COOH) between about 800 (or 810) to 1000 nm in wavelength was slightly higher than that of the spherical $Fe_3O_4$/(DSPE–PEG–COOH) or wire-like $Fe_3O_4$/(DSPE–PEG–COOH) ($Fe_3O_4$: 12.6 ± 5.9 nm in diameter and several hundred nanometers in length; synthesized in a water/ethanol mixture solution; all samples contained 0.2 mg/mL of $Fe_3O_4$) (**Figure 1(d)** [right panel]).[31]

Theoretically, the photothermal conversion efficiency of the hexagonal $Fe_3O_4$/[DSPE–PEG–COOH]) irradiated using a NIR laser may be slightly higher than that of the spherical or wire-like $Fe_3O_4$/(DSPE–PEG–COOH). However, a laser with a wavelength of 808 nm is widely used in cancer PTT, because the use of NIR (700–1000 nm) light for biomedical application is the best understood approach in the context of photon propagation through deep living tissue.

The rates of temperature increase of both spherical and wire-like $Fe_3O_4$/(DSPE–PEG–COOH) suspensions upon 808 nm laser irradiation (808 nm; 0.25 W/cm$^2$) have been found to be similar to that for the hexagonal $Fe_3O_4$/(DSPE–PEG–COOH) (**Figure 2(a)**).[31] This may be due to their comparable absorptions at around 808 nm wavelength, and the fact that their surfaces were all modified with lipid shells. The photothermal conversions of these $Fe_3O_4$ nanoparticles are rapid and highly efficient. For example, at a concentration of only 0.08 mg/mL of spherical $Fe_3O_4$ nanoparticles (coated with DSPE–PEG–COOH), the temperature increase of the suspension is ≤ 7°C over a period of 3 minutes, and 13°C over a period of 10 minutes after 808 nm laser irradiation (**Figure 2(b)**).[31] The temperature increase effect is more rapid at higher $Fe_3O_4$ nanoparticle concentrations. At concentrations of 0.8 and 8 mg/mL, the temperature increases are ≤25°C and 28°C, respectively for 10 minutes after irradiation with an 808 nm laser (**Figure 2(b)**).[31] The temperature increases are rapid in the first 5 minutes for all nanoparticle concentrations, but gradually

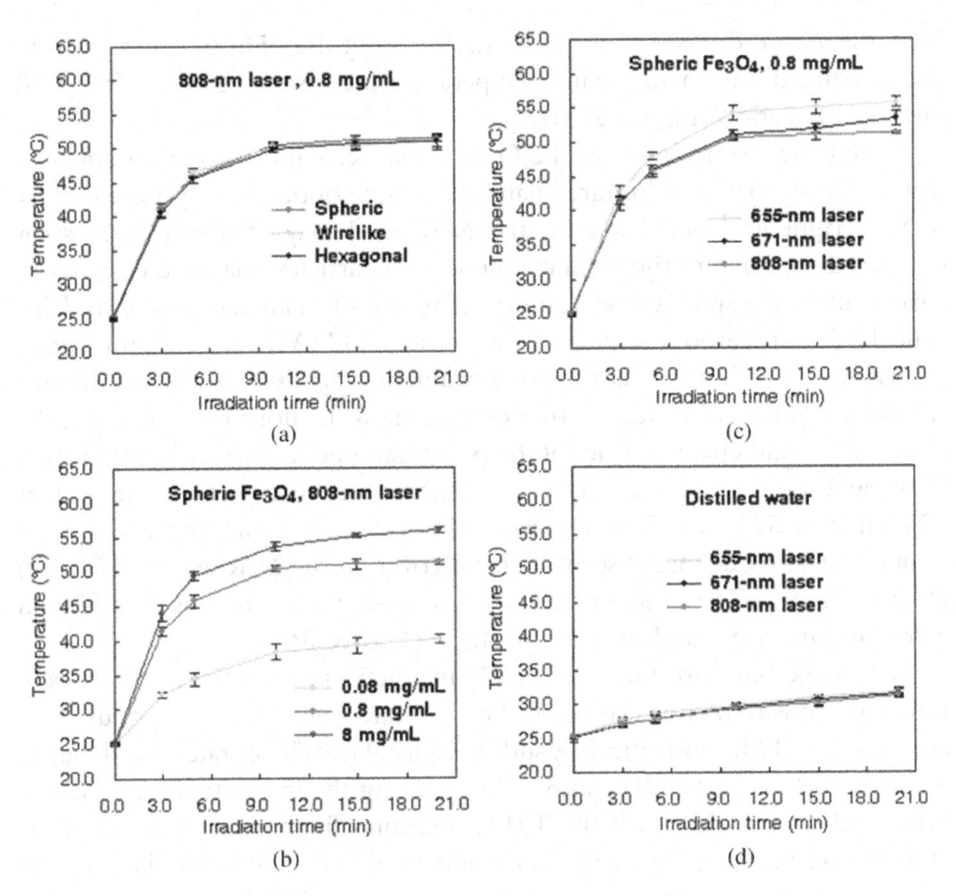


**Figure 2.** Temperature *versus* irradiation time of the aqueous suspensions of $Fe_3O_4$ nanoparticles with different concentrations and indicated shapes, including distilled water. Irradiation was done with 655, 671, and 808 nm lasers, respectively. **(a)** $Fe_3O_4$ nanoparticles with spherical, hexagonal, and wire-like shapes, all containing 0.8 mg/mL of $Fe_3O_4$ upon 808 nm laser irradiation, **(b)** Spherical $Fe_3O_4$ nanoparticles with different concentrations (0.08, 0.8 and 8 mg/mL) upon 808 nm laser irradiation, **(c)** Spherical $Fe_3O_4$ nanoparticles containing 0.8 mg/mL of $Fe_3O_4$ upon laser irradiation at 655, 671, and 808 nm, and **(d)** Distilled water upon laser irradiation at 655, 671, and 808 nm. All $Fe_3O_4$ nanoparticles were coated with DSPE–PEG–COOH. All suspension volumes were 200 $\mu$L including distilled water. Error bars were based on standard deviations. [Reused with permission from [31], Elsevier.]

level off thereafter. For the spherical $Fe_3O_4$ nanoparticle (0.8 mg/mL) solutions, irradiated at shorter irradiation wavelengths (655 nm [0.5 W/cm$^2$]; 671 nm [0.2 W/cm$^2$]; 20 minutes), the temperature increases are 2–4°C greater than those irradiated using the 808 nm laser (**Figure 2(c)**), which is the result of greater absorption at shorter wavelength.[31] The temperature increase in the above magnetic solutions is mainly produced by the $Fe_3O_4$ nanoparticles, since the temperature of distilled water irradiated using the 655, 671 and 808 nm lasers changed with a maximum temperature increase of ≤6.7°C after a 20 minutes irradiation (**Figure 2(d)**).[31]

When the diameter of the $Fe_3O_4$ nanoparticles increases from approximately 10 nm to several hundred nanometers, the optical absorption intensity of the magnetic nanoparticles in the NIR band may obviously increase as compared with that of the small magnetic nanoparticles. Liao and co-workers have synthesized spherical 440 nm diameter $Fe_3O_4$ nanoparticles using benzene-1, 3, 5-tricarboxylic acid (trimesic acid [TMA]) and sodium citrate (shortened to "citrate") as the co-coordinating agents *via* a hydrothermal reaction.[32] The results indicate that during the 3–12 hour reaction time, the absorption-band characteristics of the precipitates generated were monitored, showing a progressive red-shift with reaction time; there is a double peak at 424 nm and 521 nm (3 hour), a double peak at 497 nm and 569 nm (6 hour), and a broadening absorption band covering the NIR region (12 hour) (**Figure 3(a)**).[32] The product after a 12 hour reaction time was dark blue in color (**Figure 3(b)**) and spherical in shape (**Figure 3(c)**).[32]

The dark blue product was $Fe_3O_4$ magnetic nanoparticles; these were found to consist of polycrystalline $Fe_3O_4$ coated with an amorphous layer composed of TMA and citrate ligands (**Figure 4**).[32] The citrate ligand played a pivotal role in the Fe (II) complex by inducing the intensity increase in the NIR, and cooperation with the TMA molecules further increased the NIR absorption intensity (**Figure 5**).[32] Liao and co-workers then coated the yellow product (after a reaction time of 3 hours) and the dark blue polycrystalline $Fe_3O_4$ with mesoporous $SiO_2$ shells.[32] They further modified the $Fe_3O_4/SiO_2$ with aminopropyl-triethoxysilane (APTES), and measured its photothermal conversion. It was found that the temperature of the APTES-functionalized $Fe_3O_4/SiO_2$ solution (400 ppm [Fe]) upon NIR laser irradiation (808 nm; 2 W/cm$^2$) was increased to 44°C after 4 minutes, followed by a slow and gradual increase to 45°C during irradiation for 6 minutes.[32] In contrast, the rates of the temperature increase of the clinical resovist agent, commercial $\gamma$-$Fe_2O_3$ nanopowders and iron oxide nanoproducts after 3 hours at the same Fe concentration were significantly lower than that of the iron oxide nanoproducts after 12 hours (**Figure 6**).[32]

The photothermal conversion efficiency is not only controlled by the material components, crystallinity, morphology and other factors, as mentioned above, but also by the "solubility" of the materials in aqueous solution. If the nanoparticles are well separated from each other, and are well suspended in aqueous solution (also termed "good solubility"), these nanoparticles will exhibit high efficiency concerning photothermal conversion. In

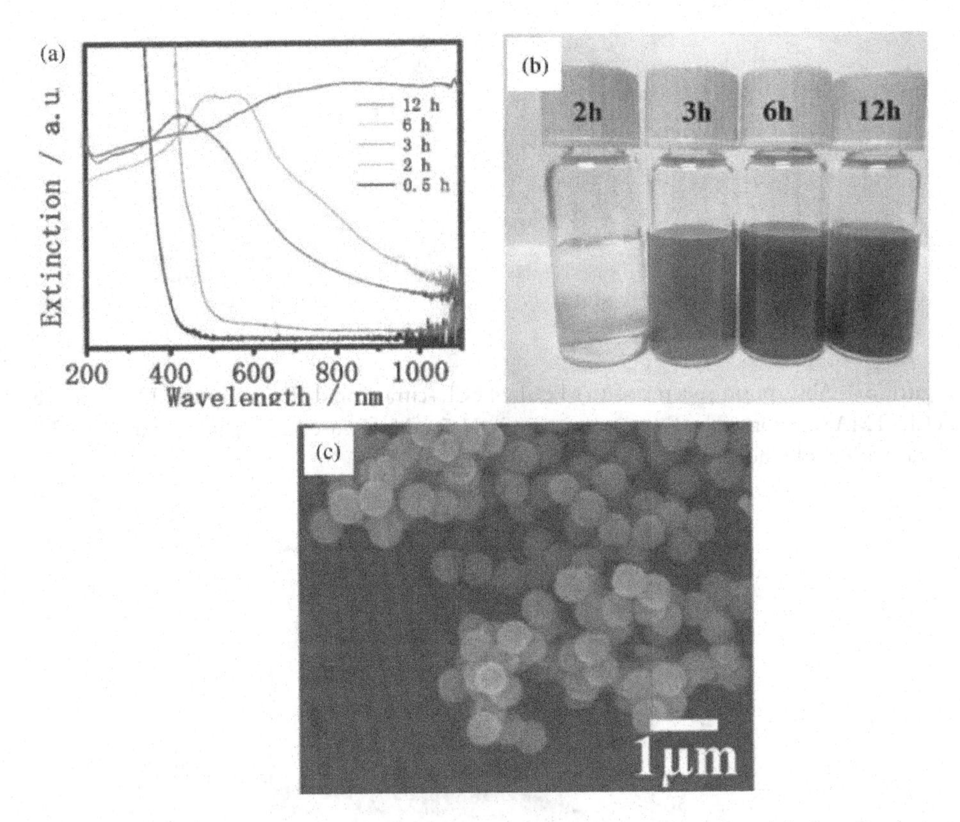


**Figure 3.** Absorption spectra **(a)** and optical image **(b)** showing the optical and solution color features, respectively, for the iron oxide nanostructures prepared with different reaction times (hydrothermal reaction). **(c)** SEM images for the iron oxide nanostructures after 12 hours. [Reused with permission from [32], Royal Society of Chemistry.]

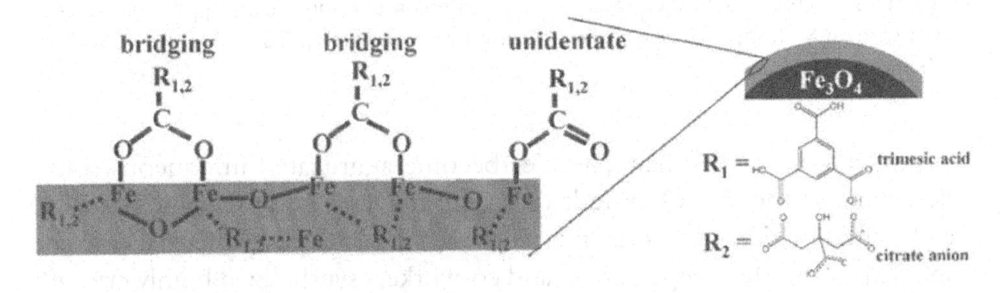


**Figure 4.** Schematic depiction representing an illustration of the amorphous layer, which consists of the Felattice–Olattice interface and the chelating ligands covering the $Fe_3O_4$ surface. [Reused with permission from [32], Royal Society of Chemistry.]

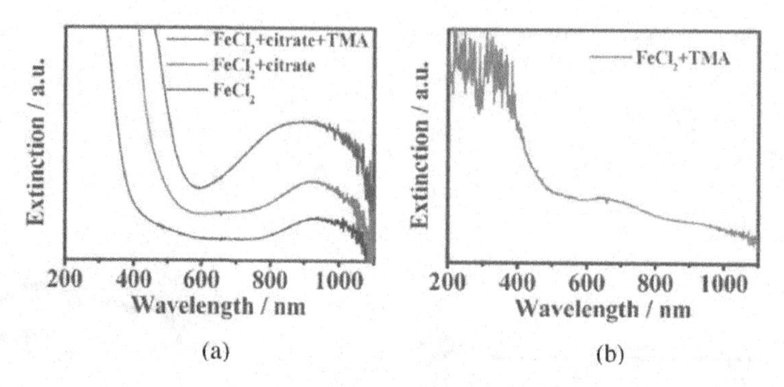


**Figure 5.** Absorption spectra of (**a**) $FeCl_2$, $FeCl_2$+citrate, and $FeCl_2$+citrate+TMA and (**b**) $FeCl_2$+TMA solutions. The Fe concentration is 34.5 mM. [Reused with permission from [32], Royal Society of Chemistry.]

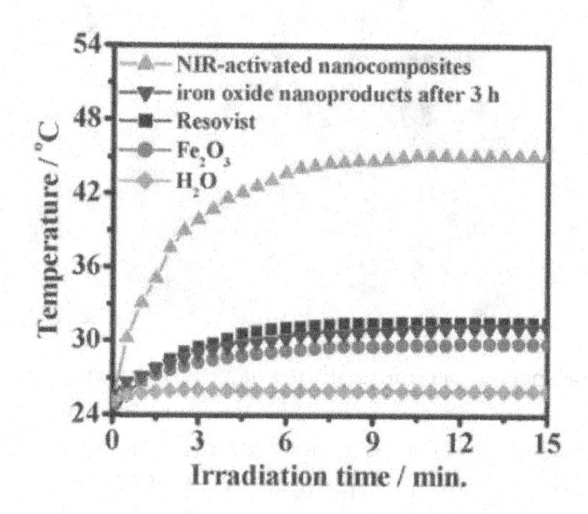


**Figure 6.** Photothermal effect (temperature *versus.* time plots) of the Resovist, commercial $Fe_2O_3$ nanopowder, iron oxide nanostructures after 3 hours, and NIR-activated nanocomposites (APTES-functionalized $Fe_3O_4/SiO_2$) with an iron concentration of 400 $ppm_{[Fe]}$ after irradiation with an 808 nm laser (2 W/cm²). [Reused with permission from [32], Royal Society of Chemistry.]

contrast, if some of the nanoparticles become aggregated in aqueous solution, not all of them can be irradiated using the laser because some particles are obscured by other particles; this will result in lower photothermal conversion efficiency in the sample. Chen and co-workers synthesized highly crystallized $Fe_3O_4$ nanoparticles that were 14 nm in diameter (**Figures 7(a) and 7(b)**) in organic solvents by thermal decomposition, and then coated them with a polysiloxane-containing diblock copolymer.[29] All of the polymer-coated highly crystallized iron oxide nanoparticles (HCIONPs) obtained were individually dispersed (**Figure 7(c)**).[29] In addition, these polymer-coated

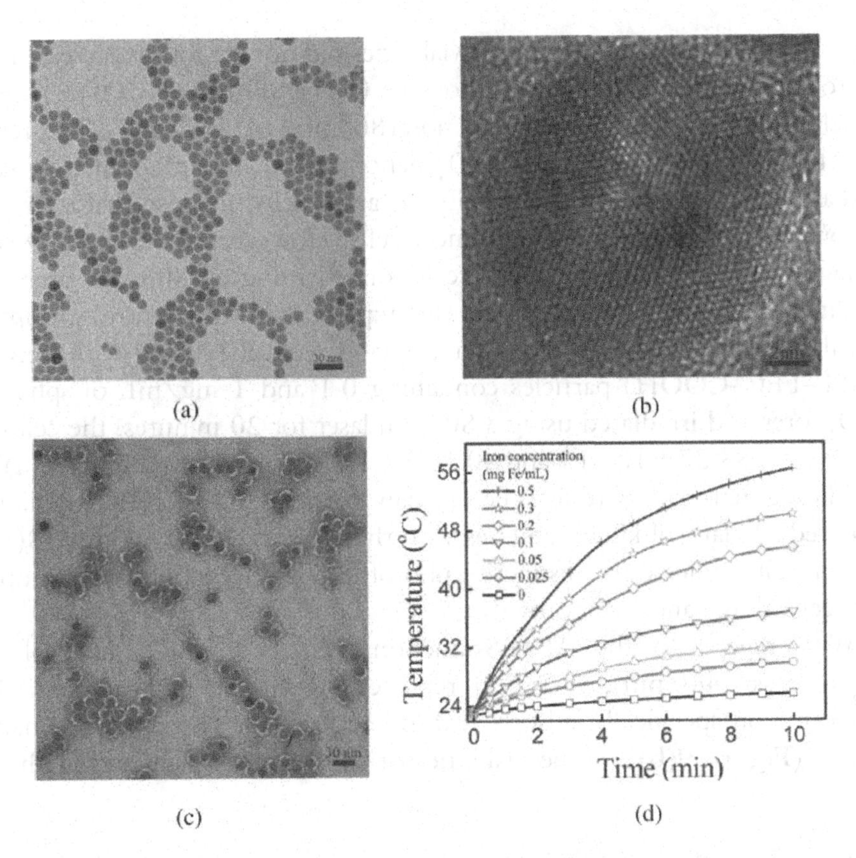


**Figure 7.** **(a)** TEM image of the as-prepared highly crystallized magnetic nanocrystals. **(b)** HRTEM image of magnetic nanocrystals. **(c)** TEM image (negative stained) of polymer-coated magnetic nanocrystals. **(d)** Measured temperatures of increasing concentrations of magnetic nanocrystals in water. All temperatures were measured during 10 minutes of illumination with a diode laser (885 nm, 2.5 W/cm$^2$). [Reused with permission from [29], Royal Society of Chemistry.]

HCIONPs are stable under various physiological conditions including phosphate buffer and cell culture media. Therefore, these highly stable nanoparticles exhibit high photothermal efficiency upon NIR laser irradiation (885 nm; 2.5 W/cm$^2$; **Figure 7(d)**).[29]

### 3.3. *PTT involving magnetic iron oxide nanoparticles in vitro*

Magnetic $Fe_3O_4$ nanoparticles have very low toxicity or are non-toxic to cells. As shown in **Figure 8(a)**,[31] when the spherical $Fe_3O_4$/(DSPE–PEG–COOH) nanoparticles containing 0.1, 0.5 and 1 mg/mL of $Fe_3O_4$ were incubated with esophageal cancer cells (Eca-109) in RPMI-1640 culture medium at

37°C for more than 1 hour, the cell viabilities did not obviously decrease relative to Eca-109 cells incubated without $Fe_3O_4$/(DSPE–PEG–COOH) nanoparticles with or without laser irradiation (808 nm; 0.25 W/cm$^2$). This result indicates that the lipid-coated $Fe_3O_4$ nanoparticles in the absence of laser irradiation have extremely low toxicity to cancer cells, and also that laser irradiation alone cannot destroy the cancer cells. However, when Eca-109 cells containing $Fe_3O_4$ nanoparticles were irradiated using an 808 nm laser, cell viability was significantly decreased as compared with the control groups as described above. For example, after the cells were incubated with $Fe_3O_4$/(DSPE–PEG–COOH) particles containing 0.1 and 1 mg/mL of spherical $Fe_3O_4$ cores and irradiated using a 808 nm laser for 20 minutes, the cell viabilities were 83.37 ± 10.01% and 40.16 ± 4.14%, respectively (**Figure 8(a)**).[31] The hexagonal and wire-like $Fe_3O_4$ nanoparticles upon NIR irradiation exhibited similar cell killing abilities with the spherical $Fe_3O_4$ (**Figure 8(b)**). These results indicate the distinctive photothermal effects of $Fe_3O_4$ nanoparticles regarding cancer cell growth.

With regard to the APTES-functionalized $Fe_3O_4/SiO_2$ nanospheres ($Fe_3O_4$ cores: ~440 nm), it has been reported that human oral carcinoma KB cells can non-specifically adsorb and/or accumulate the magnetic nanospheres (**Figure 9(b)**).[32] The viabilities of the KB cells incubated with the

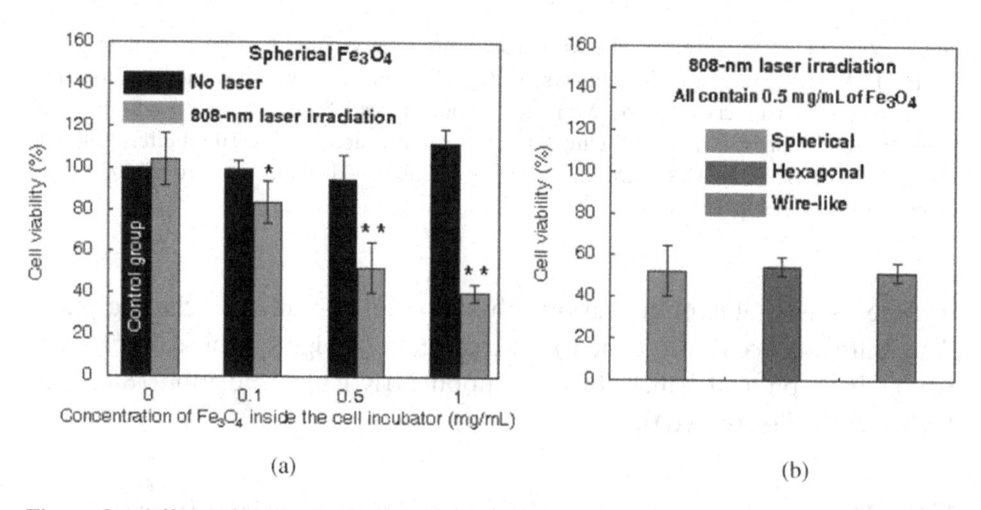


**Figure 8.** Effects of $Fe_3O_4$/(DSPE–PEG–COOH) on esophageal cancer cell viability with or without 808 nm laser irradiation. **(a)** Spherical $Fe_3O_4$ nanoparticles with different concentrations (0, 0.1, 0.5 and 1 mg/mL), with or without irradiation. **(b)** $Fe_3O_4$ nanoparticles with spherical, hexagonal, and wire-like shapes, all containing 0.5 mg/mL of $Fe_3O_4$ upon irradiation. The significance levels observed are *P <0.05 and **P <0.01 in comparison to control group values. Error bars were based on standard deviations. [Reused with permission from [31], Elsevier.]

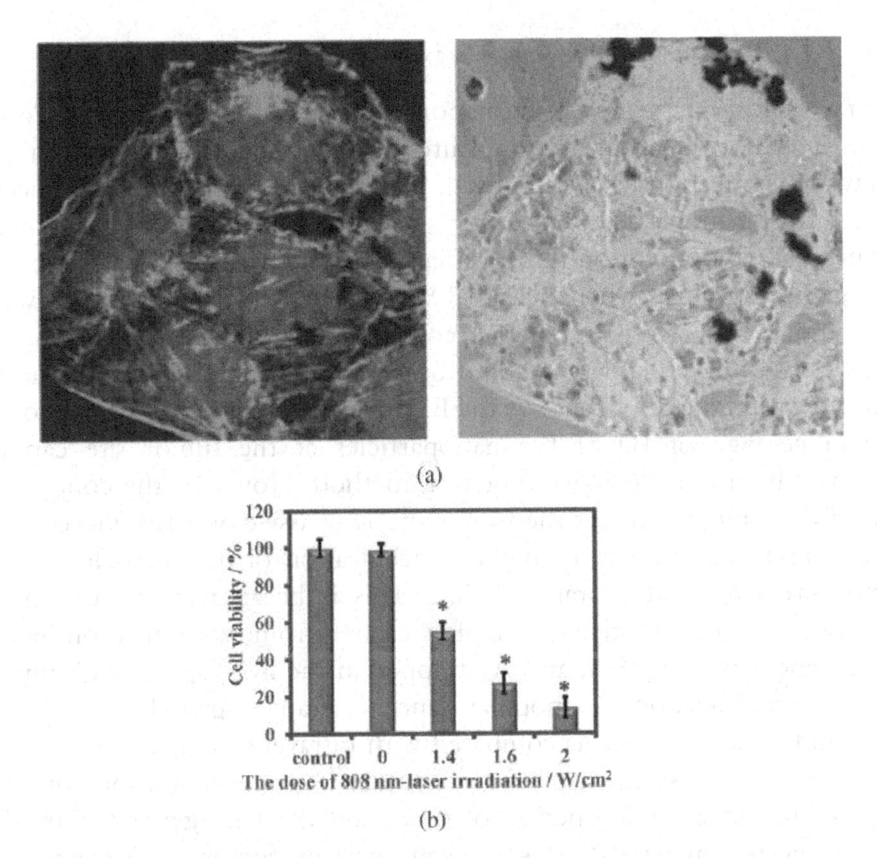


**Figure 9.** (a) Confocal laser images of KB cells treated with NIR-activated nanocomposites (APTES-functionalized $Fe_3O_4/SiO_2$): color merged multiple fluorescence (left) and brightfield + fluorescence (right). The cell cytoskeleton and nucleus were stained with Alexa 488 phalloidin (green) and DAPI (blue), respectively. (b) Viability of KB cells incubated with NIR-activated nanocomposites (120 $mg_{[Fe]}$/mL) after incubation for 1 hour and then irradiation with an 808 nm laser (15 minutes) with various power concentrations. Control indicates the cells without nanocomposite treatment and laser irradiation. *$P<0.05$ compared with control group. [Reused with permission from [32], Royal Society of Chemistry.]

magnetic nanospheres at doses of 0–120 $\mu$g/mL Fe for 24 hours are close to 100%.[32] However, when the KB cells incubated with the APTES-functionalized $Fe_3O_4/SiO_2$ nanospheres were irradiated with an 808 nm laser for 15 minutes, the cell viabilities decreased to 55% and 15% at power densities of 1.4 and 2 W/cm$^2$, respectively (**Figure 9(b)**).[32] The KB cells were killed by the heat generated from the $Fe_3O_4$ nanospheres upon laser irradiation. These results further demonstrate that the $Fe_3O_4$ nanoparticles are biocompatible materials, and that they can be used as an excellent heat source for cancer PTT.

### 3.4. *PTT in vivo of the magnetic iron oxide nanoparticles*

For *in vivo* cancer therapy using nanomaterials, including magnetic $Fe_3O_4$ nanoparticles, two administration routes namely intratumoral injection and intravenous injection have been used. For intravenous injection, the accumulation of nanoparticles at tumor sites usually relies on the enhanced permeability and retention (EPR) effect (passive targeting) and active targeting. For example, when SUM-159 tumor-bearing BALB/c mice were intravenously injected with polymer-coated highly crystallized $Fe_3O_4$ nanoparticles (14 nm in diameter), about 5.3% of the injection dose (ID) accumulated within the tumor site *via* the EPR effect at 48 hour post-injection.[29] The percentage of ID of the nanoparticles at the tumor site can be improved by using the active targeting method. However, the concentration of the nanoparticles at the tumor site *via* passive or active targeting is always significantly lower than the concentration of nanoparticles in the tumor site *via* intratumoral injection. This is because after intravenous injection many nanoparticles are rapidly cleared from the circulation by the reticuloendothelial system, and are trapped in the liver, spleen and lung.[29] Intratumoral injection is a popular cancer therapy approach used in the clinic and basic research. As compared with intravenous injection, intratumoral injection ensures that tumors contain high concentrations of heat reagents for an extended period of time, and the heat generated by the locally injected materials can specifically destroy tumors.[31] For example, when tumor-bearing mice (esophageal tumors grown from Eca-109 cells) were intratumorally injected with the $Fe_3O_4$/(DSPE–PEG–COOH) solution ($Fe_3O_4$: 8 mg/mL), a large amount of black $Fe_3O_4$ nanoparticles could be clearly observed at the tumor site even at 24 days post-injection; approximately 99.98±32.72%, 79.17±8.84% and 74.49±8.46% of $Fe_3O_4$ nanoparticles remained in the tumors for 1, 3 and 24 days post-injection, respectively (**Figure 10**).[31]

To evaluate the *in vivo* cancer treatment efficacy of PTT involving $Fe_3O_4$ nanoparticles, Chu *et al.*[31] directly injected the $Fe_3O_4$/(DSPE–PEG–COOH) solution (70 $\mu$L containing 8 mg/mL of $Fe_3O_4$) into mouse tumors (esophageal tumors) and used an 808 nm laser (0.25 W/cm$^2$) for irradiation over a period of 20 minutes every 24 hour. They found that tumor growth was significantly suppressed after 24 days of treatment (**Figure 11**). In addition, a charring spot was observed at the tumor site, indicating hemorrhage caused by the photothermal heat generated by the $Fe_3O_4$ nanoparticles. As controls, the tumors of mice injected with $Fe_3O_4$/(DSPE–PEG–COOH) solution without irradiation, or injected solely with

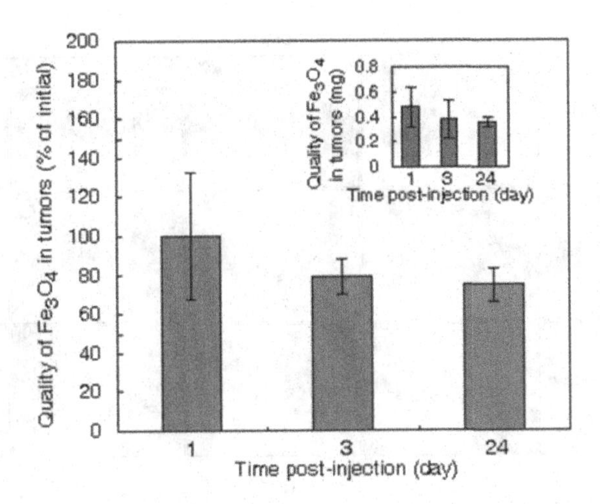


**Figure 10.** Percentage of mass retention of $Fe_3O_4$ in tumors measured by Inductively Coupled Plasma-Atomic Emission Spectrometry (ICP-AES). The tumor-bearing mice ($n = 9$) were injected with $Fe_3O_4$/(DSPE–PEG–COOH) aqueous suspensions (40 mg of $Fe_3O_4$ for each mouse) (insert: quality of $Fe_3O_4$ in tumors). Error bars were based on standard deviations according to three mice per group. [Reused with permission from [31], Elsevier.]

70 $\mu$L of Phosphate buffered saline (PBS) or PBS-dispersed DSPE–PEG–COOH (5 mg/mL), with and without irradiation, all grow markedly over time (**Figure 11**). This indicates that the magnetic nanoparticles have great potential for *in vivo* cancer PTT.

The growth inhibition of tumors treated with the magnetic iron oxide nanoparticles and NIR laser irradiation is caused by the photothermal effect of the $Fe_3O_4$ nanoparticles. For monitoring the spatial distribution of temperature in the tumor area, Liao and co-workers[32] used a thermographic camera to detect the tumor temperature in the presence or absence of iron oxide-mediated PTT. The mouse tumors (grown from KB cells) were intratumorally injected with APTES-functionalized $Fe_3O_4/SiO_2$ nanospheres (dosage: 0.5 mg/kg; $Fe_3O_4$ cores: ~440 nm) or water. The results indicated that the temperature of the tumors treated with the magnetic nanoparticles and NIR irradiation (808 nm, 2W/cm$^2$) for 10 minutes increased from approximately 34°C to 47°C.[32] In contrast, the temperature of tumor regions containing intratumorally injected $H_2O$ and exposed to NIR for 10 minutes increased from approximately 34°C to only 38°C.[32]

Chen *et al.*[29] and Shen *et al.*[30] used intravenous injection for iron oxide-mediated PTT. For example, SUM-159 tumor-bearing BALB/c mice were intravenously injected with magnetic nanoparticles (polymer-coated HCIONPs at a dose of 20 mg Fe per kg mouse body weight).[29] At 48 h post-injection, tumors were irradiated using a diode laser (885 nm; 2.5 W/cm$^2$)

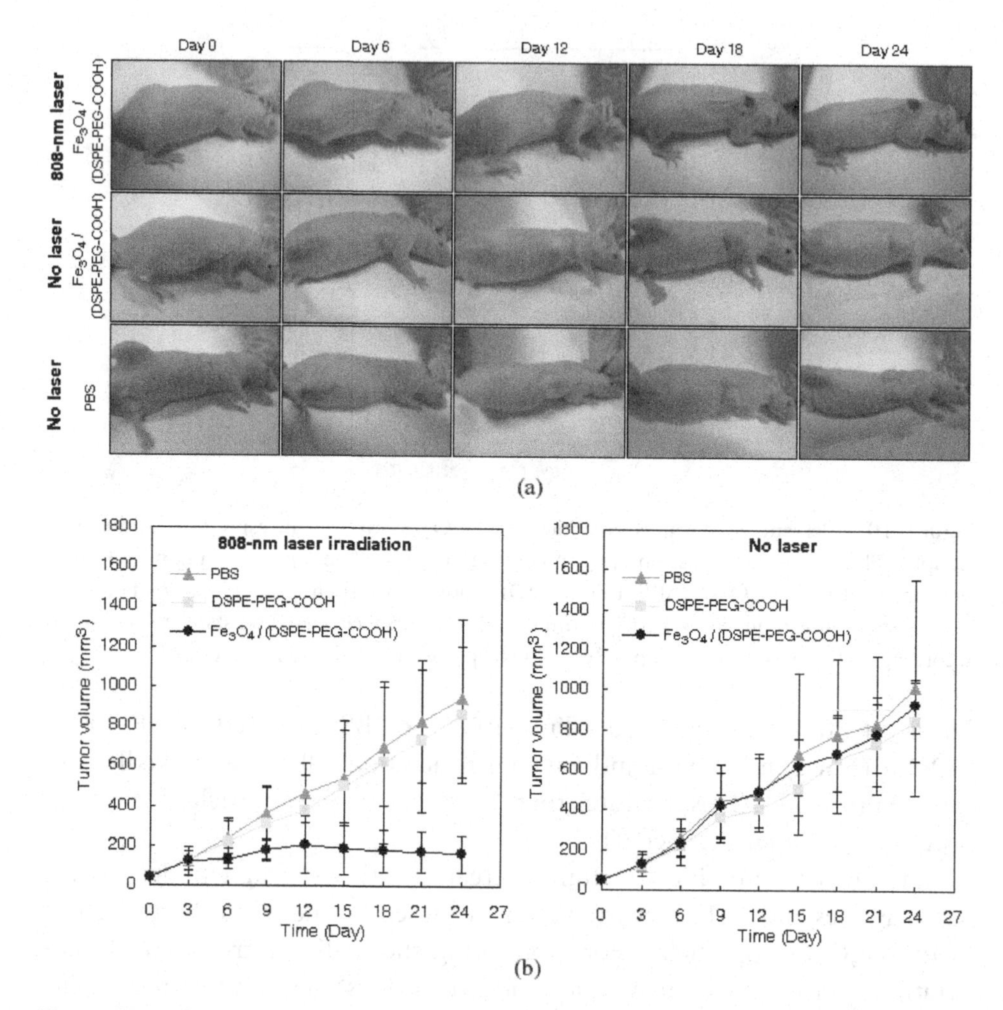


**Figure 11.** Photographs and volumes of mouse tumors treated with 70 μL of PBS-dispersed $Fe_3O_4$/(DSPE–PEG–COOH), DSPE–PEG–COOH, and PBS alone with and without 808 nm laser irradiation. **(a)** Representative photos of tumors with and without laser irradiation; **(b)** tumor growth curves with (left) and without (right) laser irradiation. Day 0 marks the tumor pre-irradiation and/or pre-injection, while day 3–24 mark the time periods from the first irradiation and/or injection. The concentrations of $Fe_3O_4$ and DSPE–PEG–COOH were 8 and 5 mg/ml, respectively. Only the $Fe_3O_4$/(DSPE–PEG–COOH) and 808 nm laser treated group showed significant tumor growth suppression compared with the control group (PBS, no laser) ($p<0.05$). Error bars were based on standard deviations. [Reused with permission from [31], Elsevier.]

for 10 minutes. The tumor surface temperature was monitored using an infrared camera. The results demonstrated that the temperature of the tumors treated with the magnetic nanoparticles and laser irradiation increased from approximately 35°C to ~60°C after irradiation for 10 minutes.[29]

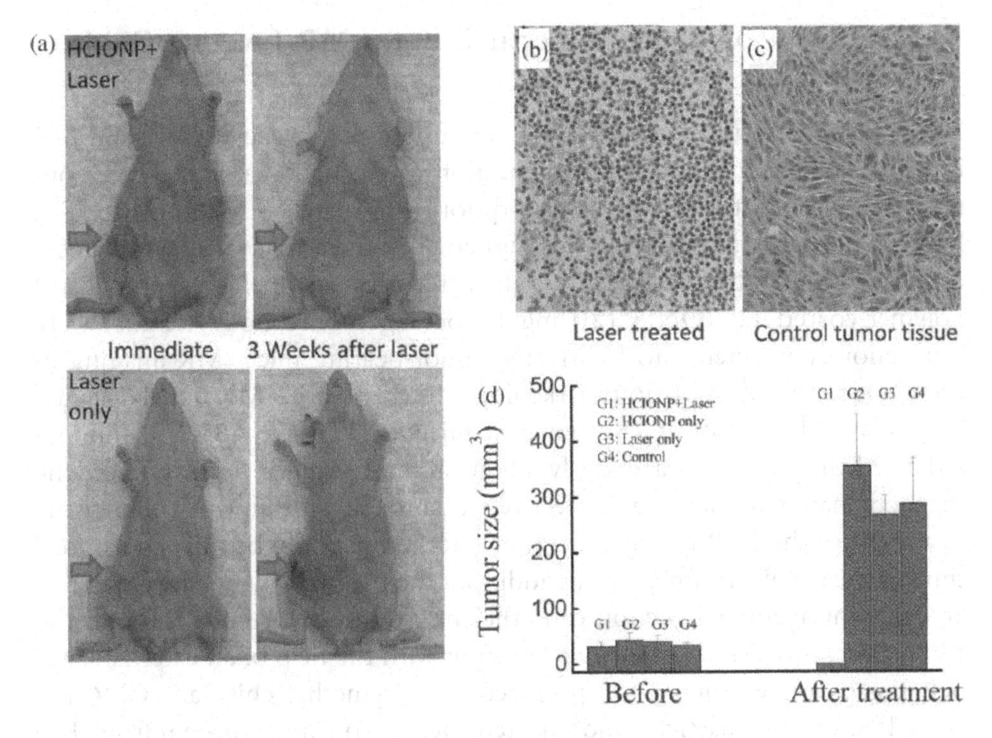


**Figure 12.** **(a)** Representative photos of SUM-159 tumor-bearing mice of both immediate and 3 weeks after laser (885 nm, 2.5W/cm$^2$) treatment. Irradiation time = 10 minutes. Arrows point to the tumor sites. H&E staining of tumor tissues from mouse treated with the polymer-coated highly crystallized iron oxide nanoparticles plus laser irradiation **(b)** and control mouse without any treatment **(c)**. **(d)** Anti-tumor efficacy of four different groups of mice before and 3 weeks post various treatments. Four groups (5 mice for each group) are magnetic nanocrystals injected mice with laser irradiation (G1), nanocrystals injected mice without laser irradiation (G2), laser treated mice without injection of nanoparticles (G3), and control mice injected with PBS (G4). Error bars are based on standard deviations. [Reused with permission from [29], Royal Society of Chemistry.]

For the mice intravenously injected with the magnetic nanoparticles and irradiated with the 885-nm laser, tumor tissue hemorrhaging was observed immediately after irradiation over a period of 10 minutes; complete tumor regression was observed after 3 weeks post-irradiation (**Figures 12(a) and 12(d)**).[29] In the control groups, the mouse tumors exhibited continuous growth leading to animal mortality within 4 weeks (**Figures 12(a) and 12(d)**). Hematoxylin and eosin (H&E) stained tumor slices revealed that the tumor tissue had obvious necrosis after treatment with the magnetic nanoparticles and laser irradiation relative to control normal tumor tissue (**Figures 12(b) and 12(c)**).[29]

## 4. Magnetic Iron Oxide Nanoparticles for MR Imaging Guided Cancer PTT

Magnetic nanoparticles can be simultaneously used for cancer PTT and MR imaging. Despite the successful application of Au nanomaterial as CT contrast, because of its high X-ray absorption, the spatial resolution in CT is lower than that in MR imaging.[32] Magnetic materials are available for cancer diagnosis due to their MR contrast characteristics. In Chen's experiments,[29] polymer-coated HCIONPs (20 mg Fe per kg mouse body weight) were intravenously injected into SUM-159 tumor-bearing mice. MR imaging of these mice showed an obvious darkening effect with a T2-MR signal decrease by ~40% in the tumor at 48 hour post-injection (**Figure 13**). This finding indicated that when intravenously administered to tumor-bearing mice, the magnetic nanoparticles could effectively accumulate within the tumor tissue as a result of the EPR effect[29]; this tumor targeting could be observed in real time by means of MR imaging. In addition, the magnetic nanoparticles delivered *via* intravenous injection can efficiently target the tumor through the EPR effect, and also accumulate in a tumor with the help of an external magnet. For example, Shen *et al.* prepared carboxymethyl chitosan (CMCTS) coated $Fe_3O_4$ nanoparticles and injected them into tumor (grown from 180 cells)-bearing mice *via* intravenous injection.[30] After maintenance of the tumor under a magnetically targeted field for 24 hours, the tumor-bearing mouse was taken for MR imaging, and the tumor was then irradiated using

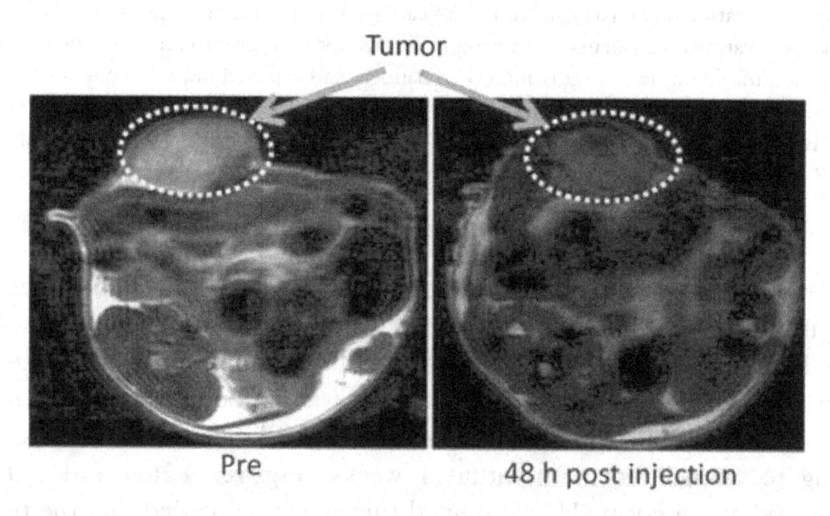


**Figure 13.** MR images of SUM-159 tumor-bearing mouse of pre (left) and 48 hour post tail-vein injection (right) of polymer-coated as-prepared highly crystallized iron oxide nanoparticles at a dosage of 20 mg Fe per kg body weight. [Reused with permission from [29], Royal Society of Chemistry.]

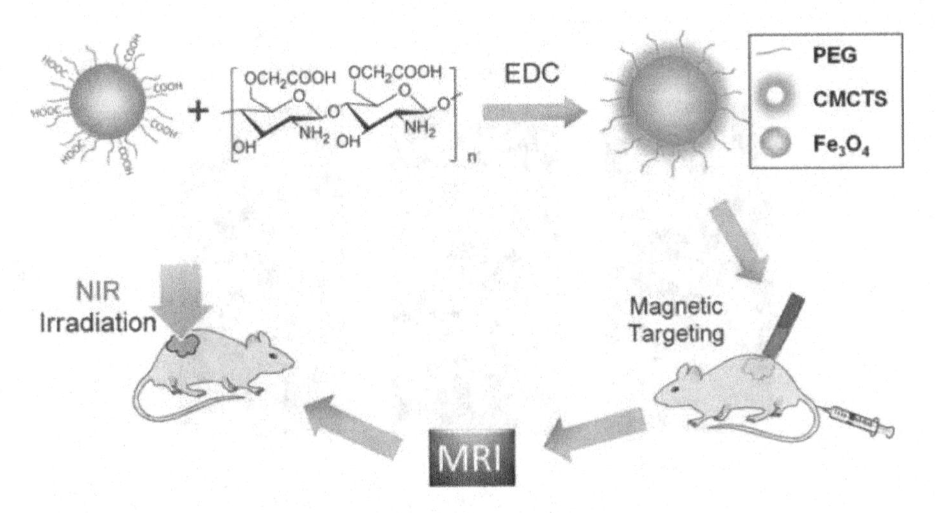


**Figure 14.** A schematic diagram showing the coating of the PEGylated $Fe_3O_4$ with CMCTS and the program of MR imaging-guided magnetically targeted hyperthermia therapy. The accumulation of $Fe_3O_4$@CMCTS is realized by placing an external magnet around the tumor region after intravenous injection. [Reused with permission from [30], Royal Society of Chemistry.]

an 808 nm laser (1.5 W/cm$^2$) (**Figure 14**).[30] MR imaging showed that the $Fe_3O_4$/CMCTS core-shell nanocomposites had accumulated in the tumor and that an external targeted magnetic field could improve the targeting efficiency (**Figure 15(a)**).[30] NIR-laser induced cancer PTT were then performed after the magnetic nanoparticles had accumulated in the tumor confirmed by the MR imaging. The mouse tumors treated with $Fe_3O_4$/CMCTS under a magnetically targeted field disappeared at about 10 days post-irradiation, whereas the mouse tumors in control groups all continued to grow (**Figure 15(b)**).[30]

## 5. Limitations of Cancer PTT and the Prospects for $Fe_3O_4$ Nanoparticle-Mediated Cancer PTT

It should be noted that although NIR laser light has strong penetration characteristics, tumors in deep tissue (such as visceral tumors) cannot usually receive sufficiently intense irradiation. To overcome this limitation, optical fibers may be used for delivering the laser light for cancer PTT. In addition, single therapy methods including PTT also have some limitations; the cancer cells usually cannot be completely killed using a single therapy approach. Multimodal therapy has frequently been used in the clinic. In recent years, therefore, cancer PTT combined with other therapeutic techniques including PDT and chemotherapy has been widely investigated.

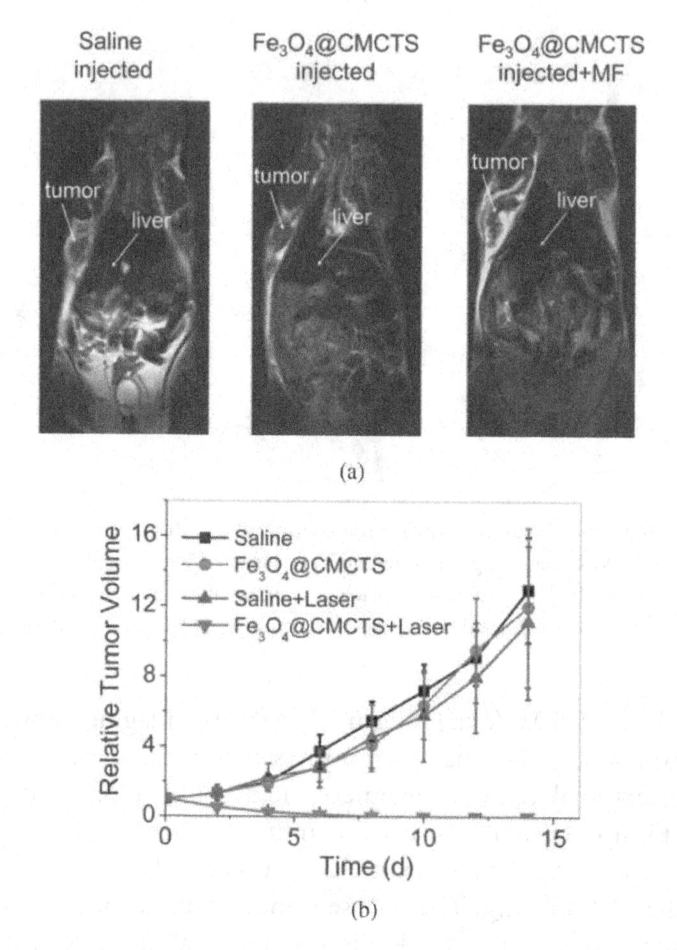


**Figure 15.** **(a)** *In vivo* T2-weighted MR images of S180 tumor-bearing mice obtained 24 hours after intravenous injection of saline (left), $Fe_3O_4$@CMCTS with MF (right), and without tumor-targeted magnetic field (middle, no MF). **(b)** Volumetric changes of S180 tumors in different groups of mice after irradiation. The relative tumor volumes were normalized to their initial sizes. For each group, 6 mice were used. Error bars were based on standard deviations. [Reused with permission from [30], Royal Society of Chemistry.]

Magnetic $Fe_3O_4$ nanoparticles have been safely used in patients, and the use of NIR light irradiation for heat generation is safer and simpler than that using an AMF. Consequently, magnetic $Fe_3O_4$ nanoparticles employed for cancer PTT, or for both MR imaging and PTT, may be applicable as a novel highly efficient cancer therapy strategy or for MR imaging guided cancer PTT. For the improvement of therapeutic efficacy against cancer, PTT using magnetic iron oxide nanoparticles should also be combined with various established cancer treatments in the future. Therefore, the use of highly crystallized $Fe_3O_4$

nanoparticles with special nanostructures including hollow, porous or polymer spheres should be encouraged for cancer PTT, since these nanostructures may exhibit the capacity to deliver a high drug payload. In addition, the NIR optical absorption coefficient of the highly crystallized $Fe_3O_4$ nanoparticles with special nanostructures may be significantly higher than that of particles with impure. Nanostructures Thus, $Fe_3O_4$ nanoparticle-mediated cancer therapy using the magnetic nanoparticles with novel or specific nanostructures can be performed using a NIR laser with only an ultra-low power density.

## References


1. Raaphorst GP. Fundamental aspects of hyperthermic biology. In *An Introduction to the Practical Aspects of Clinical Hyperthermia*, Field SB & Hand JW, eds., Taylor & Francis, London, 1990, pp. 10–54.
2. Fajardo LF. Pathological effects of hyperthermia in normal tissues. *Cancer Res* 1984;44:4826s–4835s.
3. Hegyi G, Szigeti GP, Szász A. Hyperthermia *versus* oncothermia: cellular effects in complementary cancer therapy. *Evid Based Complement Alternat Med* 2013; 672873. doi: 10.1155/2013/672873.
4. Timmerman RD, Bizekis CS, Pass HI, Fong Y, Dupuy DE, Dawson LA, Lu D. Local surgical, ablative, and radiation treatment of metastases. *CA Cancer J Clin* 2009;59:145–170.
5. Wu C, Yu C, Chu M. A gold nanoshell with a silica inner shell synthesized using liposome templates for doxorubicin loading and near-infrared photothermal therapy. *Int J Nanomed* 2011;6:807–813.
6. Hirsch LR, Stafford RJ, Bankson JA, Sershen SR, Rivera B, Price RE, Hazle JD, Halas NJ, West JL. Nanoshell-mediated near-infrared thermal therapy of tumors under magnetic resonance guidance. *Proc Natl Acad Sci* 2003;100:13549–13554.
7. Sikdar D, Rukhlenko ID, Cheng W, Premaratne M. Effect of number density on optimal design of gold nanoshells for plasmonic photothermal therapy. *Biomed Opt Express* 2013;4:15–31.
8. Chen J, Wang D, Xi J, Au L, Siekkinen A, Warsen A, Li ZY, Zhang H, Xia Y, Li X. Immuno gold nanocages with tailored optical properties for targeted photothermal destruction of cancer cells. *Nano Lett* 2007;7:1318–1322.
9. Chen J, Glaus C, Laforest R, Zhang Q, Yang M, Gidding M, Welch MJ, Xia Y. Gold nanocages as photothermal transducers for cancer treatment. *Small* 2010;6:811–817.
10. Hauck TS, Jennings TL, Yatsenko T, Kumaradas JC, Chan WCW. Enhancing the toxicity of cancer chemotherapeutics with gold nanorod hyperthermia. *Adv Mater* 2008;20: 3832–3838.
11. Alkilany AM, Thompson LB, Boulos SP, Sisco PN, Murphy CJ. Gold nanorods: their potential for photothermal therapeutics and drug delivery, tempered by the complexity of their biological interactions. *Adv Drug Deliv Rev* 2012;64:190–199.
12. Ren F, Bhana S, Norman DD, Johnson J, Xu L, Baker DL, Parrill AL, Huang X. Gold nanorods carrying paclitaxel for photothermal-chemotherapy of cancer. *Bioconjug Chem* 2013;24:376–386.
13. Wang Y, Black KCL, Li HLW, Zhang Y, Wan XCD, Liu S-Y, Li M, Kim P, Li Z-Y, Wang LV, Liu Y, Xia Y. Comparison study of gold nanohexapods, nanorods, and nanocages for photothermal cancer treatment. *ACS Nano* 2013;7:2068–2077.

14. Wang S, Huang P, Nie L, Xing R, Liu D, Wang Z, Lin J, Chen S, Niu G, Lu G, Chen X. Single continuous wave laser induced photodynamic/plasmonic photothermal therapy using photosensitizer-functionalized gold nanostars. *Adv Mater* 2013;25:3055–3061.
15. Chen H, Zhang X, Dai S, Ma Y, Cui S, Achilefu S, Gu Y. Multifunctional gold nanostar conjugates for tumor imaging and combined photothermal and chemo-therapy. *Theranostics* 2013;3:633–649.
16. Huang X, Tang S, Mu X, Dai Y, Chen G, Zhou Z, Ruan F, Yang Z, Zheng N. Freestanding palladium nanosheets with plasmonic and catalytic properties. *Nat Nanotechnol* 2011;6:28–32.
17. Moon HK, Lee SH, Choi HC. *In vivo* near-infrared mediated tumor destruction by photothermal effect of carbon nanotubes. *ACS Nano* 2009;3:3707–3713.
18. Wang L, Shi J, Zhang H, Li H, Gao Y, Wang Z, Wang H, Li L, Zhang C, Chen C, Zhang Z, Zhang Y. Synergistic anticancer effect of RNAi and photothermal therapy mediated by functionalized single-walled carbon nanotubes. *Biomaterials* 2013;34:262–274.
19. Hashida Y, Tanaka H, Zhou S, Kawakami S, Yamashita F, Murakami T, Umeyama T, Imahori H, Hashida M. Photothermal ablation of tumor cells using a single-walled carbon nanotube–peptide composite. *J Control Release* 2014;173:59–66.
20. Zhang M, Murakami T, Ajima K, Tsuchida K, Sandanayaka ASD, Ito O, Iijima S, Yudasaka M. Fabrication of ZnPc/protein nanohorns for double photodynamic and hyperthermic cancer phototherapy. *Proc Natl Acad Sci* 2008;105:14773–14778.
21. Whitney JR, Sarkar S, Zhang J, Do T, Young T, Manson MK, Campbell TA, Puretzky AA, Rouleau CM, More KL, Geohegan DB, Rylander CG, Dorn HC, Rylander MN. Single walled carbon nanohorns as photothermal cancer agents. Lasers Surg Med. 2011;43:43–51.
22. Jung HS, Kong WH, Sung DK, Lee M-Y, Beack SE, Keum DH, Kim KS, Yun SH, Hahn SK. Nanographene oxide–hyaluronic acid conjugate for photothermal ablation therapy of skin cancer. *ACS Nano* 2014;8:260–268.
23. Sahu A, Choi WI, Lee JH, Tae G. Graphene oxide mediated delivery of methylene blue for combined photodynamic and photothermal therapy. *Biomaterials* 2013;34:6 239–6248.
24. Yang K, Zhang S, Zhang G, Sun X, Lee S-T, Liu Z. Graphene in mice: ultrahigh *in vivo* tumor uptake and efficient photothermal therapy. *Nano Lett* 2010;10:3318–3323.
25. Chu M, Peng J, Zhao J, Liang S, Shao Y, Wu Q. Laser light triggered-activated carbon nanosystem for cancer therapy. *Biomaterials* 2013;34:1820–1832.
26. Chu M, Pan X, Zhang D, Peng J, Hai W. The therapeutic efficacy of CdTe and CdSe quantum dots for photothermal cancer therapy. *Biomaterials* 2012;33:7071–7083.
27. Hessel CM, Pattani VP, Rasch M, Panthani MG. Copper selenide nanocrystals for photothermal therapy. *Nano Lett* 2011;11:2560–2566.
28. Li Y, Lu W, Huang Q, Huang M, Li C, Chen W. Copper sulfide nanoparticles for photothermal ablation of tumor cells. *Nanomed* (*Lond*). 2010;5:1161–1171.
29. Chen H, Burnett J, Zhang F, Zhang J, Paholak H, Sun D. Highly crystallized iron oxide nanoparticles as effective and biodegradable mediators for photothermal cancer therapy. *J Mater Chem B* 2014;2:757–765.
30. Shen S, Kong F, Guo X, Wu L, Shen H, Xie M, Wang X, Jin Y, Ge Y. CMCTS stabilized $Fe_3O_4$ particles with extremely low toxicity as highly efficient near-infrared photothermal agents for *in vivo* tumor ablation. *Nanoscale* 2013;5:8056–8066.
31. Chu M, Shao Y, Peng J, Dai X, Li H, Wu Q, Shi D. Near-infrared laser light mediated cancer therapy by photothermal effect of $Fe_3O_4$ magnetic nanoparticles. *Biomaterials* 2013;34:4078–4088.
32. Liao M-Y, Lai P-S, Yu H-P, Lin H-P, Huang C-C. Innovative ligand-assisted synthesis of NIR-activated iron oxide for cancer theranostics. *Chem Commun* 2012;48:5319–5321.

33. Xu L, Cheng L, Wang C, Peng R, Liu Z. Conjugated polymers for photothermal therapy of cancer. *Polym Chem* 2014;5:1573–1580.
34. Huang X, Neretina S, El-Sayed MA. Gold nanorods: from synthesis and properties to biological and biomedical applications. *Adv Mater* 2009;21:4880–4910.
35. Mie G. Contributions to the optics of turbid media, particularly of colloidal metal solutions. *Ann Phys (Leipzig)* 1908;25:377–445.
36. Kerker, M. The scattering of light and other electromagnetic radiation. In *Physical Chemistry, a Series of Monographs*, Loebl EM, ed., Vol. 16, Academic Press, New York, 1969, pp. 42–50.
37. Papavassiliou GC. Optical properties of small inorganic and organic metal particles. *Prog Solid State Chem* 1979;12:185–271.
38. Bohren CF, Huffman DR. *Absorption and Scattering of Light by Small Particles*, Wiley, New York, 1983.
39. Kreibig U, Vollmer M. *Optical Properties of Metal Clusters*, Springer, Berlin, 1995.
40. Huang X, El-Sayed MA. Gold nanoparticles: Optical properties and implementations in cancer diagnosis and photothermal therapy. *J Adv Res* 2010;1:13–28.
41. Zeng S, Yong K-T, Roy I, Dinh X-Q, Yu X, Luan F. A review on functionalized gold nanoparticles for biosensing applications. *Plasmonics* 2011;6:491–506.
42. Rhodes C, Franzen S, Maria JP, Losego M, Leonard DN, Laughlin B, Duscher G, Weibel S. Surface plasmon resonance in conducting metal oxides. *J Appl Phys* 2006;100:054905.
43. Franzen S. Surface plasmon polaritons and screened plasma absorption in indium tin oxide compared to silver and gold. *J Phys Chem* C 2008;112:6027–6032.
44. Odom TW, Nehl CL. How gold nanoparticles have stayed in the light: the 3M's principle. *ACS Nano* 2008;2:612–616.
45. Maier-Hauff K, Ulrich F, Nestler D, Niehoff H, Wust P, Thiesen B. Efficacy and safety of intratumoral thermotherapy using magnetic iron-oxide nanoparticles combined with external beam radiotherapy on patients with recurrent glioblastoma multiforme. *J Neurooncol* 2011;103:317–324.
46. Maier-Hauff K, Rothe R, Scholz R, Gneveckow U, Wust P, Thiesen B. Intracranial thermotherapy using magnetic nanoparticles combined with external beam radiotherapy: results of a feasibility study on patients with glioblastoma multiforme. *J Neurooncol* 2007;81:53–60.
47. Johannsen M, Gneveckow U, Thiesen B, Taymoorian K, Cho CH, Waldöfner N, Scholz R, Jordan A, Loening SA, Wust P. Thermotherapy of prostate cancer using magnetic nanoparticles: feasibility, imaging, and three-dimensional temperature distribution. *Eur Urol* 2007;52:1653–1661.
48. Johannsen M, Gneveckow U, Taymoorian K, Thiesen B, Waldöfner N, Scholz R. Morbidity and quality of life during thermotherapy using magnetic nanoparticles in locally recurrent prostate cancer: results of a prospective phase I trial. *Int J Hyperthermia* 2007;23:315–323.
49. Jordan A, Scholz R, Maier-Hau K, Johannsen M, Wust P, Nadobny J, Schirra H, Schmidt H, Deger S, Loening S, Lanksch W, Felix R. Presentation of a new magnetic field therapy system for the treatment of human solid tumors with magnetic fluid hyperthermia. *J Magn Magn Mater* 2001;225:118–126.
50. Thiesen B, Jordan A. Clinical applications of magnetic nanoparticles for hyperthermia. *Int J Hyperthermia* 2008;24:467–474.

Chapter 5

# Nanomaterial-Based Sensors for Environmental Monitoring

*Jinhu Yang*

*Department of Chemistry, Tongji University,*
*Shanghai 200092, P. R. China*
*Research Center for Translational Medicine, East Hospital,*
*The Institute for Biomedical Engineering and Nano Science,*
*Tongji University School of Medicine, Shanghai 200092, P. R. China*

The rapid developments in nanotechnology bring new opportunities to many important fields. Nanomaterial-based novel sensors for environmental safety have attracted great attention. In this chapter, we will review recent progress in nanomaterial-based sensors for detection of environmental pollutants such as heavy metal ions, toxic gases, pesticides and industrial waste water. We will discuss the categories of the three different nanomaterials, namely, metals, carbon materials and semiconductors. Some representative samples will be provided and discussed in brief to illustrate the significant potential of nanomaterial-based sensors in the field of environmental monitoring.

## 1. Introduction

Environmental pollution has become a global issue and received special attention. Heavy metals, toxic gases, organophosphorous compounds (pesticides and insecticides) and wastewater (phenol, $H_2O_2$) are major environment pollutants produced from the industry and people activities. They can directly or indirectly damage our ecosystem and threaten environmental safety and people's health. For example, it is reported that heavy metal ions such as $Hg^{2+}$, $Pb^{2+}$, $Cd^{2+}$, etc. can accumulate in human bodies and cause some serious diseases that even develop to various cancers, if people drink or eat heavy-metal-contaminated water and food for a long time.[1] Some toxic and flammable gases also do great harm to people's health. They can affect the nervous system of humans and can cause people to lose consciousness at very low concentrations. Since people are living in a world surrounded by toxic pollutants, it is highly desired to develop advanced sensors with high sensitivity, selectivity as well as simple low-cost detection methods to monitor these toxic substances and protect ourselves from the toxic.

Technically, sensors are comprised of two functional components — the receptor and the transducer. The receptor is designed with high specificity to identify the different analytes, which is related closely to the performance of sensors. The transducer is another separate part to output sense responses in chemical or physical ways, such as electrochemical, optical, magnetic, thermal, and piezoelectric signals, depending on specific electrode materials of sensor devices associated with different detection principles. Nanotechnology — the science of controlling the structure of materials by atomic- or molecular-level manipulation and assembling of nanostructured materials as devices for potential applications — presents both opportunities and challenges. The development of nanotechnology promotes the generation of novel structured nanomaterial with advanced and unprecedented properties.[2] Nanostructured materials with various sizes, shapes and compositions often possess unique chemical, physical, optical, catalytic and electronic properties, which offer new possibilities for the development of novel sensing and monitoring technology. In the field of environmental monitoring, nanomaterial-based sensors have showed great potential in trace amount detection of pollutants, owing to their large specific surface area, high surface reactivity, high catalytic activity and strong adsorption capacity.[3–8] Therefore, the design and fabrication of nanomaterial-based chemosensors has generated great interest in a variety of scientific communities ranging from biological and environmental sciences to engineering.

In this chapter, our review will focus on recent progress in nanomaterial-based sensors for environmental pollution monitoring. The sensor devices made of different categories of nanomaterials such as metals, carbons and metal oxides will be discussed associating them with main pollutants of heavy metal ions, toxic gases, pesticide and industrial waste water detection. Some selected examples of nanomaterial-based sensors will be given to illustrate their applications in environmental monitoring, according to the above three categories of nanomaterials.

## 2. Nanomaterial-Based Sensors in the Field of Environmental Monitoring

### 2.1. *Metal nanoparticle-based sensors*

Au and Ag nanoparticles are dominant candidates for metal-based sensors, due to their high extinction coefficients and strong size- and distance-dependent optical properties which result from the collective oscillation of electrons at their surfaces, known as the unique surface plasmon resonances (SPR).[9] The SPR of Au and Ag is sensitive to nanoparticle size, shape, composition, interparticle distance and local dielectric environment.[10] On the basis of this unique property, various Au/Ag-based sensors relying on the technologies of colorimetry and surface-enhanced Raman scattering (SERS) have been constructed, which have been applied to the detection of a broad range of pollutants including heavy metal cations, aromatic compounds, inorganic pollutants, organophosphate, toxins, etc.[11]

From 1996, a series of colorimetric sensors based on the color changing of the AuNPs or AgNPs have been developed by Mirkin's group, which demonstrate the feasible, *in situ*, real-time qualitative/semiquantitative detection.[12,13] After that, colorimetric sensors for the detection of heavy metal ions and radioactive metal ($UO_2^{2+}$)[14,15] have been reported by Lu, *et al.* The developed system was not only selective and sensitive but also practical and convenient at room temperature. Taking one as an example, as illustrated in **Figure 1**, the $Pb^{2+}$ ions colorimetric sensor wherein was constructed by employing $Pb^{2+}$-specific DNAzyme and DNA-functionalized gold nanoparticles as the target recognition element and probing/signaling element, respectively.[14]

There were systematic investigations of Au nanoparticle-based colorimetric sensors for $Hg^{2+}$ ion detection, which can be classified into oligonucleotide–Au NP sensors,[16–23] oligopeptide–Au NP sensors[24–27] and functional molecule decorated Au nanoparticle sensors[28–33] in terms of the types of decorators on Au nanoparticles. For instance, T-rich nucleic acid (5'-TTCTTTCT TCCCTTGTTTGTT-3') modified Au nanoparticle sensor was developed for fast, simple and wide-range detection of $Hg^{2+}$ ions (**Figure 2**).[21] The ssDNA-treated Au nanoparticles showed a red color with the absorption band around 520 nm. After $Hg^{2+}$ ion addition, an obvious color change from red to blue caused by aggregation of Au nanoparticles occurred, which could be both detected by naked eyes and evidenced by the UV-vis spectrophotometer. The only interfering ions of $Pb^{2+}$ could be eliminated by using the mask agent of 2,6-pyridinedicarboxylic acid (PDCA). Therefore, the probe exhibited excellent selectivity, sensitivity, and low limit of detection (10 nM, 2 ppb).

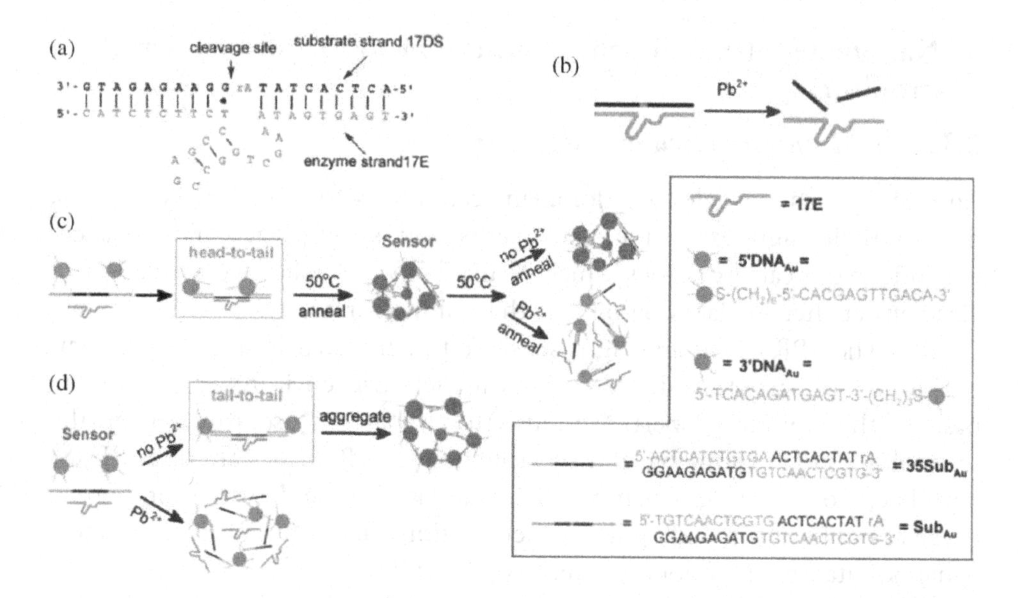


**Figure 1.** (**a**) Secondary structure of the "8–17" DNAzyme. (**b**) Cleavage of 17DS by 17E in the presence of $Pb^{2+}$. (**c**) Schematic of the previous colorimetric $Pb^{2+}$ sensor design. The three components of the sensor ($50DNA_{Au}$, $Sub_{Au}$, and 17E) can assemble to form blue-colored aggregates. Nanoparticles were aligned in a "head-to-tail" manner. A heating-and-cooling process (annealing) was required for detection. (**d**) Schematic of the new colorimetric sensor design. The nanoparticles are aligned in a "tail-to-tail" manner. In the absence of $Pb^{2+}$, nanoparticles can be assembled by the DNAzyme at ambient temperature within 10 minutes, resulting in a blue color; while in the presence of $Pb^{2+}$, $35Sub_{Au}$ is cleaved by 17E, inhibiting the assembly, resulting in a red color of separated nanoparticles. Reproduced from Ref. 14 with the permission of the author.

An oligopeptide–Au nanoparticle probe for $Hg^{2+}$ detection was prepared by Mandal *et al.* by means of *in situ* reduction and stabilization of oligopeptide on Au nanoparticles.[26] The probe showed remarkable color changes with adding $Hg^{2+}$ at ppm concentrations, i.e., from an initial red color with the SPR band at 527 nm in the absence of $Hg^{2+}$, to a purple color in the presence of 4 ppm $Hg^{2+}$ and further a blue color with a new SPR band at 670 nm in the presence of 12 ppm $Hg^{2+}$. In addition, this oligopeptide-Au nanoparticle probe displayed a good selectivity toward $Hg^{2+}$ among other metal ions such as $Pb^{2+}$, $Cu^{2+}$, $Cd^{2+}$, $Zn^{2+}$ and $Ca^{2+}$. Chen *et al.* demonstrated a pre-modification-free route to construct a new $Hg^{2+}$-detection sensor based on oligopeptide–Au nanoparticle probe. Due to the strong affinity between $Hg^{2+}$ and cysteine groups in oligopeptides (Lys-Cys-Gly-Trp-Gly-Cys), the detection system showed high ionic selectivity.[24]

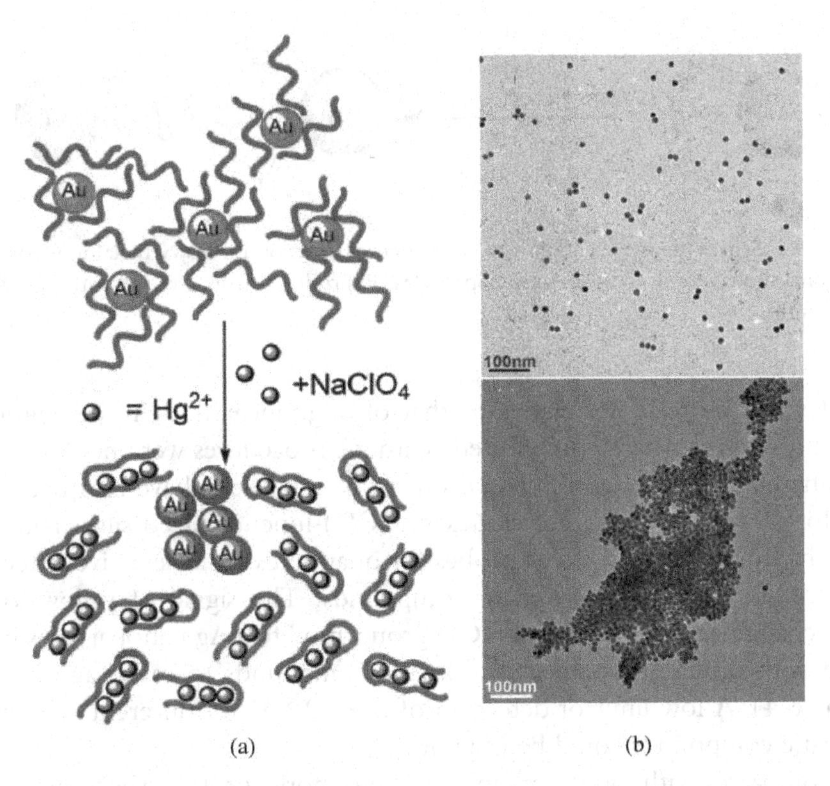


**Figure 2.** **(a)** Schematic representation of $Hg^{2+}$-stimulated aggregation of Au nanoparticles and **(b)** TEM images of nonaggregated Au nanoparticles stabilized by ssDNA in the presence $NaClO_4$ and the aggregated Au nanoparticles after addition of $Hg^{2+}$. Reproduced from Ref. 21 with permission of the author.

Functional molecules were good candidates to functionalize Au nanoparticles for colorimetric sensor construction. Functional groups containing S, O and N elements play important roles in ligands, such as in thiols, carboxyls, aminos and hydroxyls. Among these, thiol is one of the main functional group for Au nanoparticle decoration due to the strong Au–S bond. Hence, derivatives of mercapto aliphatic acid are preferred decorators for Au-based sensor devices. Tupp *et al.* developed a colorimetric sensor for heavy metal ions by utilizing 11-mercaptoundecanoic acid (MUA) as the capping agent for Au nanoparticles (13 nm) (**Figure 3**).[28] Strong SPR absorptions at 526 nm with a red color were obtained for aqueous MUA-capped Au nanoparticles. As the recognizing group, carboxyls in MUA molecules have strong chelation ability to heavy metal ions of $Pb^{2+}$, $Cd^{2+}$, $Hg^{2+}$, etc. Similar aggregation was observed when MUA-capped Au nanoparticles interacted with these heavy metal ions, resulting in the SPR band shift and color change of the system from red to blue. Functional molecule–Au nanoparticle preparation is

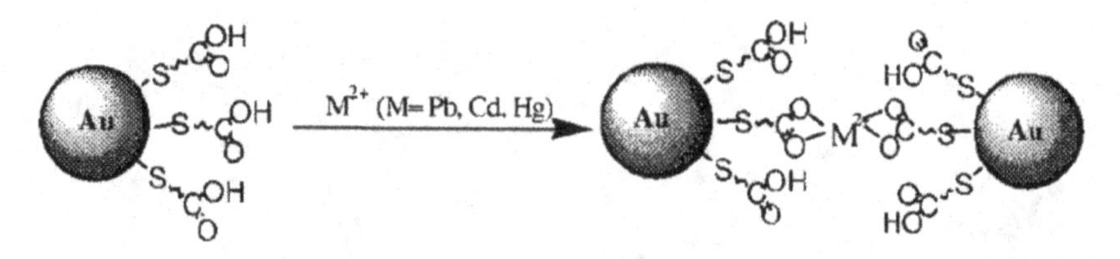


**Figure 3.** Schematic representation of colorimetric detection for heavy metal ions based on functional molecules decorated Au nanoparticles. Reproduced from Ref. 28 with permission of the author.

simpler and more facile relative to that of oligonucleotide–Au or oligopeptide–Au systems where complicated synthetic procedures were needed.

The colorimetric sensors based on Ag nanoparticles have also been well developed. For example, β-cyclodextrin (CD)-functionalized silver nanoparticles had been synthesized as probes for quantitative and sensitive detection of different isomers of aromatic compounds. The signal of yellow-to-red color change was observed when CD group-modified Ag nanoparticles interacted with different isomers of aromatic compounds to cause aggregation (**Figure 4**). A low limit of detection of $5 \times 10^{-5}$ M for different isomers of aromatic compounds could be obtained.[34]

Compared with other organic chromophoric probes, there are some advantages of nanoparticle-based colorimetric detection, based on the above description: (i) tunable novel optical properties (e.g., SPR); (ii) higher absorption extinction coefficient (ca. $10^8$ $cm^{-1}M^{-1}$ for AuNPs, three orders of magnitude higher than common organic dyes); (iii) high photostability; and (iv) ease of functionalization. Therefore, with these merits, metal nanoparticle-based colorimetric sensors are attractive and promising for environmental pollutant monitoring.

Another type of metal-nanoparticle-based sensors supported by a different SERS technique has also been developed for contaminant detection.[35–39] So far, some main toxic organic molecules such as phenol,[35] pesticides,[36] TNT[37] as well as heavy metal ions,[38,39] have been successfully detected. For example, Dasary and co-workers reported an ultrasensitive, low-cost, and miniaturized SERS probe made of cysteine–Au nanoparticles for 2,4,6-trinitrotoluene (TNT) detection. The designed SERS probe demonstrated high sensitivity and selectivity.[37] Zhang *et al.* fabricated a novel structure of Au@Ag core-shell nanoparticles as a SERS probe for detection of pesticide residues in a variety of fruit peels. The core-shell Au@Ag- nanoparticle-based probe gave a shell-thickness-dependent Raman enhancement signal. The detections of different pesticide residue such as thiocarbamate and organophosphorous compounds

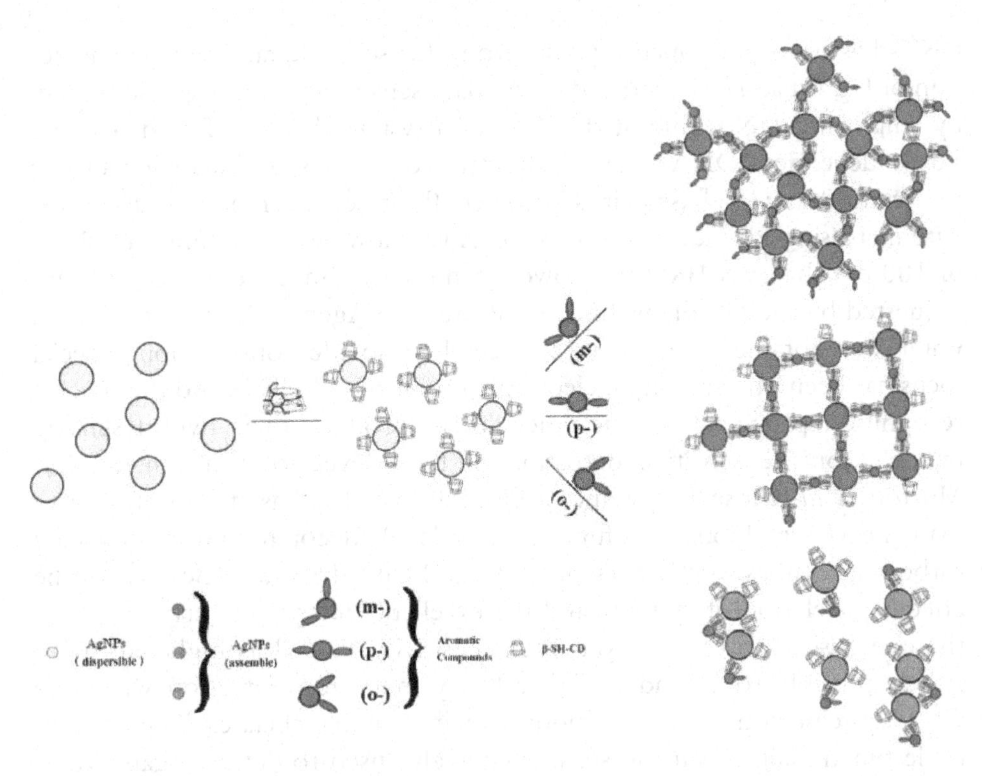


**Figure 4.** Schematic of host–guest recognition for β-CD-modified AgNPs with different aromatic compounds. Reproduced from Ref. 34 with permission of the author.

were realized with a satisfying sensitivity and simple manipulation by directly casting the particle sensors onto fruit peels.[36]

## 2.2. *Carbon-material-based sensors*

Since the discovery of carbon nanotubes (CNTs) by Iijima in 1991 and single-layer graphene by Geim *et al.* in 2004, CNTs and graphene, two most representative carbon materials, have received intense attention due to their fascinating properties of superior electric conductivity, high specific surface area, excellent mechanical flexibility and widespread applications in energy and environmental areas.[40–44] CNTs and graphene play particularly important roles in environmental monitoring with a wide range of pollutant detection, such as heavy metal ions ($Hg^{2+}$, $Pb^{2+}$, $Cu^{2+}$, $Ag^{+}$),[45–52] $H_2O_2$,[53] organophosphate pesticides,[54,55] toxic gases,[56–59] industrial wastewater,[60] etc.

For instance, some key achievements have been made using CNTs and graphene as conductive substrates to construct field-effect transistor (FET) sensors for heavy metal ion detection. A FET sensor based on single-wall

CNTs has been developed by Kim group for sensitive and selective detection of $Hg^{2+}$. Fan *et al.* reported a $Pb^{2+}$ nanosensor with improved selectivity by employing DNAzyme in the detection system.[52] It was found that the $Pb^{2+}$-independent DNAzyme was deactivated once its G-T wobble pair was replaced with a G-C base pair. The system fluorescence signal was decreased with increasing $Pb^{2+}$ ion concentration. A very low detection limit of 0.5 nM or 100 ppt that was 100 times lower than the maximum tolerance (72 nM) requested by the US Environmental Protection Agency (EPA) for drinking water was obtained, thanks to the graphene oxide contribution. Special focus has been on the unique electronic properties of CNTs and the specific recognition properties of biomolecules to produce their hybrid sensing material for the sensitive detection of trace levels of heavy metals.[61,62] Morton *et al.* chose L-cysteine, a biomolecule that has strong affinity to many heavy metal ions, as a functional molecule decorated on multi-walled carbon nanotubes (MWCNTs) for $Pb^{2+}$ and $Cu^{2+}$ detection. Because of the effective chelation of cysteine and the excellent electric property of CNTs, the composite sensor displayed high sensitivity as well as high selectivity. It was possible to extend to other heavy metal ion detection when the CNT–cysteine composite was modified with suitable chelates. The biomolecule-functionalized carbon sensors were also used to detect organic toxic pollutants. An enzyme-modified gold nanoparticle-graphene nanosheet composite probe has been designed for organophosphate pesticide detection.[55] The enzyme assembled on the Au/graphene hybrid material was induced by a PDDA link. The sensor gave an ultrasensitive detection and an ultralow detection limit of 0.1 pM, compared with other nanomaterial-based organophosphate pesticide sensors.

CNT and graphene are promising gas-sensing materials, and due to their unique properties of large surface area, strong gas adsorption capability and high electric conductivity that are favorable for high-performance gas sensing.[56–59,63,64] It was investigated that pristine CNT based sensors could give strong responses to various gases, such as $NO_2$, NO, $NH_3$, $SO_2$, etc., but poor selectivity and slow recovery.[65–67] Modi *et al.* developed a novel kind of ionization microsensors by using vertically aligned MWCNT film as the anode (**Figure 5**).[57] The sensor showed excellent performance that was independent of the influence of some system parameter, such as temperature, humidity, and gas flow. On the other hand, graphene-based gas sensors have been constructed for the detection of a range of gases including $H_2O$, $NH_3$, CO, $NO_2$ and NO.[58] These inexpensive high-sensitive sensors are expected to find more significant applications such as in counter-terrorism, besides pollutant detection in environmental monitoring. However, conventional solid-state gas

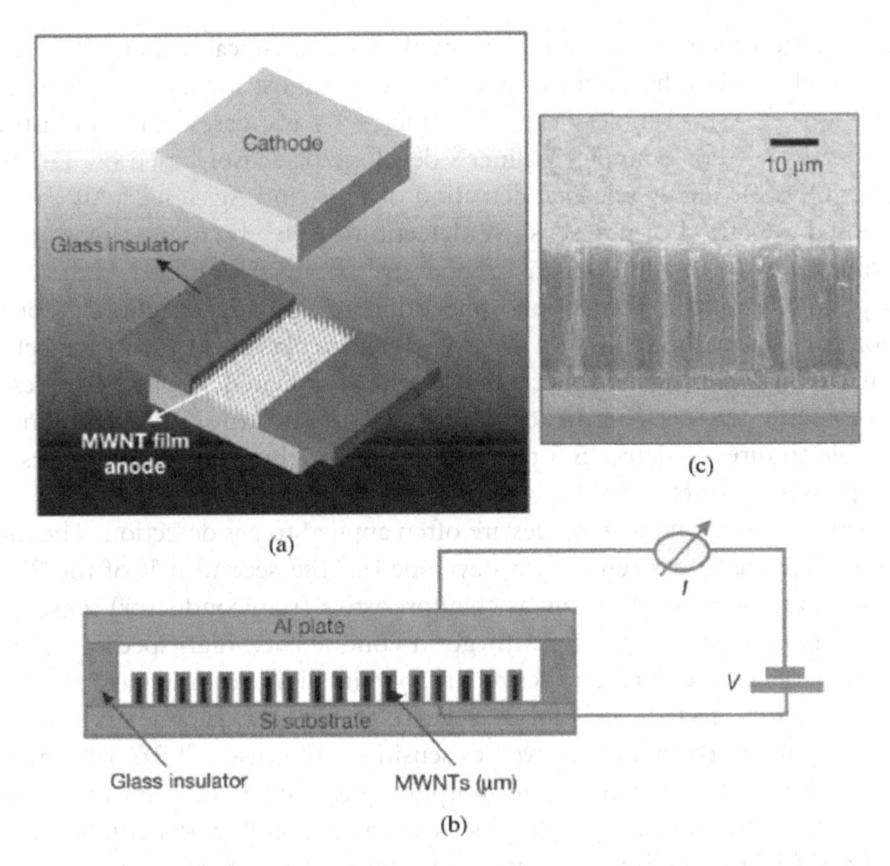


**Figure 5.** The nanotube sensor device. (**a**) Exploded view of the sensor showing MWNT film as the anode, 180 mm-thick glass insulator plates, and Al plate as cathode. (**b**) Diagram of actual test set-up. (**c**) SEM micrograph of a CVD-grown, vertically aligned MWNT film used as the anode. Reproduced from Ref. 57 with permission of the author.

sensors often have the advantages of high working temperature and long-term stability, which need special concern in future.

## 2.3. *Semiconductor-material-based sensors*

Semiconductor materials are of fundamental importance to both modern material science and engineering because of their unique energy band structure and novel electrical, optical, electrochemical, optoelectronic, catalytic, and sensing properties.[68,69] As semiconductor sensing materials, quantum dots (QDs) such as CdSe, CdTe, ZnS, etc. and oxide materials of $SnO_2$, CuO, ZnO as well as $TiO_2$ have received considerable attention.

Semiconductor QDs are extensively investigated as high-efficiency optical labels in the sensing research field due to their narrow emission bands, high fluorescence quantum yield, and photophyscial/chemical stability.[70] So far, QD-based sensors have been used to detect most pollutants including metal ions,[71–73] small molecules,[74] pesticides,[75,76] toxic gases,[77] and industrial wastewater.[78,79] For example, Willner's developed a sensor based on T-rich/C-rich QDs for highly selective detection of $Hg^{2+}$ and $Ag^{+}$ ions.[72] Ali' group proposed a method of ultrasensitive detection of $Pb^{2+}$ by using glutathione-modified QDs.[71] In addition, an antibody-free strategy with both high sensitivity and selectivity for organophosphorothioate pesticide detection has been reported. The method relies on a simple ligand-replacement turn-on mechanism coupled with the surface coordination-originated fluorescence resonance energy transfer (**Figure 6**).[76] This fluorescence turn-on chemosensor was able to directly detect 5.5 ppb of organophosphorothioate pesticides of chlorpyrifos in fruits with a detection limit of 0.1 nM.

Sensors based on metal oxides are often applied to gas detection. The first metal-oxide-based gas sensor was developed in the second half of the 20th century and are applied mainly as chemoresistive (semiconductor) sensors.[80] In principle, sensing materials with good conductivity, high specific surface area and open structure are desired to obtain high performance, because these structure factors are closely related to the key parameters such as response and recovery time as well as sensitivity of sensors.[81] To date, metal oxides with various structures and morphologies such as 1D nanostructures, mesoporous structures, hierarchical structures, and hollow structures, which are favorable for high-performance gas sensors have been employed. For

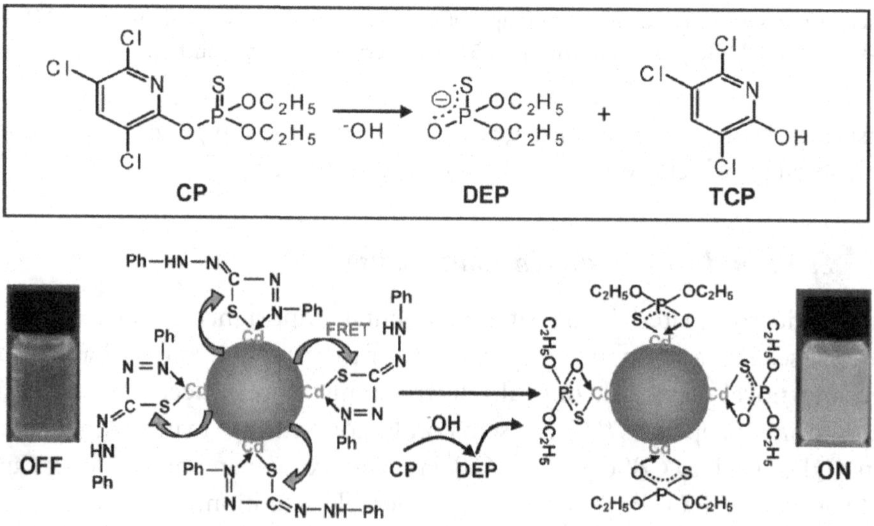

**Figure 6.** Chlorpyrifos is hydrolyzed to diethylphosphorothioate (DEP) and trichloro-2-pyridinol (TCP) in basic media. Reproduced from Ref. 76 with permission of the author.

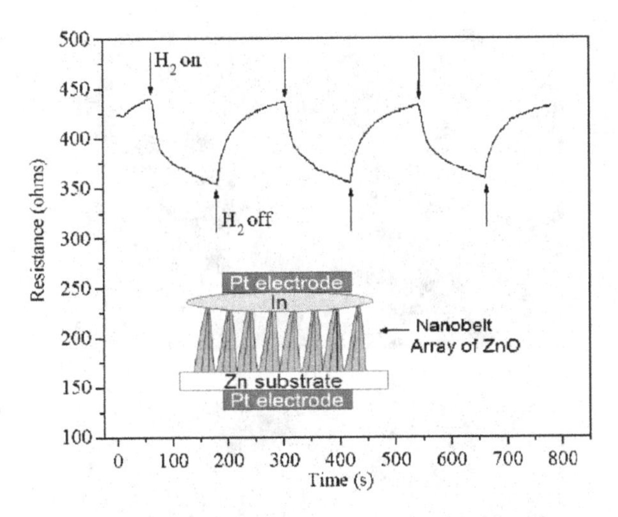


**Figure 7.** Gas sensor responses of ZnO nanobelt array grown on Zn substrate to $H_2$ at a concentration of 250 ppm. Inset is a schematic of the device based on a ZnO nanobelt array on Zn substrate. Reproduced from Ref. 85 with permission of the author.

example, the 1D nanostructures such as nanowires, nanorods, and nanotubes with opened configuration have been fabricated and used for gas sensing.[82,83] With the development of synthetic strategies, the improvement of gas sensing characteristics using 1D ZnO, $SnO_2$, $In_2O_3$, and $WO_3$ nanostructures has been investigated.[84,85] Yang *et al.* proposed a novel electrochemical route to the *in situ* fabrication of ultrathin ZnO nanobelt (<8 nm) arrays on Zn foils and investigated sensing performance of ZnO nanobelts toward $H_2$ (**Figure** 7).[85] The sensitivity of the sensor at 250 ppm $H_2$ concentration was 14%, which is higher than the common ZnO film sensors under similar working conditions. Remarkably, the sensor response and recovery were very good even at room temperature, which is ascribed to the high surface area of the ordered ZnO nanobelt arrays consisted by ultrathin ZnO nanobelts.

Mesoporous oxide materials with well-aligned pores are another kind of attractive candidates for gas sensing technology.[86–90] Owing to their high specific surface area that allows full adsorption of target gases and porous structures that facilitate gas diffusion kinetics, the oxide-based sensors with porous nanostructure often showed very high gas response and short response time.[91–94] This is similar to that of the hierarchical oxide structures which usually possessed micropores or mesopores within the architectures formed during the assembly processes. Lee and co-workers hydrothermally fabricated a novel kind of 3D flower-like $SnO_2$ hierarchical microspheres which consist of 2D $SnO_2$ nanoflakes, as shown in **Figure 8**.[95] The hierarchical $SnO_2$ spheres contained a higher surface area and mesopores (4.5–20 nm) compared to dense

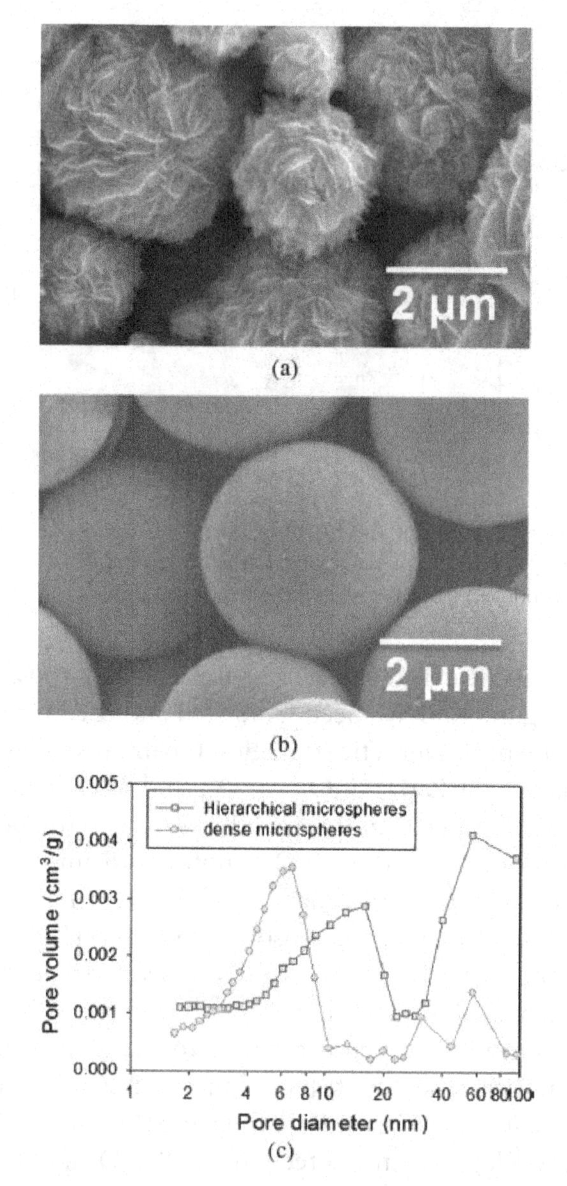


**Figure 8.** SEM images of (**a**) flower-like $SnO_2$ hierarchical microspheres and (**b**) dense $SnO_2$ microspheres, and (**c**) the pore-size distributions of hierarchical and dense $SnO_2$ microspheres determined from nitrogen adsorption–desorption isotherm. Reproduced from Ref. 95 with permission of the author.

$SnO_2$ microspheres, indicating a desirable structure for gas sensing. Ponzoni *et al.* employed the W thermal evaporation method and synthesized novel $WO_3$ nanowire networks for $NO_2$ sensing. The sensor could give a response of ~6-fold increase of resistance when exposed to 50 ppb $NO_2$ at 300°C.[96]

Hollow oxide structures have also been the main focus for enhancing the gas sensing performance. For gas sensor applications, the hollow interior and the thin and permeable shell layers are advantageous for complete electron depletion and effective gas diffusion, respectively. Thus, representative gas sensing materials have been prepared as hollow structures.[97–103] Sb-doped $SnO_2$ hollow spheres were prepared by Martinez *et al.* by using templating of polystyrene spheres and the sensor based on the $SnO_2$ hollow spheres delivered a 3–5 fold resistance changes compared to the sensors made of polycrystalline $SnO_2$ films or Sb:$SnO_2$ microporous nanoparticle films.[101] Kim *et al.* constructed the gas sensors based on $TiO_2$ hollow hemisphere for $NO_2$ gas detection, which demonstrated 2-time sensitivity relative to solid structures of plain $TiO_2$ film.[102] It is noteworthy that Yang *et al.* proposed a simple one-pot template-free method to synthesize CuO hollow spheres with hierarchical pores, that is, quasimicropores (1.0–2.2 nm), mesopores (5–30 nm), and macropores (hollow cores, 2–4 μm).[104]

As shown in **Figure 9**, the CuO hollow spheres also displayed a hierarchical architecture, namely, the primary CuO nanograins, the quasi-single-crystal

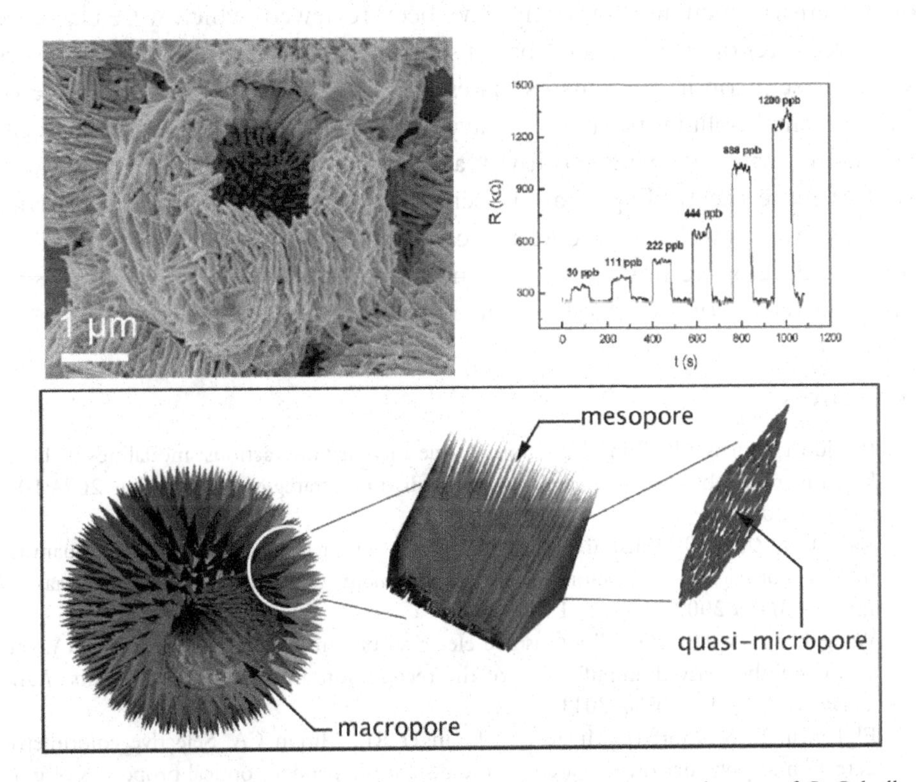


**Figure 9.** Schematic illustration to show the hierachcial structure and pores of CuO hollow spheres with remarkable sensing performance. Reproduced from Ref. 104 with permission of the author.

nanosheets assembled by nanograins, and the spheres composed of the nanosheets. With such unique hierarchical pores and architecture, the CuO hollow spheres exhibited excellent sensing performance toward $H_2S$ as gas sensing material, such as low detection limit of 2 ppb, high sensitivity at parts per billion level concentration, broad linear range, short response time of 3 seconds, and recovery time of 9 seconds. The excellent performance was ascribed to a synergetic effect of the hierarchical structure of the unique CuO spheres: the quasi-micropores offer active sites for effectively sensing, the mesopores facilitate the molecular diffusion kinetics, and the macropores serve as gas reservoirs and minimize diffusion length, while good conductivity of the quasi-single-crystal nanosheets favors fast charge transportation, which in turn contributes to the high sensitivity, quick response, and recovery of the $H_2S$ sensor.

## 3. Conclusions and Future Perspectives

In conclusion, some important advances regarding nanomaterial-based sensors for environmental monitoring have been reviewed, which were classified into three categories, i.e., metal-based sensors, carbon-material-based sensors as well as semiconductor sensors. These sensors can detect a wide range of environmental pollutants, such as heavy metal ions, toxic gases, organic molecules, etc., with satisfying sensitivity and selectivity. However, from a viewpoint of more practical application, sensor devices with enhanced sensitivity and selectivity as well as low cost and ease of operation are highly desired. In addition, developing novel sensing nanomaterials to obtain wearable sensors with long-term working stability is also necessary.

## References

1. Onyido I, Norris AR, Buncel E. Biomolecule-mercury interactions: modalities of DNA based-mercury binding mechanisms. Remediation strategies. *Chem Rev* 2004;104: 5911–5930.
2. Sadik OA, Zhou AL, Kikandi S, Du N, Wang Q, Varner K. Sensors as tools for quantitation, nanotoxicity and nanomonitoring assessment of engineered nanomaterials. *J Environ Monit* 2009;11:1782–1800.
3. Wang J, Liu G, Jan MR. Ultrasensitive electrical biosensing of proteins and DNA: carbon-nanotube derived amplification of the recognition and transduction events. *J Am Chem Soc* 2004;126:3010–3011.
4. Elghanian R, Storhoff JJ, Mucic RC, Lesinger RL, Mirkin CA. Selective colorimetric detection of polynucleotides based on the distance-dependent optical properties of gold nanoparticles. *Science* 1997;277:1078–1081.

5. Zhang J, Wang L, Pan D, Song S, Boey FYC, Zhang H, Fan C. Visual cocaine detection with gold nanoparticles and rationally engineered aptamer structures. *Small* 2008;4:1196–1200.
6. Lee J, Mahendra S, Alvarez PJJ. Nanomaterials in the construction industry: a review of their applications and environmental health and safety considerations. *ACS Nano* 2010;4:3580–3590.
7. Zhang L, Fang M. Nanomaterials in pollution trace detection and environmental improvement. *Nano Today* 2010;5:128–142.
8. Halvoroson RA, Vikesland PJ. Surface-enhanced Raman spectroscopy (SERS) for environmental analyses. *Environ Sci Technol* 2010;44:7749–7755.
9. Kelly KL, Coronado E, Zhao LL and Schatz GC. The optical properties of metal nanoparticles: the influence of size, shape, and dielectric environment. *J Phys Chem B* 2003;107: 668–677.
10. Willets KA, Van Duyne RP. Localized surface plasmon resonance spectroscopy and sensing. *Annu Rev Phys Chem* 2007;58:267–297.
11. Wang C, Yu CX. Detection of chemical pollutants in water using gold nanoparticles as sensors: a review. *Rev Anal Chem* 2013;32(1):1–14.
12. Rosi NL, Giljohann DA, Thaxton CS, Lytton-Jean AKR, Han S, Mirkin CA. Oligonucleotide-modified gold nanoparticles for intracellular gene regulation. *Science* 2006;312:1027–1030.
13. Mirkin CA, Letsinger RL, Mucic RC, Storhoff JJ. A DNA-based method for rationally assembling nanoparticles into macroscopic materials. *Nature* 1996;382:607–609.
14. Liu J, Lu Y. Accelerated color change of gold nanoparticles assembled by DNAzymes for simple and fast colorimetric $Pb^{2+}$ detection. *J Am Chem Soc* 2004;126:12298–12305.
15. Lee JH, Wang Z, Liu J, Lu Y. Highly sensitive and selective colorimetric sensors for uranyl ($UO_2^{2+}$): development and comparison of labeled and label-free DNAzyme-gold nanoparticle systems. *J Am Chem Soc* 2008;130:14217–14226.
16. Nykypanchuk D, Maye MM, van der Lelie D, Gang O. DNA-guided crystallization of colloidal nanoparticles. *Nature* 2008;451:549–552.
17. Tanaka Y, Oda S, Yamaguchi H, Kondo Y, Kojima C, Ono A. N-15-N-15 J-coupling across Hg-II: direct observation of Hg-II-mediated T-T base pairs in a DNA duplex. *J Am Chem Soc* 2007;129:244–245.
18. Xue XJ, Wang F, Liu XG. One-step, room temperature, colorimetric detection of mercury ($Hg^{2+}$) using DNA/nanoparticle conjugates. *J Am Chem Soc* 2008;130:3244–3245.
19. Oh JH, Lee JS. Designed hybridization properties of DNA-gold nanoparticle conjugates for the ultraselective detection of a single-base mutation in the breast cancer gene BRCA1. *Anal Chem* 2011;83:7364–7370.
20. Li HX, Rothberg L. Colorimetric detection of DNA sequences based on electrostatic interactions with unmodified gold nanoparticles. Proceedings of the National Academy of Sciences of the United States of America 2004;14036–14039.
21. Li D, Wieckowska A, Willner I. Optical analysis of Hg(2+) ions by oligonucleotide-gold-nanoparticle hybrids and DNA-based machines. *Angew Chem Int Ed* 2008;47: 3927–3931.
22. Wang Y, Yang F, Yang XR. Colorimetric biosensing of mercury(II) ion using unmodified gold nanoparticle probes and thrombin-binding aptamer. *Biosens Bioelectron* 2010;25:1994–1998.

23. Yu CJ, Cheng TL, Tseng WL. Effects of $Mn^{2+}$ on oligonucleotide-gold nanoparticle hybrids for colorimetric sensing of $Hg^{2+}$: improving colorimetric sensitivity and accelerating color change. *Biosens Bioelectron* 2009;25:204–210.
24. Du JJ, Sun YH, Jiang L, Cao XB, Qi DP, Yin SY, Ma J, Boey FYC, Chen XD. Flexible colorimetric detection of mercuric ion by simply mixing nanoparticles and oligopeptides. *Small* 2011;7:1407–1411.
25. Joshi JP, Lee WI, Lee KH. Ratiometric and turn-on monitoring for heavy and transition metal ions in aqueous solution with a fluorescent peptide sensor. *Talanta* 2009;78:903–909.
26. Si S, Kotal A, Mandal TK. One-dimensional assembly of peptide-functionalized gold nanoparticles: an approach toward mercury ion sensing. *J Phys Chem C* 2007;111:1248–1255.
27. Slocik JM, Zabinski JS, Phillips DM, Naik RR. Colorimetric response of peptide-functionalized gold nanoparticles to metal ions. *Small* 2008;4:548–551.
28. Kim YJ, Johnson RC, Hupp JT. Gold nanoparticle-based sensing of "spectroscopically silent" heavy metal ions. *Nano Lett* 2001;1:165–167.
29. Chen L, Lou TT, Yu CW, Kang Q, Chen LX. N-1-(2-Mercaptoethyl)thymine modification of gold nanoparticles: a highly selective and sensitive colorimetric chemosensor for $Hg^{2+}$. *Analyst* 2011;136:4770–4773.
30. Yang XR, Liu HX, Xu J, Tang XM, Huang H, Tian DB. A simple and cost-effective sensing strategy of mercury (II) based on analyte-inhibited aggregation of gold nanoparticles. *Nanotechnology* 2011;22:275503.
31. Du JJ, Hu MM, Fan JL, Peng XJ. Fluorescent chemodosimeters using "mild" chemical events for the detection of small anions and cations in biological and environmental media. *Chem Soc Rev* 2012;41:4511–4535.
32. Hung YL, Hsiung TM, Chen YY, Huang YF, Huang CC. Colorimetric detection of heavy metal ions using label-free gold nanoparticles and alkanethiols. *J Phys Chem C* 2010;114:16329–16334.
33. Liu DB, Qu WS, Chen WW, Zhang W, Wang Z, Jiang XY. Highly sensitive, colorimetric detection of mercury(II) in aqueous media by quaternary ammonium group-capped gold nanoparticles at room temperature. *Anal Chem* 2010;82:9606–9610.
34. Chen X, Parker SG, Zou G, Su W, Zhang Q. Beta-cyclodextrin-functionalized silver nanoparticles for the naked eye detection of aromatic isomers. *ACS Nano* 2010;4:6387–6394.
35. Li D, Fossey JS, Long Y. Portable surface-enhanced Raman scattering sensor for rapid detection of aniline and phenol derivatives by on-site electrostatic preconcentration. *Anal Chem* 2010;82:9299–9305.
36. Liu B, Han G, Zhang Z, Liu R, Jiang C, Wang S, Han M. Shell thickness-dependent Raman enhancement for rapid identification and detection of pesticide residues at fruit peels. *Anal Chem* 2012;84:255–261.
37. Dasary SSR, Singh AK, Senapati D, Yu H, Ray PC. Gold nanoparticle based label-free SERS probe for ultrasensitive and selective detection of trinitrotoluene. *J Am Chem Soc* 2009;131:13806–13812.
38. Zamarion VM, Timm RA, Araki K, Toma HE. Ultrasensitive SERS nanoprobes for hazardous metal ions based on trimercaptotriazine-modified gold nanoparticles. *Inorg Chem* 2008;47:2934–2936.
39. Wang Y, Irudayaraj J. A SERS DNAzyme biosensor for lead ion detection. *Chem Commun* 2011;47:4394–4396.

40. Wang XL, Li G, Chen Z, Augustyn V, Ma XM, Wang G, Dunn B, Lu YF. High-performance supercapacitors based on nanocomposites of $Nb_2O_5$ nanocrystals and carbon nanotubes. *Adv Energy Mater* 2011;1:1089–1093.
41. Chen Z, Augustyn V, Wen J, Zhang YW, Shen MQ, Dunn B, Lu YF. High-Performance Supercapacitors Based on Intertwined CNT/$V_2O_5$ Nanowire Nanocomposites, *Adv Mater* 2011;23:791–795.
42. Liu CG, Yu ZN, Neff D, Zhamu A, Jang BZ. Graphene-based supercapacitor with an ultrahigh energy density. *Nano lett* 2010;10:4863–4868.
43. Kim TY, Lee HW, Stoller M, Dreyer DR, Bielawski CW, Ruoff RS, Suh KS. High-performance supercapacitors based on poly(ionic liquid)-modified graphene electrodes. *ACS Nano* 2011;5:436–442.
44. Zhu YW, Murali S, Stoller MD, Ganesh KJ, Cai WW, Ferreira PJ, Pirkle A, Wallace RM, Cychosz KA, Thommes M, Su D, Stach EA, Ruoff RS. Carbon-based supercapacitors produced by activation of graphene. *Science* 2011;332:1537–1541.
45. Zhang L, Li T, Li B, Li J, Wang E. Carbon nanotube-DNA hybrid fluorescent sensor for sensitive and selective detection of mercury(II) ion. *Chem Commun* 2010;46:1476–1478.
46. Gao X, Xing G, Yang Y, Shi X, Liu R, Chu W, Jing L, Zhao F, Ye C, Yuan H, Fang X, Zhao Y. Detection of trace Hg(2+) via induced circular dichroism of DNA wrapped around single-walled carbon nanotubes. *J Am Chem Soc* 2008;130: 9190–9191.
47. Gong J, Zhou T, Song D, Zhang L. Monodispersed Au nanoparticles decorated graphene as an enhanced sensing platform for ultrasensitive stripping voltammetric detection of mercury(II). *Sens Actuators, B* 2010;150:491–497.
48. Zhao X, Kong R, Zhang X, Meng H, Liu W, Tan W, Shen G, Yu R. Graphene-DNAzyme based biosensor for amplified fluorescence "Turn-On" detection of $Pb^{2+}$ with a high selectivity, *Anal Chem* 2011;83:5062–5066.
49. Sudibya HG, He Q, Zhang H, Chen P. Electrical detection of metal ions using field-effect transistors based on micropatterned reduced graphene oxide films. *ACS Nano* 2011;5:1990–1994.
50. Liu M, Zhao H, Chen S, Yu H, Zhang Y, Quan X. Label-free fluorescent detection of Cu(II) ions based on DNA cleavage-dependent graphene-quenched DNAzymes *Chem Commun* 2011;47:7749–7751.
51. Wang B, Luo B., Liang M., Wang A, Wang J, Fang Y, Chang Y, Zhi L. Chemical amination of graphene oxides and their extraordinary properties in the detection of lead ions. *Nanoscale* 2011;3:5059–5066.
52. Wen Y, Peng C, Li D, Zhuo L, He S, Wang L, Huang Q, Xu Q, Fan C. Metal ion-modulated graphene-DNAzyme interactions: design of a nanoprobe for fluorescent detection of lead (II) ions with high sensitivity, selectivity and tunable dynamic range. *Chem Commun* 2011;47(3):6278–6280.
53. Wang J, Musameh M, Lin Y. Solubilization of carbon nanotubes by Nafion toward the preparation of amperometric biosensors. *J Am Chem Soc* 2003;125:2408–2409.
54. Mauter MS, Elimelech M. Environmental applications of carbon-based nanomaterials. *Environ Sci Technol* 2008;42:5843–5859.
55. Wang Y, Zhang S, Du D, Shao Y, Li Z, Wang J, Engelhard MH, Li J, Lin Y. Self-assembly of acetylcholinesterase on a gold nanoparticles-graphene nanosheet hybrid for organophosphate pesticide detection using polyelectrolyte as a linker. *J Mater Chem* 2011;21: 5319–5325.

56. Qi P, Vermesh O, Grecu M, Javey A, Wang Q, Dai H. Toward large arrays of multiplex functionalized carbon nanotube sensors for highly sensitive and selective molecular detection. *Nano Lett* 2003;3:347–351.
57. Modi A, Koratkar N, Lass E, Wei B, Ajayan PM. Miniaturized gas ionization sensors using carbon nanotube. *Nature* 2003;424:171–174.
58. Schedin F, Geim AK, Morozov SV, Hill EW, Blake P, Katsnelson MI, Novoselov KS. Detection of individual gas molecules adsorbed on graphene. *Nat Mater* 2007;6: 652–655.
59. Li W, Geng X, Guo Y, Rong J, Gong Y, Wu L, Zhang X, Li P, Xu J, Cheng G, Sun M, Liu L. Reduced graphene oxide electrically contacted graphene sensor for highly sensitive nitric oxide detection. *ACS Nano* 2011;5:6955–6961.
60. Goh MS, Pumera M. Number of graphene layers exhibiting an influence on oxidation of DNA bases: analytical parameters. *Anal Bioanal Chem* 2011:399:127–131.
61. Morton J, Havens N, Mugweru A, Wanekaya AK. Detection of trace heavy metal ions using carbon nanotube-modified electrodes. *Electroanalytical* 2009;21:1597–1603.
62. Janegitz BC, Marcolino-Junior LH, Campana-Filho SP, Faria RC, Fatibello-Filho O. Anodic stripping voltammetric determination of copper(II) using a functionalized carbon nanotubes paste electrode modified with crosslinked chitosan. *Sens Actuators B Chem* 2009;142:260–266.
63. Wei L, Shizhen H, Wenzhe C. An MWCNT-doped $SNO_2$ thin film $NO_2$ gas sensor by RF reactive magnetron sputtering. *J Semicond* 2010;31:024006.
64. Penza M, Rossi R, Alvisi M, Signore MA, Cassano G, Dimaio D, Pentassuglia R, Piscopiello E, Serra E, Falconieri M. Characterization of metal-modified and vertically-aligned carbon nanotube films for functionally enhanced gas sensor applications. *Thin Solid Films* 2009;517:6211–6216.
65. Moon S-IL, Paek K-K, Lee Y-H, Park H-K, Kim J-K, Kim S-W Ju B-K. Bias-heating recovery of MWCNT gas sensor. *Mater Lett* 2008;62:2422–2425.
66. Goldoni A, Petaccia L, Gregoratti L, Kaulich B, Barinov A, Lizzit S, Laurita A, Sangaletti L, Larciprete R. Spectroscopic characterization of contaminants and interaction with gases in single-walled carbon nanotubes. *Carbon* 2004;42:2099–2112.
67. Valentini L, Cantalini C, Armentano I, Kenny J M, Lozzi L, Santucci S. Highly sensitive and selective sensors based on carbon nanotubes thin films for molecular detection. *Diam Relat Mater* 2004;13:301–1305.
68. Zhang T, Dong W, Keeter-Brewer M, Konar S, Njabon RN, Tian ZR. Site-specific nucleation and growth kinetics in hierarchical nanosyntheses of branched ZnO crystallites. *J Am Chem Soc* 2006;128:10960–10968.
69. Lu CH, Qi LM, Yang JH, Wang XY, Zhang DY, Xie JL, Ma JM. One-pot synthesis of octahedral $Cu_2O$ nanocages via a catalytic solution route. *Adv Mater* 2005;17:2562–2567.
70. Han M, Gao X, Su JZ, Nie S. Quantum-dot-tagged microbeads for multiplexed optical coding of biomolecules. *Nat Biotechnol* 2001;19:631–635.
71. Ali EM, Zheng Y, Yu H, Ying JY. Ultrasensitive $Pb^{2+}$ detection by glutathione-capped quantum dots. *Anal Chem* 2007;79:9452–9458.
72. R Freeman, T Finder and I Willner, Multiplexed Analysis of $Hg^{2+}$ and $Ag^{+}$ Ions by Nucleic Acid Functionalized CdSe/ZnS Quantum Dots and Their Use for Logic Gate Operations, *Angew Chem Int Ed* 48 2009; 7818–7821.

73. Long Y, Jiang D, Zhu X, Wang J Zhou F. Trace $Hg^{2+}$ Analysis via Quenching of the Fluorescence of a CdS-Encapsulated DNA Nanocomposite. *Anal Chem* 2009;81: 2652–2657.
74. Koneswaran M, Narayanaswamy R. Mercaptoacetic acid capped CdS quantum dots as fluorescence single shot probe for mercury(II). *Sens Actuators B* 2009;139:91–96.
75. Vinayaka AC, Thakur MS. Focus on quantum dots as potential fluorescent probes for monitoring food toxicants and foodborne pathogens. *Anal Bioanal Chem* 2010;397:1445–1455.
76. Zhang K, Mei Q, Guan G, Liu B, Wang S, Zhang Z. Ligand replacement-induced fluorescence switch of quantum dots for ultrasensitive detection of organophosphorothioate pesticides. Anal Chem 2010;82:9579–9586.
77. Ma Q, Cui H, Su X. Highly sensitive gaseous formaldehyde sensor with CdTe quantum dots multilayer films. *Biosens Bioelectron* 2009;25:839–844.
78. Li H, Han C. Sonochemical synthesis of cyclodextrin-coated quantum dots for optical detection of pollutant phenols in water. *Chem Mater* 2008;20:6053–6059.
79. Goldman ER, Medintz IL, Whitley JL, Hayhurst A, Clapp AR, Uyeda HT, Deschamps JR, Lassman ME, Mattoussi H. A hybrid quantum dot-antibody fragment fluorescence resonance energy transfer-based TNT sensor. *J Am Chem Soc* 2005;127:6744–6751.
80. Baker AR, Firth JG. *Min Eng* 1969;128:237–244.
81. Lee J-H. Gas sensors using hierarchical and hollow oxide nanostructures: overview. *Sens Actuators B* 2009;140:319–336.
82. Comini E, Bratto C, Faglia G, Ferroni M, Vomiero A, Sberveglieri G. Quasi-one dimensional metal oxide semiconductors: preparation and characterization and application as chemical sensors. *Prog Mater Sci* 2009;54:1–67.
83. Kolmakov A, Moskovits M. Chemical sensing and catalyst by one-dimensional metal oxide nanostructures. *Annu Rev Mater Res* 2004;34:151–180.
84. Xia Y, Yang P, Sun Y, Wu Y, Mayers B, Gates B, Yin Y, Kim F, Yan H. One dimensional nanostructures: synthesis, characterization, and applications. *Adv Mater* 2003;15:353–389.
85. Yang JH, Liu GM, Lu J, Qiu YF, Yang SH. Electrochemical route to the synthesis of ultrathin ZnO nanorod/nanobelt arrays on zinc substrate. *Appl Phys Lett* 2007,90:103109.
86. Ciesla U, Schüth F. Ordered mesoporous materials. *Micropor MesoporMater* 1999;27:131–149.
87. Yang P, Zhao D, Margolese DI, Chmelka BF, Stucky GD. Generalized syntheses of large-poremesoporous metal oxides with semicrystalline frameworks. *Nature* 1998;396:152–155.
88. Shon JK, Kong SS, Kim YS, Lee J-H, Park WK, Park SC, Kim JM. Solvent free infilteration method for mesoporous $SnO_2$ using mesoporous silica templates. *Micropor Mesopor Mater* 2009;120:441–446.
89. Shimizu Y, Hyodo T, Egashira M. Mesoporous semiconducting oxides for gas sensor application. *J Eur Ceram Soc* 2004;24:1389–1398.
90. Shimizu Y, Jono A, Hyodo T, Egashira M. Preparation of large mesoporous $SnO_2$ powders for gas sensor application. *Sens Actuators B* 2005; 108:56–61.
91. Devi GS, Hyodo T, Shimizu Y, Egashira M. Synthesis of mesoporous $TiO_2$-based powders and their gas-sensing properties. *Sens Actuators B* 2002;87:122–129.
92. Hyodo T, Shimizu Y, Egashira M. Gas-sensing properties of ordered mesoporous $SnO_2$ and effects of coating thereof. *Sens Actuators B* 2003;93:590–600.

93. Yang J, Hidajat K, Kawi S. Synthesis of nano-$SnO_2$/SBA-15 composite as a highly sensitive semiconductor oxide gas sensor. *Mater Lett* 2008;62:1441–1443.
94. Waitz T, Wagner T, Sauerwald T, Kohl C-D, Tiemann M. Ordered mesoporous $In_2O_3$: synthesis by structure replication and application as a methane gas sensor. *Adv Funct Mater* 2009; 19:653–661.
95. Kim H-R, Choi K-I, Lee J-H, Akbar SA. Highly sensitive and ultra-fast responding gas sensors using self-assembled hierarchical SnO2 spheres. *Sens Actuators B* 2009;136:138–143.
96. Ponzoni A, Comini E, Sberveglieri G, Zhou J, Deng SZ, Xu NS, Ding Y, Wang ZL. Ultrasensitive and highly selective gas sensors using three dimensional tungsten oxide nanowire networks. *Appl Phys Lett* 2006; 88:203101.
97. Hyodo T, Sasahara K, Shimizu Y, Egashira M. Preparation of macroporous $SnO_2$ films using PMMA microspheres and their sensing properties to NO*x* and $H_2$. *Sens Actuators B* 2005;106:580–590.
98. Tan Y, Li C, Wang Y, Tang J, Ouyang X. Fast-response and high sensitivity gas sensors based on $SnO_2$ hollow spheres. *Thin Solid Films* 2008; 516:7840–7843.
99. Zhang J, Wang S, Wang Y, Wang Y, Zhu B, Xia H, Guo X, Zhang S, Huang W, Wu S. $NO_2$ sensing performance of $SnO_2$ hollow-sphere sensor. *Sens Actuators B* 2009;135:610–617.
100. Zhong Z, Yin Y, Gates B, Xia Y. Preparation of mesoscale hollow spheres of $TiO_2$ and $SnO_2$ by templating against crystalline arrays of polystyrene beads. *Adv Mater* 2000;12:206–209.
101. Martinez CJ, Hockey B, Montgomery CB, Semancik S. Porous tin oxide nanostructured microspheres for sensor applications. *Langmuir* 2005;21:7937–7944.
102. Kim I-D, Rothschild A, Yang D-J, Tuller HL. Macroporous $TiO_2$ thin film gas sensors obtained using colloidal templates. *Sens Actuators B* 2008;130:9–13.
103. Wang WW, Zhu YJ, Yang LX. ZnO–$SnO_2$ hollow spheres and hierarchical nanosheets: hydrothermal preparation, formation mechanism, and photocatalytic properties. *Adv Funct Mater* 2007;17:59–64.
104. Qin Y, Zhang F, Chen Y, Zhou YJ, Li J, Zhu AW, Luo YP, Tian Y, Yang JH. Hierarchically porous CuO hollow spheres fabricated via a one-pot template-free method for high-performance gas sensors. *J Phys Chem C* 2012;116:11994–12000.

# Chapter 6

# Janus Nanostructures and Their Bio-medical Applications

*Yilong Wang* and Feng Wang†*

**Research Center for Translational Medicine, East Hospital,*
*The Institute for Biomedical Engineering & Nano Science,*
*Tongji University School of Medicine*
*Shanghai, 200092, P. R. China*
*†Department of Physics and TcSUH*
*University of Houston*
*Houston, TX77204, USA*

Nanotechnology has great potential in bio-medicine, leading to tremendous efforts in gene transfection, drug delivery, cell targeting and hyperthermia. As researchers gain familiarity with homogeneous nanomaterials, more attention is drawn to Janus nanostructure, an emerging new material with two or more distinct properties. With the uniqueness in asymmetry, Janus nanomaterials offer more possibilities for assembling multifunctional carriers. In this chapter, the synthesis of ternary Janus system of silica, $Fe_3O_4$ and polystyrene is introduced. A relatively easy method of combining miniemulsion and sol-gel is employed. The reaction mechanism is discussed with controllable morphology obtained. Dual surface functionalities are explored by selectively conjugating targeting group and anti-cancer drug. *In vitro* drug release profiles indicate a pH-responsive fashion. Cell study proves that the Janus nanocomposites successfully combine targeting and controlled drug release, achieving higher drug efficacy. Future endeavors will focus on structure novelty as well as fabrication simplification and scaling up. With its potential for higher integration, Janus nanostructures can well-serve as a versatile platform for biomedical applications.

# 1. Background

Nanotechnology has become an important means to help medical researchers and doctors addressing key clinical issues in diagnosis as well as treatment, including drug delivery, hyperthermia, imaging and cell targeting, preferably in a combined fashion. Nano-carrier systems have been well developed to achieve isolated functionalities. However, driven by the need for integrated functionalities, more research is focused on nano-carrier systems to achieve drug delivery, imaging, cell targeting and hyperthermia simultaneously, which is referred to as "multifunctional nano-carriers."

Such nano-carriers must have, but not be limited to, the following features to achieve multifunctionality.

1. The surface functional groups of the nano-carriers are crucial to further manipulation for conjugation of biological moieties such as drugs and targeting ligands. Besides, surface properties determine the colloidal stability of the nano-carrier systems.
2. The nano-carrier needs to have imaging capability, either for fluorescent imaging or magnetic resonance imaging (MRI).
3. The nano-carrier has to be nontoxic for clinical application.

An idealized multifunctional nano-carrier designed for diagnostic and therapeutic purpose is illustrated in **Figure 1**.

## 1.1. *Introduction to Janus particles*

Through the extensive efforts devoted to constructing nano-carrier systems to achieve multifunctionalities toward clinical application,[2] the vast majority of the approaches currently being used concentrate on particle surface functionalization with drugs,[3] DNA,[4] RNA,[5] peptide,[6] antibodies,[7] and imaging probes.[8] One major limitation in this approach is that a nanoparticle normally assumes a symmetrical geometry, offering limited possibilities for surface conjugation. Furthermore, the potential interaction of multiple components on a single carrier could result in undesired effects. It is not uncommon that the design and assembly of a single surface symmetrical carrier involves complicated reaction and purification steps because of the difficulties in the structural and chemical arrangements of the functional components.

Therefore, the critical question appears to be developing multi-surface nanostructures that can best utilize the intrinsic properties of nanomaterials. Being different from isotropic materials, the fundamentals of which

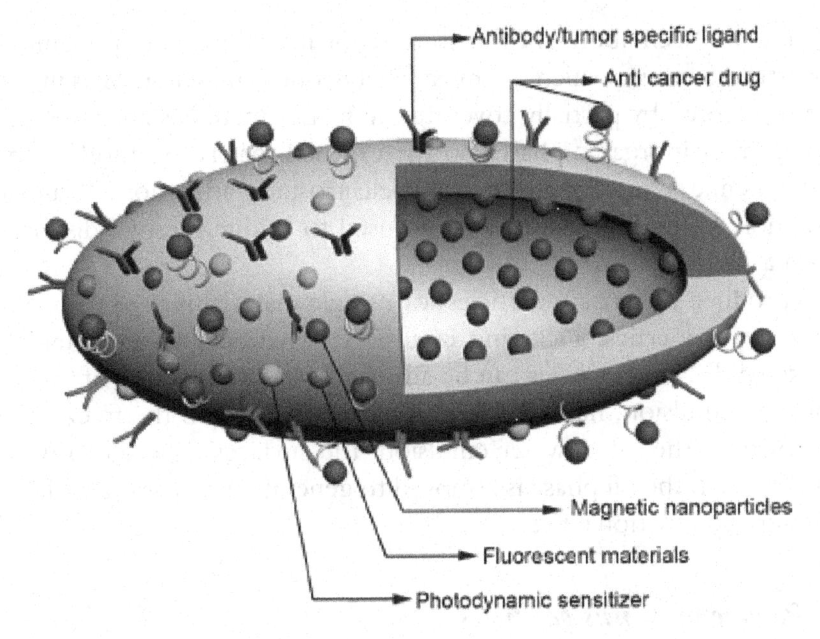


**Figure 1.** Schematic illustration of an idealized nano-carrier for cancer treatment.[1]

have been well-understood by scientists, anisotropic materials offer new possibilities, inspiring researchers to explore ways of fabrication as well as related application.[9] Successful attempts have been made to generate anisotropic microparticles,[10] but it becomes challenging when it is being attempted at the nanoscale level. Janus nanoparticles, named after the Roman god who has two distinct faces, are multiple-surfaced structures.[11] The asymmetry in shape, composition, and surface chemistry ideally suits the assembly of multiple components on a single-particle system.[12] The multiple surfaces with different functional groups makes the selective conjugation of targeting ligands, imaging probes, or drugs convenient.[13] Therefore, Janus nanoparticles have earned the reputation of "truly multifunctional entities."[9]

## 1.2. *Inorganic Janus particles*

Generally inorganic Janus nanoparticles are divided into two categories:[11]

1. One nanoparticle consisting of two or more phases, with each component expressing different optical, electrical or magnetic properties.
2. The nanoparticle composed of one material phase but possessing distinct chemical properties on both sides.

There has been abundant literature reporting fabricating inorganic Janus nanoparticles, among which the most flexible one is masking. Masking is usually implemented by partially covering the matrix materials and subsequently growing other inorganic phases on the exposed matrix. Methods employed include chemical functionalization,[14,15] electrostatic attraction[16,17] and metal evaporation.[18] Besides, ligands are also used to bridge the two distinct parts to form a Janus structure.[19]

To produce Janus nanoparticles with single material phase but two different surface properties, Pickering emulsion was developed by Granick and coworkers.[11,20] Nanoparticles can be adsorbed onto the surface of droplets to stabilize an emulsion, namely, Pickering emulsion. Upon the freezing of the oil droplets in the oil-in-water emulsion, the surfaces exposed to water are modified. Then the oil phase is removed to generate nanoparticles with asymmetric surface functionalities.

## 1.3. *Polymeric Janus particles*

Currently the phase separation is the most widely used method in producing polymeric Janus nanoparticles. This relatively complex approach involves the change in affinity between monomers and polymers, among polymers or between the generated nanoparticles and the aqueous phase.[21] One common approach is growing polymers through emulsion polymerization on the presynthesized polymer seeds of another kind. On the other hand, regarding the copolymer blocks ABC, the middle block B can connect with the two ends A and C to form a two-module Janus nanocomposite. Phase separation is usually low in yield and the morphology control is rather difficult.[11]

The masking approach also finds its way in polymeric Janus particle fabrication. The key is to functionalize only partial, either by masking or conducting interface modification. Subsequent growth of other polymers would occur on the exposed surfaces.[22,23]

Microfluid is also used to prepare Janus particles. Monomers are dissolved in the dispersed phase, which is injected perpendicular to the immiscible continuous phase. Droplets are formed in the stream for polymerization.[24] Microfluid is perhaps the only method that can be easily scaled up for mass production.[11] However, the sizes produced from microfluid approach are a lot bigger than those prepared by phase separation and masking.

## 1.4. *Polymeric-inorganic Janus composites*

The polymer-inorganic nanocomposites combine advantages from both organic and inorganic portion. They have attracted extensive interests due to the potential application as catalysts, functional coatings and biosensors.[9] They are also preferable in biomedical application because inorganic components such as silica and iron oxides offer further opportunities for bio-conjugation as well as the capability of MRI and hyperthermia. Polystyrene/silica was investigated most among the model structures, possibly because both can be easily fabricated in a controllable fashion.

Selective coating via masking remains quite useful in fabrication polymer-inorganic Janus nanocomposites.[25,26] Phase separation represents another important approach in generating polymer-inorganic composites, which often occurs in the miniemulsion process. Miniemulsion has been attracting increasing attention in encapsulation of inorganic solids to generate Janus structures. In the miniemulsion process, individual droplets serve as independent reactors and mass transfer is suppressed by the co-stabilizer. This nature allows inorganic particles to be directly dispersed in the monomer droplets. Janus structures can be obtained upon polymerization. Professor Donglu Shi, Dr. Feng Wang from University of Cincinnati, USA, and their collaborator Dr. Yilong Wang from Tongji University, China, have synthesized a series of polystyrene/iron oxides/silica Janus nanocomposites with tunable structures and surface properties in the past few years. The mechanism behind has also been explored, which is presented in the next section.

## 2. Preparation of Janus Composite Nanostructures via Miniemulsion

### 2.1. *Janus polystyrene/silica composites via miniemulsion polymerization of monomer based on locally surface-modified inorganic particles*

Initially, polystyrene/silica system was most frequently used to establish new methodology. Xia and coworkers[27] prepared the asymmetric dimers of polystyrene and silica on the micrometer scale via a template-directed self-assembly process. In 2005, Bourgeat Lami and coworkers[28] first reported the preparation of PS/$SiO_2$ Janus composites via a multi-step emulsion polymerization process. In that case, although finally uniform Janus pair could be obtained when the number of formed polymer latexes were adjusted to be the same as the inorganic particle seeds, it is not easy and convenient to get the ideal 1:1 number ratio in several other experiments.

In 2008, Wang and Shi *et al.*[29] reported a facile and novel chemical method to synthesize the polystyrene/silica Janus pair through a miniemulsion polymerization of styrene monomer in the presence of locally surface-modified silica seeds. In this process, the strength of both miniemulsion

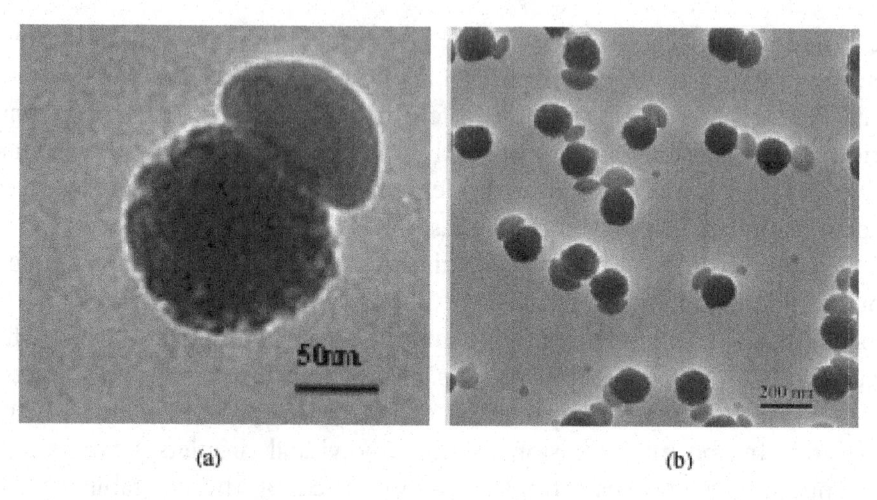


**Figure 2.** TEM images of mushroom-like silica/polystyrene nanocomposite particles: (**a**) single pair and (**b**) multiple pairs.[29]

polymerization and locally hydrophobic surface property of silica particles were taken into full use. Formation of composite droplets with a monomer droplet partially linked on the silica surface as a result of strong hydrophobic interaction between the monomer molecules and the modified area of the silica surface is critical for the synthesis. Then, Janus structures were obtained after the monomer was polymerized.

Transmission electron microscopic (TEM) observation reveals that Janus composites consisted of a silica particle and a polystyrene semi-sphere. Such unique semi-sphere shape of polystyrene nodules comes from the nucleation and growth of styrene monomer from the surface of modified silica particles (**Figure 2**). The isolated polystyrene latexes could be removed through repeated centrifugation and wash procedures. The locally surface-modified silica particles and as-synthesize Janus composite particles are both amphiphilic, so that they preferentially exist at the interface in the water–oil dual-phase system.

This facile method could be easily scaled up and was followed by several research groups in the preparation of Janus polystyrene/silica composites.[30]

## 2.2. *Janus $Fe_3O_4$@silica/Polystyrene composites via phase separation during the combing process of miniemulsion polymerization and sol-gel reaction*

Although the aforementioned method is fairly suitable for polystyrene/silica system, it is not feasible for more complicated ternary composite systems, such as polymer/inorganic 1/inorganic 2 systems due to the multi-step process. Phase separation between inorganic and organic components during the reaction had been recognized as an effective and facile method to control the composites' morphology. Montagne and coworkers[31] have synthesized Janus morphology by phase separation between the oleic acid-modified magnetic core and polystyrene chains in emulsion polymerization. Yin and co-workers[32] have prepared the eccentric shape $Fe_3O_4$@silica@polystyrene-coreshell structures by phase separation in a seed emulsion polymerization using γ-methacr yloxypropyltrimethoxysilane (MPS)-modified $Fe_3O_4$@silica as seeds. However, both counterparts of the composite particles exhibited hydrophobic surface properties, which greatly reduced the asymmetric characteristic of the whole composites. Later, Wu *et al.*[33] reported preparation of Janus polystyrene/silica composite particles through the phase separation approach. MPS played an important role in the formation of Janus composites via phase separation and different monomer/TEOS ratio could influence the final size and structure. However, there is seldom report about the synthesis of Janus magnetic composites composed of polymer/silica components via phase separation approach. Therefore, there is a critical need to develop novel synthesis route to synthesize such ternary magnetic composites with anisotropic surface property and shape via phase separation.

Wang and coworkers[34] have synthesized Janus magnetic composite particles (JMCPs) with high anisotropy and magnetite content consisting of iron oxide, silica, and polystyrene by phase separation in a combining miniemulsion polymerization with a sol–gel technique. This approach eliminates the multistep procedures needed in the preparation of composites with magnetic contents. The morphologies of the magnetic composites can be well controlled in the facile process by changing the reaction parameters. The one-pot process based on oleic-acid-modified iron oxide nanoparticles can be described as shown in **Figure 3**: (1) The miniemulsion droplets system was produced by a typical miniemulsifying process when a water phase and an oil phase are mixed. Surfactant SDS is dissolved in DI water to form the water phase. The mixture of hydrophobic St monomer, costabilizer, oleic acid-modified $Fe_3O_4$ nanoparticles (OAIOs), TEOS, and the coupling agent is an oil phase; (the miniemulsion droplet enlarged in stage (b). (2) Based on the formed miniemulsion droplet system, the polymerization of St monomer is initiated mainly in the droplets rather than in the monomer-swollen micelles according to the miniemulsion reaction mechanism. In the initial period of polymerization, part of the St monomer in the miniemulsion droplets is transformed into the polystyrene chains, leading to heterogeneous intermediate droplets as shown in stage c as a result of the reduced affinity between the

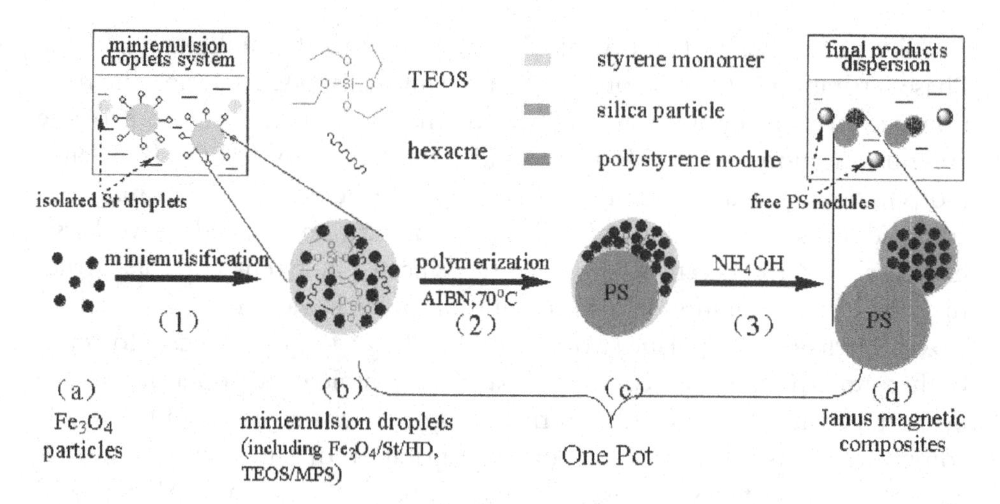


**Figure 3.** Schematic diagram showing the formation of the Janus $Fe_3O_4$@silica/polystyrene composite particles by the one-pot process combining the *in situ* miniemulsion polymerization of the St monomer and the sol–gel reaction of TEOS in the same miniemulsion droplet containing an iron oxide core.[34]

polystyrene chains and other hydrophobic components, especially with OAIOs. The mixture of St monomer, OAIOs, and TEOS tends to concentrate and be located eccentrically on the surfaces of the PS particles (stage c). (3) After $NH_4OH$ is added, TEOS starts to hydrolyze and condense to form hydrophilic silica chain. The addition of $NH_4OH$ may decrease the hydrophobicity of the OAIOs through the transformation of oleic acid molecules to oleate at high pH values. Thus OAIOs may be more compatible with TEOS molecules. As a result, the formed silica chains tend to hybridize with iron oxide particles, inducing the formation of iron oxide/silica hybrid particles. The pair consisting of a growing PS nodule and an iron oxide/silica hybrid particle is prepared by the bridging effect of the coupling agent MPS molecules. Finally, after most of the St and TEOS are dispensed, Janus composite particles are obtained (stage d).

TEM observation and EDX analysis had characterized the microstructure and chemical composition of the Janus composite particles (**Figure 4**). The Janus particles are composed of polystyrene and silica, with the silica portion loaded with magnetite core. As shown in the EDX spectra obtained from isolated area of composite, there is no signal from the Fe or Si element in the neat region of the composite but relatively strong peaks from Si and Fe elements from the darker counterpart (inset of **Figure 4**). This is different from the results from Hatton and co-workers[35] regarding sonochemical synthesis, possibly because of different phase-separation behaviors.

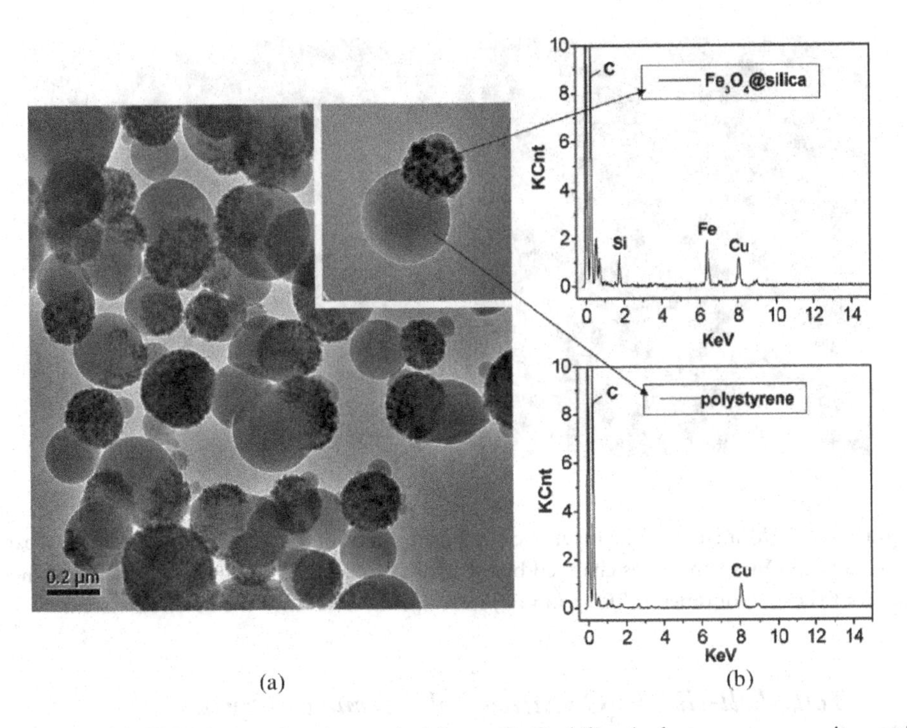


**Figure 4.** (**a**) TEM image showing typical Janus $Fe_3O_4$/silica/polystyrene composite particles and (**b**) EDX elemental analysis of an as-synthesized Janus magnetic composite particle as shown in the inset of a.[34]

In this study, it was found that phase separation could be controlled by the addition of $NH_4OH$ to trigger the hydrolysis and polycondensation of TEOS with MPS molecules leading to the formation of a silica chain. The OAIOs tend to be hydrophilic as a result of the transformation of oleic acid to oleate on its surface with $NH_4OH$ addition. The nearly complete phase separation between OAIO/polystyrene and the hydrophobic-to-hydrophilic transformation of the iron oxide particles enhance the formation of the Janus structures and $Fe_3O_4$/silica hybrid. When $NH_4OH$ was added much earlier, changed from 60 min to 10 min, no Janus structure could be found. The presence of the spherical composite particles with dispersed iron oxide in the matrix composed of polystyrene and silica (**Figure 5(a)**) indicates that the silica molecules in the early stage of polymerization can interrupt further phase separation between the iron oxide and polymer. In addition, introducing divinylbenzene (DVB) into the polymerization also generated symmetric particles due to the incomplete phase separation between OAIOs and polystyrene induced by the addition of DVB.

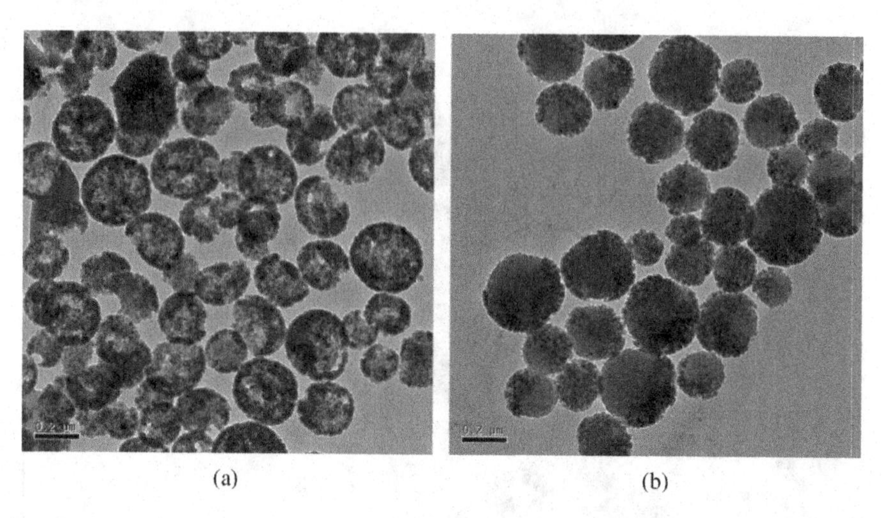

(a) (b)

**Figure 5.** TEM images of the magnetic composite particles when the degree of phase separation between OAIO and PS is changed by **(a)** adding $NH_4OH$ 10 minutes after the polymerization of the St monomer at 70°C and **(b)** introducing DVB into the reaction system.[34]

## 2.3. *Yolk-shell-like $Fe_3O_4$@silica/polystyrene composites via phase separation during the combing process of miniemulsion polymerization and sol–gel reaction*

Using the above-mentioned combining process of miniemulsion polymerization and sol–gel reaction, Wang *et al.*[36] prepared yolk-shell magnetic composite structures (YSMCs) containing an $Fe_3O_4$/silica hybrid shell and a polystyrene. It is worth noting that such yolk-shell structure can be obtained via one-pot ternary phase separation of the inorganic and organic components during the reaction for preparation of Janus magnetic composites in the absence of MPS as the coupling agent. Being a totally different methodology from the template-assisted selective etching approach, this method provided a facile way to get yolk-shell composite nanoparticles for the first time. The combined process of miniemulsion polymerization and sol–gel reaction based on pre-made OAIOs is as shown in **Figure 6**. After phase separation, silica chains are coated on the OAIOs as a result of the lowest systematic energy. The $Fe_3O_4$/silica hybrid shell forms outside the oil droplets as the *in situ* templates. The St monomers diffuse inwardly to the PS particles and TEOS outwardly to the inorganic shell. As the St monomers and TEOS precursors in the droplets are almost exhausted, YSMCs with a PS core and an incomplete $Fe_3O_4$/silica hybrid shell are prepared (stage d). The interior space of the composites is created through volume shrinkage of the PS chains, while the diameter of the inorganic shell is kept unchanged once it forms.

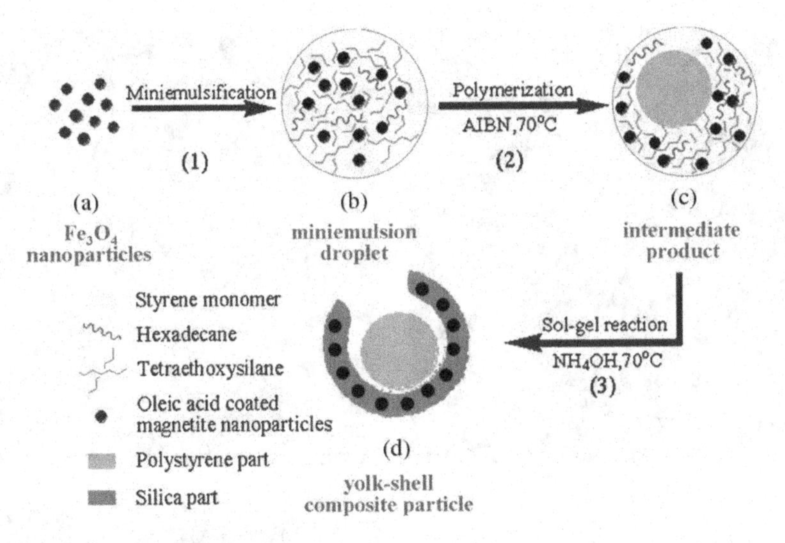


**Figure 6.** Schematic illustration showing the formation of the yolk-shell magnetic nanocomposites via a combined process of miniemulsion polymerization and sol–gel reaction.[36]

The microstructure and chemical compositions of the typical YSMCs are studied by TEM, SEM, and EDX analysis (**Figure** 7). The YSMCs are composed of a neat core and a dark shell with magnetite embedded (**Figure 7(a)**). There is an obvious interior space between the core and the shell (**Figure 7(b)**). In **Figure** 7(**c**), signals of Fe, Si and C elements are involved in the composites. The specific location of the polymeric component is identified after being etched with toluene for 5 hours at room temperature. The entire core disappears and hollow $Fe_3O_4$/silica hybrid structures are obtained (**Figure** 7(**d**)). These results show that each YSMC is composed of a polystyrene core and an $Fe_3O_4$/silica hybrid shell.

## 2.4. *Dual surface-functionalized superparamagnetic Janus nanocomposites (SJNCs) of $Fe_3O_4$@silica/polystyrene via miniemulsion polymerization and sol–gel reaction*

Targeting at biomedical application, Wang *et al.*[37] successfully fabricated a superparamagnetic Janus nanocomposite (SJNCs) of polystyrene/$Fe_3O_4$@$SiO_2$ with both surface bearing functional groups using previous developed technique. The nanocomposites are synthesized by a one-pot combination of miniemulsion followed by sol–gel reaction. Briefly, the oil phase (styrene monomers, tetraethoxysilane (TEOS), hexadecane, oleic acid capped iron oxides (OAIOs)) is emulsified with the deionized water to generate a miniemulsion system with sodium dodecyl sulfate as emulsifier. Polymerization of

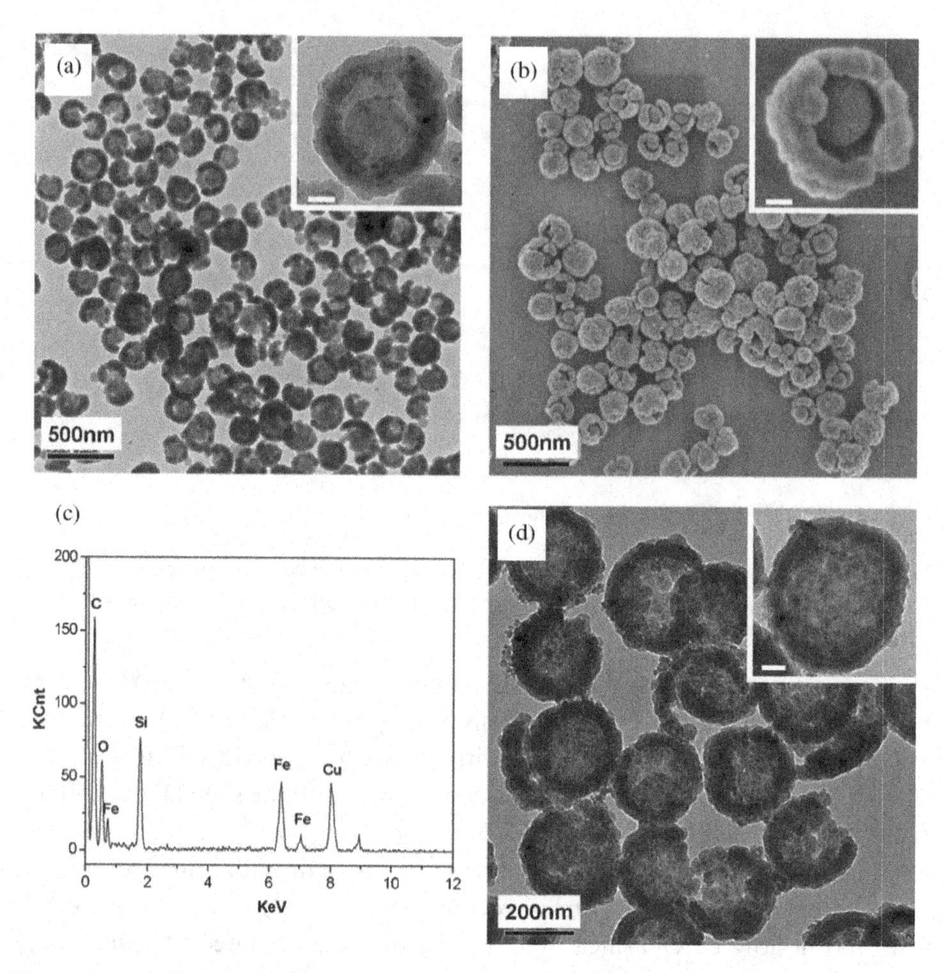


**Figure 7.** (**a**) TEM; (**b**) SEM images; (**c**) EDX spectrum of the typical yolk-shell magnetic composites, and (**d**) TEM image of the hollow $Fe_3O_4$/silica hybrids obtained from YSMCs after the PS core is etched by toluene, all inset scale bars are 50 nm.[36]

styrene is initiated by 4,4'-azobis(4-cyanovaleric acid) (ACVA) followed by silica formation via sol–gel reaction (**Figure 8**).

TEM image (**Figure 9(a)**) of SJNCs clearly indicates a structure consisted of a polystyrene core covered by a half shell of silica embedded with $Fe_3O_4$ nanoparticles. The PS core has a diameter of about 200 nm, and the $Fe_3O_4$@ Silica hybrid shell has a thickness of around 100 nm, adding up to a diameter of about 300 nm for one SJNC. As shown in the EDS spectra (**Figure 9(b)**), the core gives out mainly signal from C and O with minor signals of Si and Fe coming from the hybrid shell at the back of the core. On the other hand, stronger signals from Si and Fe are obtained from the shell, indicating that the core of SJNCs is composed of pure polymer, which is polystyrene in this

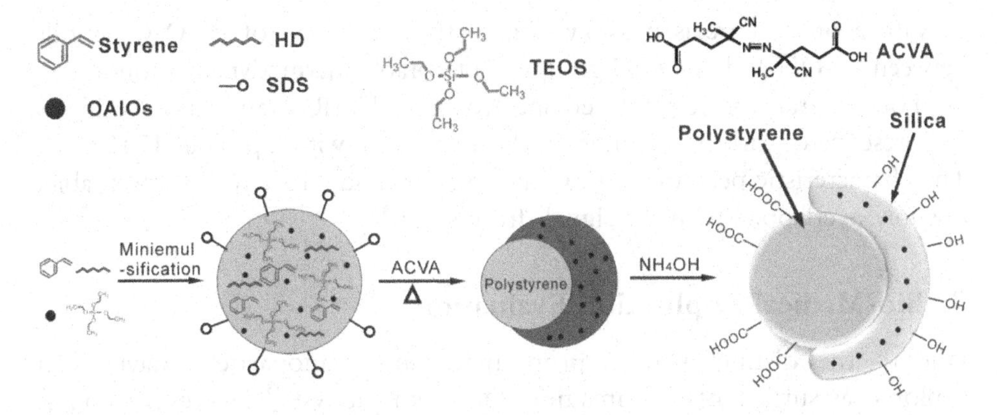


**Figure 8.** Schematic illustration of the formation of polystyrene/$Fe_3O_4$@silica superparamagnetic Janus nanocomposites (SJNCs).[37]

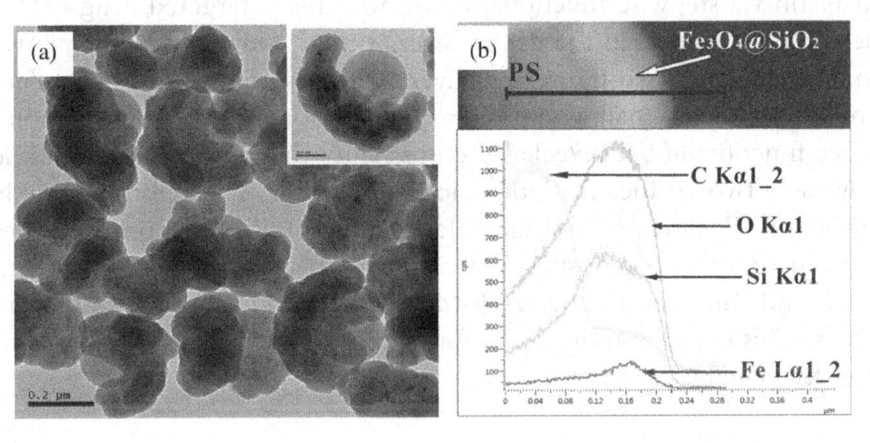


**Figure 9.** **(a)** TEM images of the dual functionalized polystyrene/$Fe_3O_4$@silica Janus nanoparticles, the scale bar is 200 nm (inset image is the magnified image of one nanocomposite particle, scale bar is 100 nm), and **(b)** elemental mapping of one nanocomposite particle.[37]

case, while the shell is composed of iron oxide and silica. It is quite consistent with the previous report[34] that iron oxides are located in silica rather than in polystyrene. It is also proposed that phase separation between inorganic and organic components during polymerization is crucial to the formation of SJNCs. Once polystyrene chains are formed, their low affinity toward St monomers drives away the other organic phase (TEOS, St monomers and OAIOs) eccentrically. The presence of $NH_4OH$ initiates sol–gel reaction of TEOS as well as the transformation of oleic acid to oleate. Consequently, OAIOs have become hydrophilic and are incorporated in the silica rather than PS after condensation. The observation that silica forms a half shell along the curvature of PS core is possibly due to the formation of -Si-O-C- bonds between -Si-OH and –COOH groups at a favorable thermodynamic condition.

Fourier Transformed Infrared Spectroscopy (FTIR) examination confirms the presence of carboxyl groups on the PS surface with a peak at 1712 $cm^{-1}$. The characteristic peak of -Si-O-C- bond at around 1100 $cm^{-1}$ is observable, possibly overlapped with the signals from silica.[37]

## 3. Bio-Medical Application Evaluation

Due to the distinct surface properties, Janus nanoparticles have been explored as surfactants,[38,39] imaging,[11,40,41] and catalyst.[42] However, reports of Janus nanoparticles on biomedical application have been very limited. Xu et al.[13c] fabricated Au/$Fe_3O_4$ nanoparticles and conjugated Herceptin and Platin via stepwise functionalization to achieve targeted drug delivery. The conjugates released the drugs faster at mild acidic environment. Wang *et al.*[43] conjugated two fluorescent dyes to both surfaces of the silica/polystyrene asymmetric nanocomposites. In addition to the confirmation of surface functionality, it revealed the possibility that this material can selectively carry two entities for biomedical application. Through the same line of thought, Wang *et al.*[37] produced Janus dual surface functionalized superparamagnetic nanocomposites and conjugated the surfaces with targeting ligands and anti-cancer drug. Following cell experiments showed increased efficacy. This opens another route for targeted drug delivery with emphasis on surface chemistry.

### 3.1. *Surface conjugation and drug release*

Folic acid is essential for cell proliferation. Its uptake is mediated by the protein known as the folate receptor,[44] which facilitates cellular internalization after binding to folic acid via receptor-mediated endocytosis.[44,45] Although present in almost all tissues, folate receptors are significantly overexpressed in malignant tissues,[44,46] making folic acid suitable as the targeting moiety toward tumor cells.[47] Poly(ethylene glycol) bis-amine with ($NH_2$-PEG-$NH_2$) was linked to folic acid via one of the two amino ending groups. The product was then conjugated to the polystyrene surfaces through carbodiimide-mediated coupling with the carboxyl groups on polystyrene surfaces **(Figure 10A)**.

The anti-cancer drug doxorubicin (DOX) was selected to study the drug delivery performance of SJNCs. DOX, with the trade name of Adriamycin,

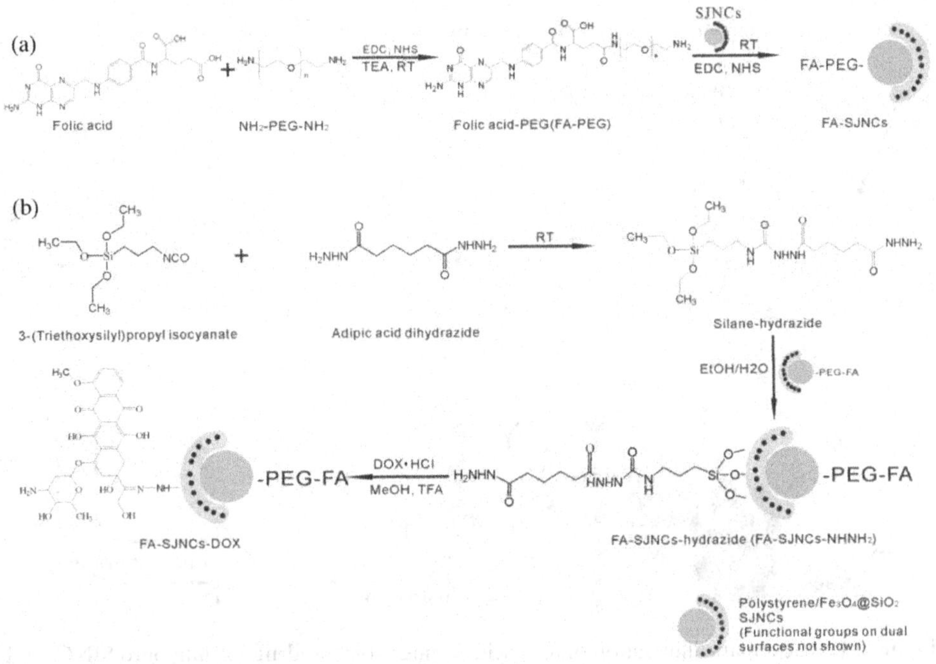

**Figure 10.** Schematic drawing of **(a)** folic acid functionalization on polystyrene core surface and **(b)** DOX loading onto the hybrid shell surface.[37]

was discovered in the 1950s. DOX interacts with DNA by intercalation and inhibition of macromolecular biosynthesis, thus preventing DNA chains from replication.[48] However, DOX has many adverse effects, including cardiotoxicity which is fatal. It was proposed that hydrazone bonds be used to conjugate doxorubicin to the hybrid shell surfaces. A hydrazone bond is relatively stable at physiological condition (pH 7.2–7.4) but will hydrolyze at endocytic condition (pH 4.5–6.5). In such manner, highly controlled release of doxorubicin could be achieved, thus significantly reducing the side effects.[3,49]

As shown in **Figure 10**, folic acid and DOX are conjugated onto dual surfaces separately without interference. The DOX loading capacity is determined to be 2.08 ± 0.2 wt% using the subtraction method. It was hypothesized upon internalization of the conjugates by tumor cells via folate receptor-mediated endocytosis, drug comes off faster in endocytic compartments (**Figure 11**) to minimize undesired release during bloodstream circulation.

The pH dependence of DOX release behaviors were monitored *in vitro* in buffer solutions with different pH values. At pH 7.4, which was similar to the physiological condition, accumulative DOX release after 195 hours was about 25.1% (**Figure 12**). On the other hand, at pH 5.0 and 6.0, which

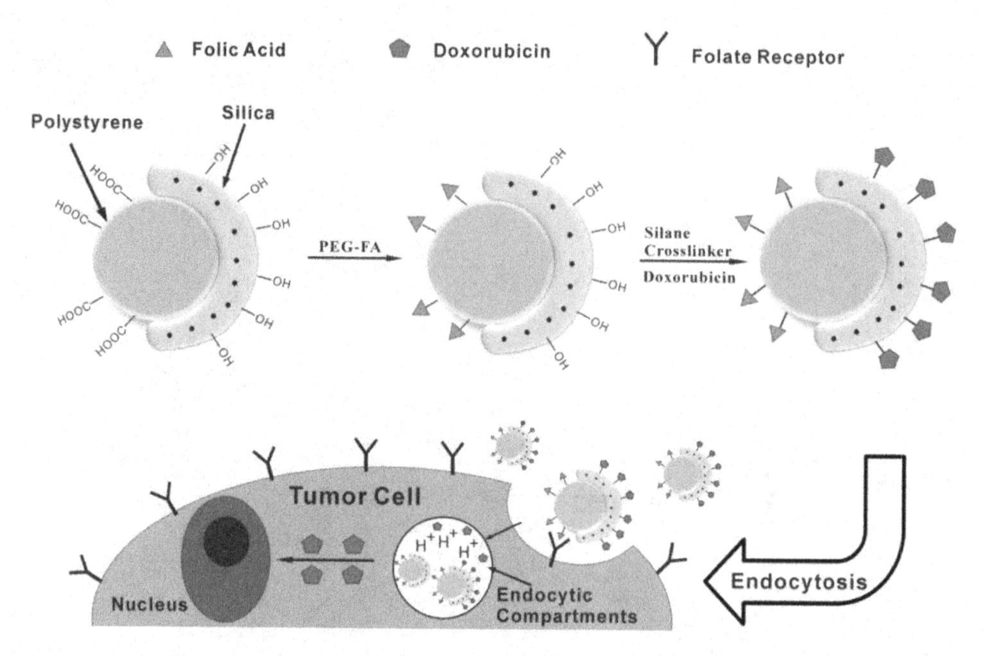


**Figure 11.** Schematic illustration of folic acid conjugation and drug loading onto SJNCs and subsequent cell internalization through endocytosis.[37]

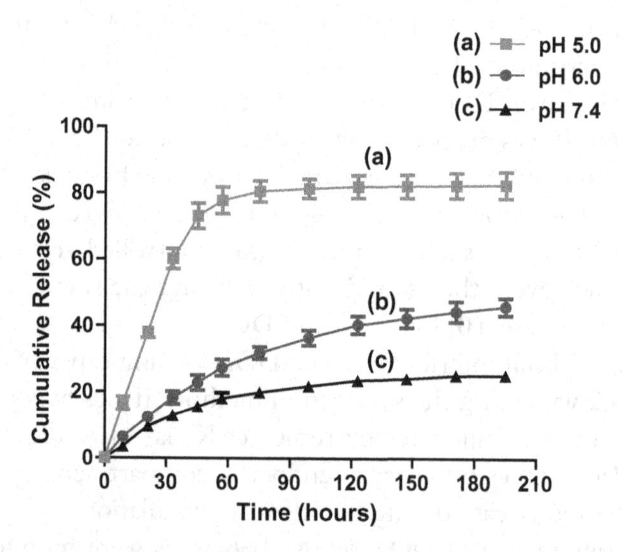


**Figure 12.** DOX release profiles from DOX conjugated SJNCs at (a) pH 5.0 and (b) pH 6.0 and (c) pH 7.4.[37]

simulated the endocytic compartment environment, DOX were release much faster, reaching 82.6% and 47.1% after 195 hours, respectively. Apparently, as the concentration of $H^+$ increases, hydrazone bonds hydrolyze at a faster rate, releasing more drug from the carriers.

### 3.2. *Cell experiments*

Targeting effect and pH-triggered DOX release were combined and evaluated against the human breast cell line MDA-MB-231 overexpressing folate receptors. SJNCs with or without drugs were incubated with cells for 4 hours in Hank's Balanced Salts buffer solution (HBSS) before the cells were washed and incubated for another 48 hours in fresh cell culture media. Finally cell viability was evaluated. The SJNCs bearing both drug and folic acid is defined as the targeted group.
Two control experiments were designed to test the targeting effect:

(1) The nontargeted group. The drug-bearing nanocomposites without folic acid (SJNCs-DOX) were subjected to the same procedure.
(2) The competition group. The drug-bearing and folic-acid-conjugated nanocomposites (FA-SJNCs-DOX) were tested in a free folic acid competition experiment. Free folic acid was dissolved in the HBSS buffer to make a concentration of 1 mM. Then FA-SJNCs-DOX were dispersed in this buffer and incubated with cells.

It is obvious that all curves show similar inhibitor-response characteristic (**Figure 13**). However, the competition group exhibits less cytotoxicity at the same dosage compared with targeted group. Similar trend is observed with the nontargeted group. The $IC_{50}$ values of three groups were calculated. As shown in **Figure 14**, the $IC_{50}$ value of targeted FA-SJNCs-DOX conjugates is calculated at $255.3 \pm 55.1$ ugml$^{-1}$. Significantly lower than that of the nontargeted group ($IC50 = 781.2 \pm 163.0$ ugml$^{-1}$, $p < 0.001$) and the free FA competition group ($IC50 = 1030.2 \pm 416.1$ ugml$^{-1}$, $p < 0.05$), this value gives strong evidence supporting successful cell targeting of the nanocomposites. It is therefore deduced that more targeted FA-SJNCs-DOX conjugates are able to enter tumor cells via folate-receptor-mediated endocytosis. Once inside the endocytic compartments, DOX is released at a much faster rate due to the hydrolysis of hydrazone bonds, thus significantly enhancing the efficacy of DOX. Meanwhile, the cardiac side effects of DOX can be greatly reduced.

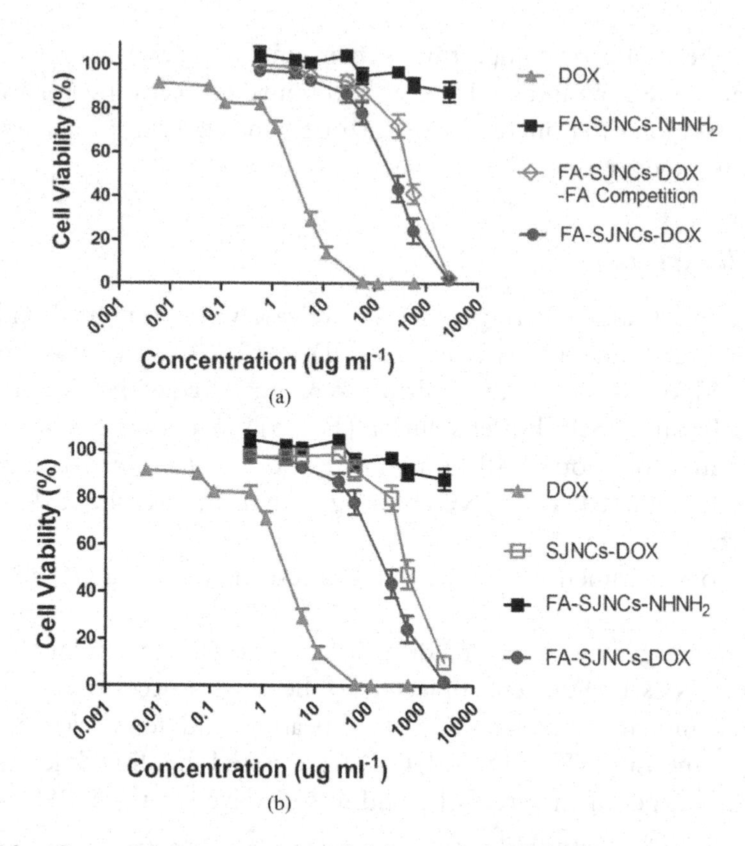


**Figure 13.** Cytotoxicity assay curves of (**a**) targeted FA-SJNCs-DOX versus targeted FA-SJNCs-DOX in free folic acid competition and (**b**) targeted FA-SJNCs-DOX versus non-targeted SJNCs-DOX against MDA-MB-231 cell line. Cytotoxicity assay curves for DOX and nanocomposites bearing folic acid but none drug (FA-SJNCs-$NHNH_2$) are displayed as reference in both graphs.[37]

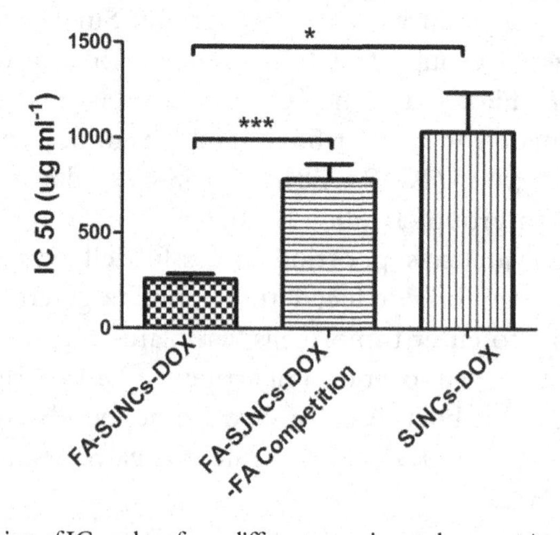


**Figure 14.** Comparsion of $IC_{50}$ values from different experimental groups. $*p < 0.05$, $***p < 0.001$.[37]

## 4. Summary

Researchers are developing more Janus structures with unique properties to offer abundant possibilities. Future efforts will focus on both structure novelty as well as simplification of fabricating process. With its potential for higher integration, Janus particles well suit the requirements as a versatile platform for bio-medical applications.

## References


1. Shi DL. Integrated multifunctional nanosystems for medical diagnosis and treatment. *Adv Funct Mater* 2009;19(21):3356–3373.
2. Gindy ME, Prud'homme RK. Multifunctional nanoparticles for imaging, delivery and targeting in cancer therapy. *Expert Opin Drug Del* 2009;6(8):865–878.
3. Lee CH, Cheng SH, Huang IP, Souris JS, Yang CS, Mou CY, Lo LW. Intracellular pH-responsive mesoporous silica nanoparticles for the controlled release of anticancer chemotherapeutics. *Angew Chem Int Edit* 2010;49(44):8214–8219.
4. Kievit FM, Veiseh O, Bhattarai N, Fang C, Gunn JW, Lee D, Ellenbogen RG, Olson JM, Zhang MQ. PEI-PEG-chitosan-copolymer-coated iron oxide nanoparticles for safe gene delivery: synthesis, complexation, and transfection. *Adv Funct Mater* 2009;19(14): 2244–2251.
5. Liu G, Xie J, Zhang F, Wang ZY, Luo K, Zhu L, Quan QM, Niu G, Lee S, Ai H, Chen XY. N-Alkyl-PEI-functionalized Iron oxide nanoclusters for efficient siRNA delivery. *Small* 2011; 7(19):2742–2749.
6. Ojea-Jimenez I, Garcia-Fernandez L, Lorenzo J, Puntes VF. Facile preparation of cationic gold nanoparticle-ioconjugates for cell penetration and nuclear targeting. *ACS Nano* 2012;6(9):7692–7702.
7. Cho HS, Dong Z, Pauletti GM, Zhang J, Xu H, Gu H, Wang L, Ewing RC, Huth C, Wang F, Shi D. Fluorescent, superparamagnetic nanospheres for drug storage, targeting, and imaging: a multifunctional nanocarrier system for cancer diagnosis and treatment. *ACS Nano* 2010; 4(9)5398–5404.
8. Quach AD, Crivat G, Tarr MA, Rosenzweig Z. Gold nanoparticle-quantum dot-polystyrene microspheres as fluorescence resonance energy transfer probes for bioassays. *J Am Chem Soc* 2011;133(7):2028–2030.
9. Lattuada M, Hatton TA. Synthesis, properties and applications of Janus nanoparticles. *Nano Today* 2011;6(3):286–308.
10. (a) Serra CA, Chang ZQ. Microfluidic-assisted synthesis of polymer particles. *Chem Eng Technol* 2008;31(8):1099–1115. (b) Dendukuri D, Pregibon DC, Collins J, Hatton TA, Doyle PS. Continuous-flow lithography for high-throughput microparticle synthesis. *Nature Materials* 2006;5(5):365–369. (c) Dendukuri D, Doyle PS. The synthesis and assembly of polymeric microparticles using microfluidics. *Adv Mater* 2009;21(41):4071–4086.
11. Hu J, Zhou SX, Sun YY, Fang XS, Wu LM. Fabrication, properties and applications of Janus particles. *Chem Soc Rev* 2012;41(11):4356–4378.
12. Hu SH, Gao XH. Nanocomposites with spatially separated functionalities for combined imaging and magnetolytic therapy. *J Am Chem Soc* 2010;132(21):7234–7237.

13. (a) Roh KH, Martin DC, Lahann J. Biphasic Janus particles with nanoscale anisotropy. *Nat Mater* 2005;4(10):759–763. (b) Roh KH, Yoshida M, Lahann J. Compartmentalized, multiphasic nanocolloids with potential applications in drug delivery and biomedical imaging. *Materialwiss Werkst* 2007;38(12):1008–1011. (c) Xu CJ, Wang BD, Sun SH. Dumbbell-like Au-$Fe_3O_4$ nanoparticles for target-specific platin delivery. *J Am Chem Soc* 2009;131(12):4216–4217.
14. Perro A, Reculusa S, Pereira F, Delville MH, Mingotaud C, Duguet E, Bourgeat-Lami E, Ravaine S. Towards large amounts of Janus nanoparticles through a protection-deprotection route. *Chem Commun* 2005;(44):5542–5543.
15. Bae C, Moon J, Shin H, Kim J, Sung M M. Fabrication of monodisperse asymmetric colloidal clusters by using contact area lithography (CAL). *J Am Chem Soc* 2007;129(46): 14232–14239.
16. Nagle L, Ryan D, Cobbe S, Fitzmaurice D. Templated nanoparticle assembly on the surface of a patterned nanosphere. *Nano Lett* 2003;3(1):51–53.
17. McConnell MD, Kraeutler MJ, Yang S, Composto RJ. Patchy and multiregion Janus particles with tunable optical properties. *Nano Lett* 2010;10(2):603–609.
18. Correa-Duarte MA, Salgueirino-Maceira V, Rodriguez-Gonzalez B, Liz-Marzan LM, Kosiorek A, Kandulski W, Giersig M. Asymmetric functional colloids through selective hemisphere modification. *Adv Mater* 2005;17(16):2014–2018.
19. Zhang L, Zhang F, Dong WF, Song JF, Huo QS, Sun HB. Magnetic-mesoporous Janus nanoparticles. *Chem Commun* 2011;47(4):1225–1227.
20. (a) Zhang J, Jin J, Zhao HY. Surface-initiated free radical polymerization at the liquid-liquid interface: a one-step approach for the synthesis of amphiphilic Janus silica particles. *Langmuir* 2009;25(11):6431–6437. (b) Hong L, Jiang S, Granick S. Simple method to produce Janus colloidal particles in large quantity. *Langmuir* 2006;22(23):9495–9499.
21. (a) Chen KM, Zhu Y, Zhang YF, Li L, Lu Y, Guo XH. Synthesis of magnetic spherical polyelectrolyte brushes. *Macromolecules* 2011;44(3):632–639. (b) Misra A, Urban MW. Acorn-shape polymeric nano-colloids: synthesis and self-assembled films. *Macromol Rapid Commun* 2010;31(2):119–127.
22. (a) Nakahama K, Kawaguchi H, Fujimoto K. A novel preparation of nonsymmetrical microspheres using the Langmuir-Blodgett technique. *Langmuir* 2000;16(21):7882–7886. (b) Zhang SY, Li Z, Samarajeewa S, Sun GR, Yang C, Wooley KL. Orthogonally dual-clickable Janus nanoparticles via a cyclic templating strategy. *J Am Chem Soc* 2011;133(29):11046–11049.
23. Bradley M, Rowe J. Cluster formation of Janus polymer microgels. *Soft Matter* 2009;5(16):3114–3119.
24. Lone S, Kim SH, Nam SW, Park S, Joo J, Cheong IW. Microfluidic synthesis of Janus particles by UV-directed phase separation. *Chem Commun* 2011;47(9):2634–2636.
25. Feyen M, Weidenthaler C, Schuth F, Lu AH. Regioselectively controlled synthesis of colloidal mushroom nanostructures and their hollow derivatives. *J Am Chem Soc* 2010;132(19):6791–6799.
26. Gong JA, Zu XH, Li YH, Mu W, Deng YL. Janus particles with tunable coverage of zinc oxide nanowires. *J Mater Chem* 2011;21(7):2067–2069.
27. Yin YD, Lu Y, Xia YN. A self-assembly approach to the formation of asymmetric dimers from monodispersed spherical colloids. *J Am Chem Soc* 2001;123(4):771–772.
28. Reculusa S, Poncet-Legrand C, Perro A, Duguet E, Bourgeat-Lami E, Mingotaud C, Ravaine S. Hybrid dissymmetrical colloidal particles. *Chem Mater* 2005;17(13):3338–3344.

29. Qiang W, Wang Y, He P, Xu H, Gu H, Shi D. Synthesis of asymmetric inorganic/polymer nanocomposite particles via localized substrate surface modification and miniemulsion polymerization. *Langmuir* 2008;24(3):606–608.
30. (a) Liu B, Zhang C, Liu J, Qu X, Yang Z. Janus non-spherical colloids by asymmetric wet-etching. *Chem Commun (Camb)* 2009;(26):3871–3873. (b) Ge X, Wang M, Yuan Q, Wang H. The morphological control of anisotropic polystyrene/silica hybrid particles prepared by radiation miniemulsion polymerization. *Chem Commun (Camb)* 2009;(19):2765–2767.
31. Montagne F, Mondain-Monval O, Pichot C, Elaissari A. Highly magnetic latexes from submicrometer oil in water ferrofluid emulsions. *J Polym Sci Pol Chem* 2006;44(8): 2642–2656.
32. Ge J, Hu Y, Zhang T, Yin Y. Superparamagnetic composite colloids with anisotropic structures. *J Am Chem Soc* 2007;129(29):8974–8975.
33. Lu W, Chen M, Wu L M. Onestep synthesis of organic-inorganic hybrid asymmetric dimer particles via miniemulsion polymerization and functionalization with silver. *J Colloid Interf Sci* 2008;328(1):98–102.
34. Wang YL, Xu H, Ma YS, Guo FF, Wang F, Shi DL. Facile one-pot synthesis and morphological control of asymmetric superparamagnetic composite nanoparticles. *Langmuir* 2011;27(11):7207–7212.
35. Teo BM, Suh SK, Hatton TA, Ashokkumar M, Grieser F. Sonochemical synthesis of magnetic Janus nanoparticles. *Langmuir* 2011;27(1):30–33.
36. Wang YL, Wang F, Chen BD, Xu H, Shi DL. Facile one-pot synthesis of yolk-shell superparamagnetic nanocomposites via ternary phase separations. *Chem Commun* 2011;47(37):10350–10352.
37. Wang F, Pauletti GM, Wang J, Zhang J, Ewing RC, Wang Y, Shi D. Dual surface-functionalized Janus nanocomposites of polystyrene/Fe(3)O(4)@SiO(2) for simultaneous tumor cell targeting and stimulus-induced drug release. *Adv Mater* 2013;25(25):3485–3489.
38. Xu Q, Kang X, Bogomolni RA, Chen S. Controlled assembly of Janus nanoparticles. *Langmuir* 2010;26(18):14923–14928.
39. Kim JW, Lee D, Shum HC, Weitz DA. Colloid surfactants for emulsion stabilization. *Adv Mater* 2008;20(17):3239–3243.
40. Kim SH, Sim JY, Lim JM, Yang SM. Magnetoresponsive microparticles with nanoscopic surface structures for remote-controlled locomotion. *Angew Chem Int Edit* 2010;49(22):3786–3790.
41. Yin SN, Wang CF, Yu ZY, Wang J, Liu SS, Chen S. Versatile bifunctional magnetic-fluorescent responsive Janus supraballs towards the flexible bead display. *Adv Mater* 2011;23(26):2915–2919.
42. Seh ZW, Liu SH, Zhang SY, Shah KW, Han MY. Synthesis and multiple reuse of eccentric Au@TiO2 nanostructures as catalysts. *Chem Commun* 2011:47(23):6689–6691.
43. Wang Y, Xu H, Qiang W, Gu H, Shi D. Asymmetric composite nanoparticles with anisotropic surface functionalities. *Journal of Nanomaterials* 2009;2009.
44. Lee RJ, Low PS. Folate as a targeting device for proteins utilizing folate receptor-mediated endocytosis. *Methods Mol Med* 2000;25:69–76.
45. (a) Stella B, Arpicco S, Peracchia MT, Desmaele D, Hoebeke J, Renoir M, D'Angelo J, Cattel L, Couvreur P. Design of folic acid-conjugated nanoparticles for drug targeting. *J Pharm Sci-Us* 2000;89(11):1452–1464. (b) Leamon CP, Low PS. Delivery of

macromolecules into living cells — a method that exploits folate receptor endocytosis. *P Natl Acad Sci USA* 1991;88(13):5572–5576. (c) Antony AC. The biological chemistry of folate receptors. *Blood* 1992;79(11):2807–2820. (d) Rothberg KG, Ying YS, Kolhouse JF, Kamen BA, Anderson RGW. The glycophospholipid-linked folate receptor internalizes folate without entering the clathrin-coated pit endocytic pathway. *J Cell Biol* 1990;110(3):637–649.

46. (a) Buist MR, Molthoff CFM, Kenemans P, Meijer CJLM. Distribution of Ov-Tl-3 and Mov18 in normal and malignant ovarian tissue. *J Clin Pathol* 1995;48(7):631–636. (b) Ross JF, Chaudhuri PK, Ratnam M. Differential regulation of folate receptor isoforms in normal and malignant-tissues in-vivo and in established cell-lines — physiological and clinical implications. *Cancer* 1994;73(9):2432–2443. (c) Garinchesa P, Campbell I, Saigo PE, Lewis JL, Old LJ, Rettig WJ. Trophoblast and ovarian-cancer Antigen-Lk26 — sensitivity and specificity in immunopathology and molecular-identification as a folate-binding protein. *Am J Pathol* 1993;142(2):557–567. (d) Salazar MD, Ratnam M. The folate receptor: what does it promise in tissue-targeted therapeutics? *Cancer Metast Rev* 2007;26(1):141–152.
47. (a) Kono K, Liu MJ, Frechet JMJ. Design of dendritic macromolecules containing folate or methotrexate residues. *Bioconjugate Chem* 1999;10(6):1115–1121. (b) Fei XN, Liu Y, Li C. Folate conjugated chitosan grafted thiazole orange derivative with high targeting for early breast cancer cells diagnosis. *J Fluoresc* 2012;22(6):1555–1561. (c) Shen AJ, Li DL, Cai XJ, Dong CY, Dong HQ, Wen HY, Dai GH, Wang PJ, Li YY. Multifunctional nanocomposite based on graphene oxide for in vitro hepatocarcinoma diagnosis and treatment. *J Biomed Mater Res A* 2012;100A(9):2499–2506. (d) Mahajan S, Koul V, Choudhary V, Shishodia G, Bharti AC. Preparation and in vitro evaluation of folate-receptor-targeted SPION-polymer micelle hybrids for MRI contrast enhancement in cancer imaging. *Nanotechnology* 2013;24(1):015603. (e) Hou J, Zhang Q, Li X, Tang Y, Cao MR, Bai F, Shi Q, Yang CH, Kong DL, Bai G. Synthesis of novel folate conjugated fluorescent nanoparticles for tumor imaging. *J Biomed Mater Res A* 2011;99A(4):684–689. (f) Barz M, Canal F, Koynov K, Zentel R, Vicent MJ. Synthesis and in vitro evaluation of defined HPMA folate conjugates: influence of aggregation on folate receptor (FR) mediated cellular uptake. *Biomacromolecules* 2010;11(9):2274–2282. (g) Saul JM, Annapragada A, Natarajan JV, Bellamkonda RV. Controlled targeting of liposomal doxorubicin via the folate receptor in vitro. *J Control Release* 2003;92(1–2):49–67.
48. (a) Pigram WJ, Fuller W, Hamilton LD. Stereochemistry of intercalation — interaction of daunomycin with DNA. *Nature-New Biol* 1972;235(53):17-&. (b) Momparler RL, Karon M, Siegel SE, Avila F. Effect of adriamycin on DNA, RNA and protein-synthesis in cell-free systems and intact cells. *Cancer Res* 1976;36(8):2891–2895. (c) Fornari FA, Randolph JK, Yalowich JC, Ritke MK, Gewirtz DA. Interference by doxorubicin with DNA unwinding in Mcf-7 breast-tumor cells. *Mol Pharmacol* 1994;45(4):649–656. (d) Lown JW. Anthracycline and anthraquinone anticancer agents — current status and recent developments. *Pharmacol Therapeut* 1993;60(2):185–214.
49. (a) Lee JE, Lee DJ, Lee N, Kim BH, Choi SH, Hyeon T. Multifunctional mesoporous silica nanocomposite nanoparticles for pH controlled drug release and dual modal imaging. *J Mater Chem* 2011;21(42):16869–16872. (b) Yang XQ, Grailer JJ, Pilla S, Steeber DA, Gong SQ. Tumor-targeting pH-responsive and stable unimolecular micelles as drug nanocarriers for targeted cancer therapy. *Bioconjugate Chem* 2010;21(3):496–504. (c) Lu

DX, Wen XT, Liang J, Gu ZW, Zhang XD, Fan YJ. A pH-sensitive nano drug delivery system derived from pullulan/doxorubicin conjugate. *J Biomed Mater Res B* 2009;89B(1):177–183. (d) Wuang SC, Neoh KG, Kang ET, Leckband DE, Pack DW. Acid-sensitive magnetic nanoparticles as potential drug depots. *Aiche J* 2011;57(6):1638–1645. (e) Chang YL, Meng XL, Zhao YL, Li K, Zhao B, Zhu M, Li YP, Chen XS, Wang JY. Novel water-soluble and pH-responsive anticancer drug nanocarriers: doxorubicin-PAMAM dendrimer conjugates attached to superparamagnetic iron oxide nanoparticles (IONPs). *J Colloid Interf Sci* 2011;363(1):403–409.

# Chapter 7

# Nanomaterial-Involved Optical Imaging/ Spectroscopy Methods for Single-Molecule Detections in Biomedicine

*Yao Qin*

*East Hospital, The Institute of Biomedical Engineering and Nano Science, Tongji University School of Medicine, Shanghai 201203, P. R. China*

Single-molecule detection (SMD) techniques can provide information of the structures, functions, interactions of various molecules at single-molecule level, which can be difficult or impossible to obtain using conventional ensemble averaging techniques. Since traditional single-molecule techniques including scanning probe microscopy methods, organic fluorescent dye assisted optical imaging/spectroscopy methods and electrochemical methods have been well reviewed elsewhere, herein, we will pay special attention to one of the most cutting-edge aspects in recent progress of SMD techniques: nanomaterial-involved optical imaging/spectroscopy single-molecule detection methods in biomedicine. The related preparing methods of nanomaterial probes, the instruments and analysis methods for SMD and the current application assays in biomedicine will be covered in current chapter.

## 1. Introduction

The dream voiced by Richard Feynman [1-2] that to manipulate and control matter on an atomic and molecular scale has enlightened the birth of single-molecule science and the development of single-molecule detection (SMD) techniques. Probing single molecules and subsequently modifying their behavior has been an ultimate primary goal. In contrast with ensemble methods, which only yield the average value of a property parameter of the interested system, single-molecule measurements have the following specialties that conventional ensemble averaging techniques do not possess:

first, they provide information such as conformational states, conformational dynamics, and the relationship between the conformational states and functions of a single molecule; second, they can be used to track the distributions and time trajectories of the objective molecules which would be suited to study the intermediates and follow time-dependent pathways of biochemical reactions that are difficult or impossible to synchronize at the ensemble level.[3–5]

As Taekjip Ha said in Ref. 6, life essentially is the subtle interplay of large numbers of proteins, DNAs, and RNAs, which interact via enzymatic reactions and complex signaling pathways. These biomolecules have multiple conformational states and can exist in multiple association states with other molecules. Exploring the detailed dynamically conformational states of these biological molecules using SMD methods will help understanding the complex, multistep biochemical processes of life directly.[6] After almost two decades of development, great advances have been made in SMD techniques including scanning probe microscopy techniques5, optical imaging and spectroscopy[3–4] and the latest electrochemical nanohole sensors.[7–8] The obtained experimental data have demonstrated that it is possible to make observations on the dynamic behavior of single molecules, to determine their mechanisms of action and to explore heterogeneity among different molecules within a population, under physiological conditions.

Although originating from the same hypothesis of Richard Feynman, for a long time nanoscience and single-molecule science advanced separately and barely had any intersections with each other. Scientists in nanoscience area were busy designing various physical or chemical methods for synthesizing nanomaterials with every possible composition, shape, and structure. On the other hand, researchers who dedicate to single-molecule imaging were bothered by the unstable fluorescence signals of organic dye molecules. Until recently, the brilliant spectral properties and photostability of semiconductor nanocrystals (also known as quantum dots (QDs)) enlighten people to replace traditional organic dye molecules with these inorganic nanomaterials and introduce various functional nanomaterials to other SMD methods. Since excellent reviews on traditional SMD methods using scanning probe techniques,[5] and fluorescence imaging and spectroscopy[6] already exist elsewhere. Herein, we will pay special attention to one of the most cutting-edge aspects in recent progress of single-molecule detection techniques: nanomaterial-inspired single-molecule detections in biomedicine. In this chapter, nanomaterial (mainly QDs and noble metal nanoparticles)-involved optical imaging/spectroscopy techniques for SMD including the related preparing methods of nanomaterial probes, the instruments and analysis methods for SMD detections, and the current application assays of nanomaterial-inspired SMD will be covered.

# 2. Quantum Dots Serve as High-Performance Fluorescent Single Molecule Tags

## 2.1. *Preparation of QD probes for SMD*

QDs are usually semiconductor crystals of a few nanometers in diameter. The compositions (core elements are from the periodic groups II–IV, IV–VI, or III–V, e.g., CdTe, CdSe, PbSe, GaAs, GaN, InP, and InAs and shell composition are usually semiconductors with wide-band gap energy such as ZnS) and sizes (usually from 2 nm to 10 nm ) of QDs determine their emission wavelength (ranging from 400 to 1350 nm).[9–10]

Due to their excellent optical properties including high quantum yields, high molar extinction coefficients (ca. 10–100 times higher than that of organic dyes), broad absorption with narrow (25–40 nm half width), symmetric photoluminescence spectra between UV to near-infrared regions, large effective stock shifts, and high resistance against photo-bleaching and photo- and chemical-degradation,[11–12] QDs have now become the new generation of fluorescent tags which are promising to replace organic dyes and fluorescent proteins for single molecule analysis. The developments of nanomaterial synthesis methods have allowed precise control of QDs' sizes and structures by control of synthesis conditions.[13] These synthesized QDs were then functionalized using different modification methods in order to improve the biocompatibility and reduce the biotoxicity as biomolecule tags. For example, thiol (–SH)-containing molecules are often used to anchor functional groups on QDs surfaces[14] and the hydrophilic ends such as carboxyl (–COOH) groups make QDs water-soluble.[15] Oligomeric phosphine,[16] dendrons,[17] and peptides[18] are alternative choice to change the surface properties. In contrast to these covalent modification approaches, coating or encapsulation by a layer of amphiphilic polymers such as diblock[19] and triblock copolymer,[20] phospholipid micelles,[21] and polysaccharides.[22]

For probing biological molecules and live cells, the last step is to functionalize those biocompatible QDs' surface with recognition molecules (e.g., DNA oligonucleotides, RNA, peptide, antibody, etc.). One of the most frequently used modification method is combining streptavidin-coated QDs with biotinylated oligonucleotides,[23] proteins,[24] and antibodies.[25] These repeatable modification methods make QDs become useful multipotent fluorescent probes, despite their relatively large size after the formation of functional layers.

## 2.2. *Internalization of QD-based probes*

For *in vivo* applications, once the functionalization issues are addressed, the next key step is internalization, that is, to deliver the conjugated QDs to the cytoplasm.[26] Several attempts have been done to deliver QDs in the cells, such as using lipid or polymer-mediated endocytosis[27] and peptide-mediated endocytosis.[28] However, the main issue of QDs delivered in the above-mentioned way is that, depending on the cell type and on the QD surface coating, the QDs may be trapped in endosomes for a long time.[28–29] Pinocytosis process activated by a transient increase of the osmotic pressure in the cell culture medium is an alternative way[30]and after pinocytosis, the QD particles are released in the cytosol for single molecule or single cell detections.[31–33] QDs can also be delivered by physical methods, such as microinjection,[34–35] and electroporation.[35–36] Microinjection allows a control on the cell region where the nanoparticles are injected and so far is the only way to deliver QDs into the nucleus.[37] Electroporation has been successfully utilized to deliver QDs in fish keratocytes and to study the cytoplasmic flow inside the migrating cells.[36]

## 2.3. *Measurement methods for single molecule tracking*

In single fluorophore imaging, classical epi-fluorescence microscopes, equipped with a high numerical aperture objective and an ultrasensitive charge-coupled device (CCD) camera is necessary. The fluorescent objects (organic fluorophores or QDs) appear on the CCD camera as bright spots, whose size is determined by the diffraction limit. Provided a sufficient number of photons are collected, the center of mass of each spot can be determined with sub-nanometer precision.[38–39] Although in SMD method, working with a small amount of QDs limits the photon density of the fluorescent spots, it helps to improve the signal-to-noise ratio.

In living cells or under physiological conditions, tracking QDs is technically more difficult due to the high background noises from the cell autofluorescence and the surroundings. To address this problem, experiments are often performed with QDs emitting in the deep red wave band, where the auto-fluorescence of the organism is negligible.[34] Another problem in cell-imaging comes from the intrinsic thickness of the sample. As most of the protein-conjugated QDs have to be tracked in a three-dimensional space, many techniques have been developed in the past years to acquire stacks of images and reconstruct the sterical structure of the cell. These newly developed methods include 3D camera-based techniques either by encoding Z information in their point spread function[40–41] or follow the Z position with multiple cameras or image planes[42–43] and the most popular scanning confocal microscopy technique.[44–48]

Recently, a method developed by Wells *et al.*[49] using four overlapping confocal volume elements to improved the confocal tracking technique and has proved to be capable of following individual QD-labeled signaling molecules in three dimensions, on and inside living cells for minutes. In brief, their method uses four fiber optics as spatial filters to simultaneously monitor four points in the sample (**Figure 1**), and uses pulsed excitation and time-correlated single

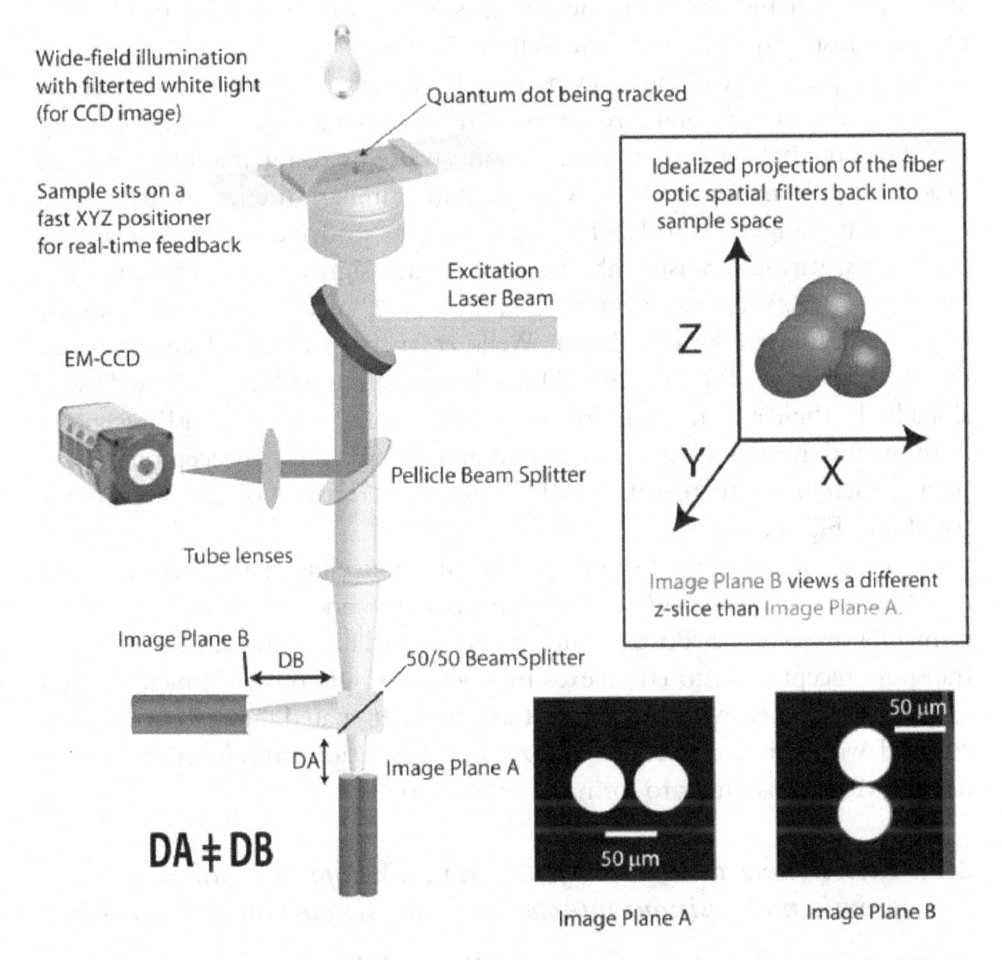


**Figure 1.** Schematic of the 3D tracking apparatus. Four fiber optics serve as spatial filters to examine four nearly diffraction limited spots in the sample simultaneously. Due to the spacing and arrangement of the fibers, these four spots form a 3D tetrahedron in the sample space. While tracking, active feedback of a XYZ piezo stage is used to keep counts on all four detectors equal and as large as possible. A small fraction (~8%) of the emission from the QD receptor being tracked is sent to an EM-CCD camera, which enables contextual information for the position of the molecule in the cell during 3D tracking. [Reproduced with permission from Ref. 49.]

photon counting to record the arrival time of every photon with ~400 ps accuracy, thus can follow small, dim objects such as single QDs as well as single organic dyes with time-resolved analysis capabilities.

## 2.4. *Application examples of QD-based single molecule tracking*

Together with the continuously improving measurement techniques in SMD, QDs have demonstrated their attractiveness as single-molecule probes in a series of recent biological/medical studies, such as membrane dynamics,[50–53] cellular substructures,[54] and drug delivery mechanisms[55] etc.

Single particle tracking in living cells have been first conducted to study the dynamics of membrane receptors. Those early studies include exploring the diffusion behavior of transmembrane proteins using micrometer sized beads[56] or gold nanoparticles.[57] More recently, similar experiments have been carried out using QDs to label proteins to help studying membrane surface topologies, various transmembrane process and signal transduction mechanism.[50–53] For example, as mentioned in the above "measurement methods for single molecule tracking" section, Wells *et al.* have invented an improved confocal technique for tracking QD-labeled signaling molecules three dimensionally. In their recent work, followed individual QD-labeled IgE receptors both on and inside rat mast cells and obtained trajectories of receptors on the plasma membrane to reveal the 3D, nano-scale features of the cell surface topology (**Figure 2**).

In terms of the application of QDs in studying cellular substructures using SMD methods, Lowe *et al.* recently reported an work[54] revealing the mechanism of how Nuclear Pore Complex (NPC) facilitates the translocation of transport receptor-cargo complexes by tracking single protein-functionalized QD cargos. In their work, an amino-PEG polymer coated fluorescent QD was coupled with the snurportin-1 IBB/Z-domain fusion protein via a bifunctional SMCC crosslinker to form the QD cargo.

## 2.5. *Restrictions and QDs as fluorescent SMD probes and other promising luminous nanoparticles for single molecule tracking*

Despite the mentioned advantages of QDs as SMD probes above, there are also restrictions. The major one is the potential toxicity of II–IV QDs in practical biological and medical applications. To address this problem, great effort has been made to develop alternative fluorescent nanoparticles with low toxic components.

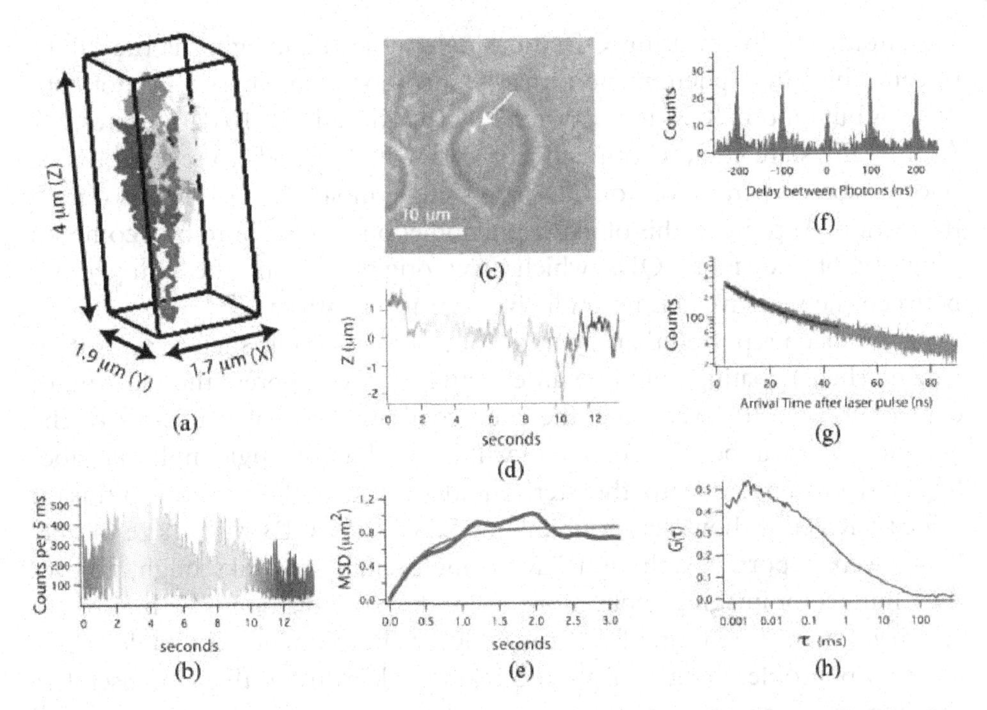


**Figure 2.** **(a)** 3D trajectory of a single QD labeled IgE-FcεRI on an unstimulated mast cell at 37°C. A rainbow color scheme is used to denote the passage of time. **(b)** The counts used for 3D tracking and feedback. **(c)** A CCD image showing the receptor location relative to the mast cell. **(d)** The Z-position of this receptor, showing over 4 μm of Z-motion that would be missed in CCD-based tracking methods. **(e)** The mean squared displacement (MSD) (blue) and fit (red). The motion is highly compartmentalized and is fit using Equation 3 of the supplementary material. **(f)** A photon pair correlation measurement derived from this ~14 second long trajectory that shows fluorescence photon anti-bunching, indicating these trajectories are from individual object molecules. **(g)** A histogram of photon arrival times with respect to the excitation laser pulse (red) and exponential fit (black) which yields a 16 ns fluorescence lifetime **(g)** An autocorrelation analysis of photon detection rate obtained during this trajectory. The decay in the correlation curve is dominated by QD blinking dynamics. [Reproduced with permission from Ref. 49.]

Recently, cadmium-free InP and $CuInS_2$ nanoparticles have attracted people's attentions as candidates of nontoxic QDs, because for one thing they do not contain toxic elements and another they possess a broad emission window from the visible to near-infrared region, with high quantum yield.[58–61] Bio-conjugated InP- and $CuInS_2$-based QDs are promising to be single molecule tracking probes in the near future. Silicon QDs are also attractive candidates as SMD probes for their relatively low toxicity and full color with water dispersion.[62–65] Recently, two-photon and three-photon excitation have been successfully observed in silicon QDs.[66]

Another problem using QDs for single molecule imaging is the "fluorescence blinking" phenomenon, that is, the fluorescent emission is not continuous but randomly switch between ON (bright) and OFF (dark) states.[67–71] As the OFF state could continued from a few milliseconds to sever-al hours, one will lose the trajectory of QD-based single molecule movements during this period. At present, this blinking phenomenon is thought to be a common property of individual QDs which may originate from the competition between the radiative and nonradiative relaxation pathways.[72]

It has been reported that accurate control of inorganic shell (e.g., CdS or CdZnS shells) coating could be an effective way to suppress the blinking of core QDs.[73–75] However, after the shell coating, the final diameters of the complex QDs can be 15–20 nm and will be too big for single molecule labeling and imaging, due to the steric hindrance, and the reduced diffusion velocity. Recently, however, nonblinking CdSe-ZnSe QDs with a size of only 8 nm were reported, although having multi-emission peaks might restrict their multicolor imaging applications.[76] Besides, nanodiamonds with nitrogen vacancy centers were found lately to be highly photoluminescent while exhibiting no photobleaching and photoblinking, which make them be useful as imaging probes for super-resolution optical microscopy in the near future.[77–78]

## 3. Metal Nanoparticles Serve as Near-Field Enhancement Substrates for SMD

### 3.1. *Noble metal nanoparticles serve as enhanced Raman substrates for SMD*

#### 3.1.1. *Fundamental mechanisms of surface enhanced Raman scattering*

First discovered in 1970s,[79] surface-enhanced Raman scattering (SERS) re-emerged recently as one of the most promising chemical and biochemical techniques after the reports of the single-molecule SERS (SM-SERS) of electronically resonant dyes.[80–81] Based on previous studies in bulk samples, the enhancement of the Raman scattering efficiency of the target molecules on plasmonic metallic surface with nanoscale roughness is attributed to two fundamental effects. The main one is concerned with the enhancement of the electromagnetic (EM) fields localized at the edges of metallic particles after excitation of surface plasmons at resonance conditions, which can lead to giant EM enhancement up to factors of ($10^{11}$–$10^{12}$).[82] The second contribution deals with a charge-transfer (CT) process wherein an electron can be transferred from an excited metal state to a vibrational energy level within the target molecule yielding an enhancement of $10\sim10^3$.[83–84] Obviously, for ensemble averaging SERS assays, the EM fields enhancement contribute much more to the overall Raman signal intensity, compared to the CT-enhancement contribution. However, at the single-particle level, recent experimental studies[85–86] reveal that there are some special phenomenon such as "blinking SERS" (the spectra show considerable time-dependent fluctuations in both signal intensities and frequencies) in SM-SERS that cannot be explained by the EM mechanism alone, indicating that chemical enhancement or related mechanism may also contribute much in SM-SERS.[87]

So far, according to extensive experimental and theoretical efforts, in addition to the vibronic CT coupling model[88] that describes the basic selection rule and energetic requirements for CT process, other microscopic models that emphasize the role of surface defects[89–91] and tunneling in the metal-molecule-metal bridge[92-94] have also been proposed. Nevertheless, the specific function models of these SERS mechanisms are still need to be explored and clarified in further SM-SERS studies.

As plasmonic metallic nanostructures are signal amplifier themselves, complex additional signal amplification facilities are not necessary, which makes nanostructures-based SM-SERS a sensitive, facile and promising detection method for SMD.

### 3.1.2. *Preparation of SERS substrates/nanotags*

As stable and reproducible SERS substrate are vital for the detection method, SERS substrate fabrication has now been a fast developing branch of nanotechnology. Various plasmonic metallic nanostructures with carefully designed shape, architecture and size have been reported. For *in vitro* analytical SM-SERS applications, substrates from a single nanoparticle,[95] nanoparticle dimmer[96] to a nanoparticle-on-a-mirror (NPOM) junction,[97] from nanowires[98] to nanotube arrays,[99] with material components from gold,[95,97,100] silver,[99,101] to their compounds,[102] are fabricated for SM-SERS studies.

But when it comes to *in vivo* or bio-medical SM-SERS use, the the metal nanoparticles usually need to be processed using various encapsulation and conjugating strategies to ensure the bio-compatibility, stability and targeting functions of the SERS substrate/nanotags. For example, Doering *et al.*[103] have synthesized a core–shell structuredAu@$SiO_2$ SERS nanotag contains three key components: a gold nanocore for optical enhancement, a reporter molecule for spectroscopic signature, and a silica shell for protection and conjugation (see **Figure 3**). This design work led the bio-analytical applications of gold nanomaterials using SERS measurements, although later it was found that the $SiO_2$ coating existed nonspecific binding limitations. Qian *et al.*'s

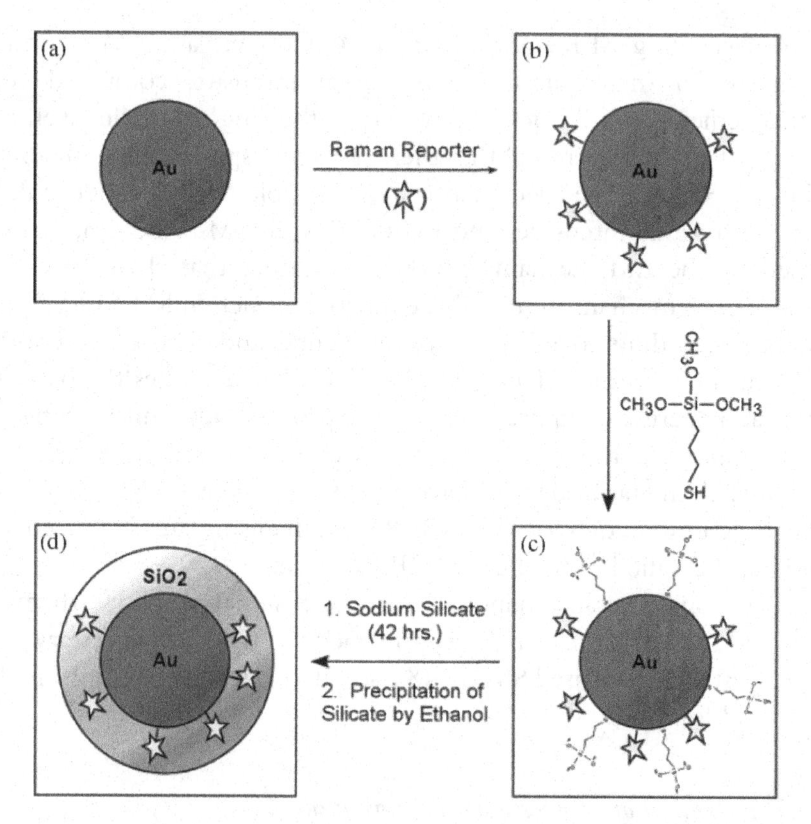


**Figure 3.** Schematic illustration of the core-shell nanoparticle structure and the procedure for preparing silica-coated SERS-active gold colloids. **(a)** Colloidal gold particles in the size range of 55–65 nm, optimized for surface Raman enhancement at 632–647-nm excitation; **(b)** gold particles with an adsorbed Raman reporter; **(c)** gold particles with both the reporter and mercaptopropyltrimethoxysilane (a common coupling agent); and **(d)** silica-coated gold particles with a Raman spectroscopic reporter embedded at the core-shell boundary. [Reproduced with permission from Ref. 103.]

work[104] reported an improved design of gold nanoparticle-based three-layer SERS tag using PEG-SH bifunctional polymer to replace silica as the shell component, which was proved to be stable and with minimal nonspecific binding and negligible cytotoxicity in intracellular delivery studies.[105]

F. Gentile *et al.*[106] recently designed a hierarchical structure which integrates several functions such as super-hydrophobicity, concentrating and selecting molecules and SERS-based single molecule sensitivity. The construction procedure and the multifunctional mechanism are shown in **Figure 4**. First, a regular hexagonal lattice of Si cylindrical micro-pillars was fabricated. Then the top of each pillar was coated with a nanoporous Si film, which would assure the capability of selecting those molecules whose size is

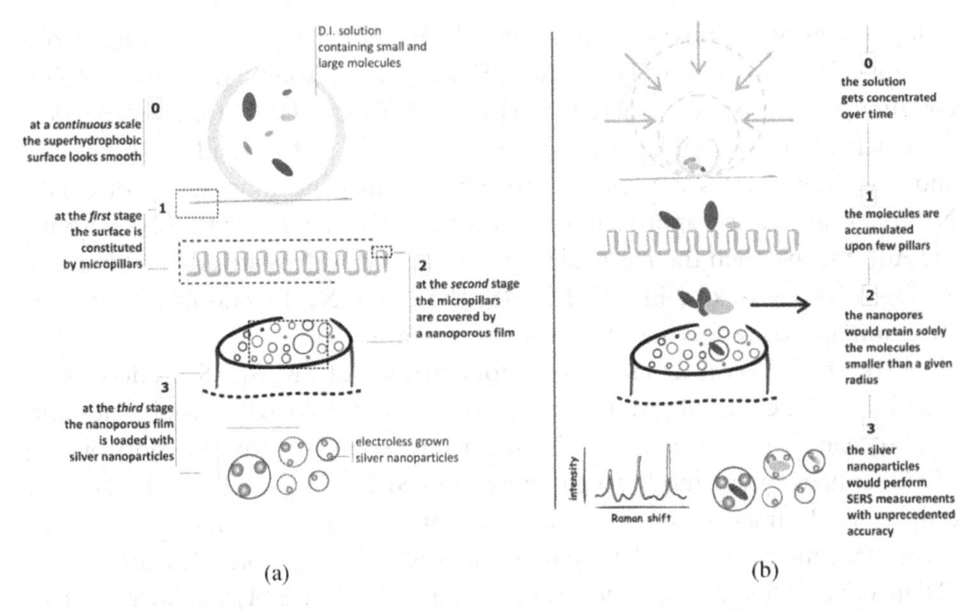


**Figure 4.** The cartoon illustrates the device. (**a**) Description of the different scales of the device. At a continuous scale the substrates looks smooth; at the micro (meso) scale, the pillars contribute to the superhydrophobicity of the substrate; at the nanoscale, the porous matrix and the silver nano-particles come into play. (**b**) Description of the functions the device is capable of, toward the sensing of few molecules. [Reproduced with permission from Ref. 106.]

strictly smaller than the pore size (4 nm in diameter). At last, a randomly distributed silver nanoparticles was infiltrated through the pores, thus providing the device with superior SERS sensitivities.

### 3.1.3. *Current studies in nanostructure-based single-molecule Raman scattering*

First reported by Nie *et al.* in 1997,[80] Single-molecule surface enhanced Raman scattering (SM-SERS) nowadays is still in its full swing in developing more efficient and reproducible SM-SERS substrates, clarifying the specific function mode of enhancement mechanism and extend its applications in bioanalytical assays.

Among various nanostructures, the NPOM configuration is believed to be almost as effective for SERS as the well-studied system of two closely coupled Au or Ag NPs, and more simple to construct. Recently using a Au nanosphere on a Au thin film (AuNS–AuTF) system combining with the bianalyte technique, Li *et al.*[97] have proved of the SM-SERS detection capability of the NPOM conguration in viarous single molecule analysis. For example, the two different conformations of big biomolecule oxytocin have been identified from the SM-SERS spectrum. They also advised that although the AuNS–AuTF conguration can be constructed via two approaches: (I) TF-functionalization approach where a monolayer of probe molecules is first formed on the AuTF, and then the AuNSs are allowed to adsorb onto the monolayer; or (II) a NP-functionalization approach where probe molecules are first adsorbed onto the AuNSs, and then the molecule -functionalized AuNSs are adsorbed on the AuTF, in order to obtain a SM-SERS, the latter NP-functionalization construction approach was adopted (shown in **Figure 5**).

Park *et al.*[84] recently established four groups of SM-SERS models using 4-amino -benzenethiol (ABT), 4-methylbenzenethiol (MBT) molecules and Au/Ag nanoparticles-on-a-Au-film junctions to investigate the contribution of CT enhancement mode to the measured SERS signals. They reveals that only a small fraction of the molecules at the junction has a significant CT-enhancement of $10–10^3$, while the rest of the molecules are nearly CT-inactive. They also claimed that overall (CT and EM) enhancement of $10^6–10^8$ is sufficient to observe the SM-SERS of an electronically off-resonant molecule, disproving the wide spread belief that a minimum enhancement of $\sim 10^{14}$ is required for detection of SM-SERS.

Stranahan *et al.*[107] presented the first super-resolution optical images of SM-SERS hotspots of Rhodamine 6G (R6G) molecules adsorbed on silver colloid aggregates and explored the interaction between single molecules and nanoparticle hot spots. Using point spread function fitting, they mapped the centroid position of SM-SERS with ~10 nm resolution, revealing the existence of multiple hot spots within a single diffraction-limited spot. They also observed the typical “blinking and wandering” of SERS centroid (shown in **Figure 6**) which is typical for SM-SERS and not exist in ensemble averaging

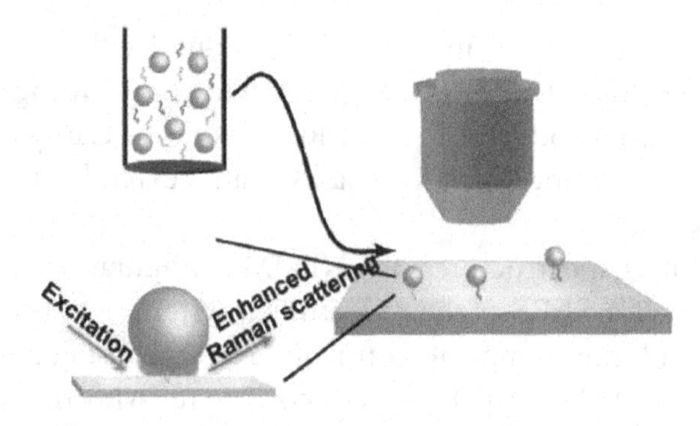


**Figure 5.** Schematic showing the SM-SERS experimental process using the bianalyte method. [Reproduced with permission from Ref. 97.]

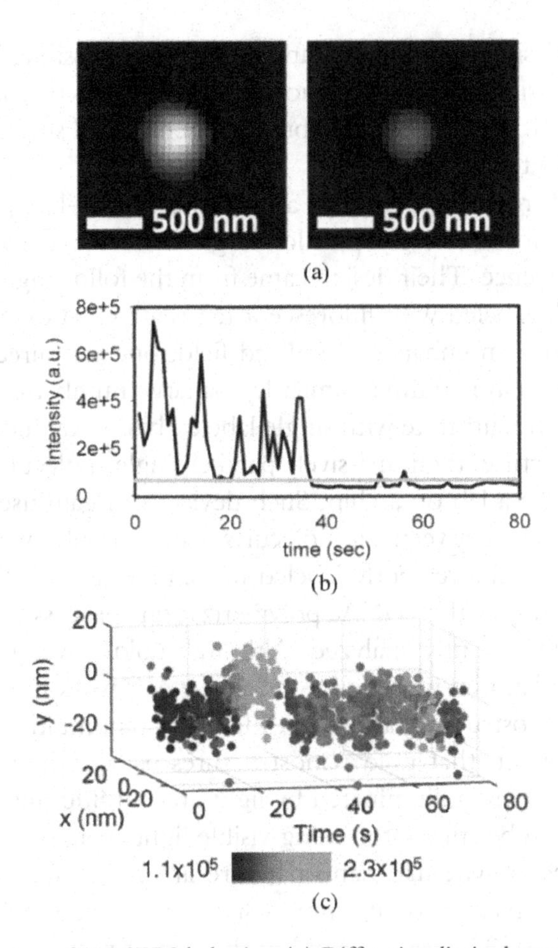


**Figure 6.** Illustration of SM-SERS behavior. **(a)** Diffraction-limited spots of a sample R6G-labeled nanoparticle aggregate when SERS is present (left) versus when only nanoparticle emission is observed (right). **(b)** Integrated SERS intensity over 80 spectra(1 s acquisition for 80 s) plotted versus time. **(c)** Diffusion trajectory showing how the position ($x$ and $y$) and intensity (color scale, arbitrary units) of the SERS centroid changes with time. [Adapted with permission from Ref. 107.]

SERS measurements. According to their experimental analysis, the "blinking and wandering" events were ascribed to the diffusion of a single molecule on the surface of the nanoparticle, which leads to changes in coupling between the scattering dipole and the optical near field of the nanoparticle.

## 3.2. *Metal nanoparticles serve as enhanced fluorescent substrates for SMD*

The plasmon field in the surface of nanoparticles (in most cases, it is noble metal nanoparticles) not only contributes a lot to Raman enhancement, which makes the single-molecule Raman detection possible, but also greatly enhance the emission intensity of fluorophores close to the nanoparticle surface, which leads to a new analytical branch of enhanced single molecule fluorescence (ESMF) technique.

Saito *et al.*[108] recently fabricated a platinum bowtie-like plasmonic nanostructure arrays for massively parallel single-molecule detection based on enhanced fluorescence. Their design came from the following thoughts: When a single molecule labeled with fluorescent tag is attached on a nanostructure, which can produce an enhanced localized field, one can directly observe the molecule behaviors in real time. Similarly, if a large number of nanostructures are arranged on the substrate with single labeled biological molecules attached on each nanostructure, then, massively parallel single-molecule analysis can be performed just like a lab on a chip. Such devices can gain useful information on kinetic processes between biomolecules. For example, when polymerase, template DNA and fluorescently labeled nucleotides are immobilized on the nanostructure arrays, the DNA polymerization process can be directly observed and subsequently analyzed. Although gold is well known as a plasmonic material, platinum was chosen in Saito *et al.*'s work to fabricate bowtie-like plasmonic nanostructure arrays in order to obtain multicolor dye labeling effects for the reason that gold nanostructures usually show ultrahigh local field enhancement when illuminated by light around 800 nm, while platinum nanostructures can be stimulated using visible light around 500 nm.

The prepared bowtie-like nanostructure arrays are shown in **Figure** 7. A very thin titanium layer about 5 nm exists between the Pt triangles. A single dye molecule (Atto532) was attached in the gap of each bowtie-like triangle pair via selective immobilization chemistry method. At last, a high fluorescence enhancement factor (ratio of fluorescence intensities of a single Atto532 molecule with and without the nanostructure) up to 30 was obtained.

Although Saito *et al.* mentioned above expect an attractive use of their nanostructures in dynamic bioanalysis in single-molecule scale, they did not demonstrate it in their work. A recent work by Chowdhury *et al.*[109] indeed have proved that the intrinsic single molecule emission of the enzyme cofactors flavin mononucleotide (FMN) can be enhanced by silver island films (SIFs). More importantly, the blinking phenomenon was also significantly reduced. Their study may encourage the future use of metallic nanostructures to study cofactors for monitoring enzyme binding reactions without the need of extrinsic labeling of the molecules.

It is notable that the luminous intensity of fluorophores can be enhanced only when there is a certain distance between the metal surface and fluorescent molecules, otherwise the fluorescence would be completely quenched.

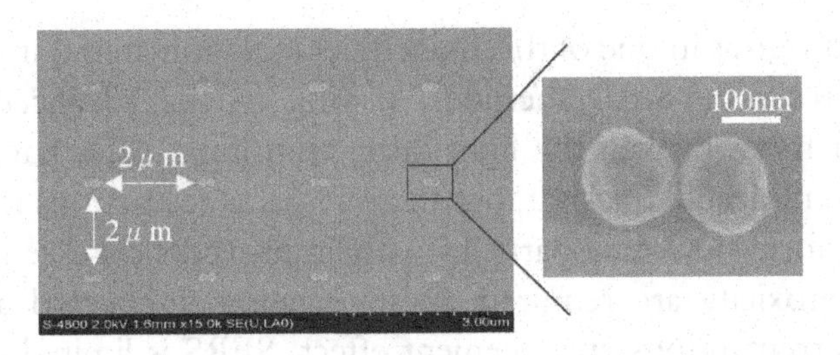


**Figure 7.** Scanning electron microscopy (SEM) images of a Pt bowtie nanostructure array fabricated by electron beam lithography and dry etching techniques. [Reproduced from Ref. 108.]

From this point of view, it is more accurate to call this method "plasmon field enhanced fluorescence" rather than "metal enhanced fluorescence". A lately reported work by Sugawa *et al.*[110] further emphasize this. Moreover, their work extends the variety of substrate materials from traditional noble metals to relatively cheap copper. In brief, ordered array of highly pure Cu nanostructures were prepared by thermally depositing copper on the upper hemispheres of two-dimensional silica colloidal crystals. The local electric fields due to the LSPR of the Cu nanostructures drastically enhanced the fluorescence signals of a porphyrin derivative. By controlling the distance between metal nanostructures and fluorescent molecules to tens of nanometers can efficiently suppress the quenching effect. The plasmonic Cu nanostructures are expected to show better enhancement effects towards fluorescence signals of biomolecules if the gap distance can be further manipulated to an appropriate thickness and are promising for SMDs.

According to the papers published on Web of Science, the related work on plasmonic nanostructure ESMF detections are far less than those on QDs-based SMD and those on SM-SERS method. It is still a relatively new analytical method. There is much to do in synthesizing high-efficient plasmonic nanostructure substrates and optimized fluorophores for preparing enhanced fluorescent probes in biomedical applications. Also as in SERS research, the enhancement mechanism need to be further clarified.

## 4. Conclusions

Being an important analysis means to uncover the ultimate secrets of life in details, single molecule detection (SMD) methods have aroused significant interests of many disciplines. As in a similarly microscale, nanoscience sheds a new light on SMD development. The combination of nanomaterials which possess remarkable optical, electrical and mechanical properties with single molecule techniques generate a new branch of biomedical detections: Nanomaterial-involved SMD methods. This chapter has delivered a brief view of the recent progress in one of this branch field: Nanomaterial-involved optical imaging/spectroscopy SMD methods. Owing to the advanced microscope apparatus which has broken the light diffraction limit, QDs have found their stride in single molecule tracking, but there remains restrictions to current QD probes. Therefore, new nanoparticles with improved fluorescence properties and lower biotoxicity are required. Despite being discovered as early as in 1970s and its tremendous enhancement effect, SERS is limited by the reproducibility of its substrate and the ambiguous enhancement mechanisms. Hence, at present, fabricating more efficient substrates with better reproducibility to clarify the real enhancement model is the biggest challenge for single molecule SERS detections. By contrast, plasmonic nanostructure ESMF detection is a much younger method. There is large space in preparing corresponding substrates and probes and exploring the enhancement mechanisms in this area.

## References

1. Feynman RP. There's Plenty of Room at the Bottom. *Sci Eng* 1960;23(5):22–36.
2. Feynman RP. *Miniaturization*, H D Gilbert, ed., Reinhold, New York, 1961.
3. Moerner WE, Orrit M. Illuminating single molecules in condensed matter. *Science* 1999;283(5408):1670–1676.
4. Weiss S. Fluorescence spectroscopy of single biomolecules. *Science* 1999;283(5408): 1676–1683.
5. Gimzewski JK, Joachim C. Nanoscale science of single molecules using local probes. *Science* 1999;283(5408)1683–1688.
6. Ha T. Single-molecule fluorescence methods for the study of nucleic acids. *Curr Opin Struct Biol* 2001;11(3):287–292.
7. Kasianowicz JJ, Brandin E, Branton D, Deamer DW. Characterization of individual polynucleotide molecules using a membrane channel. *Proc Natl Acad Sci USA* 93 1996;(24): 13770–13773.
8. Wei R, Gatterdam V, Wieneke R, Tampe R, Rant U. Stochastic sensing of proteins with receptor-modified solid-state nanopores. *Nat Nanotechnol* 2012;7(4):257–263.
9. Bruchez M, Jr, Moronne M, Gin P, Weiss S, Alivisatos AP. Semiconductor nanocrystals as fluorescent biological labels. *Science* 1998;281(5385):2013–2016.
10. Michalet X, Pinaud FF, Bentolila LA, Tsay JM, Doose S, Li JJ, Sundaresan G, Wu AM, Gambhir SS, Weiss S. Quantum dots for live cells, in vivo imaging, and diagnostics. *Science* 2005;307(5709):538–544.
11. Alivisatos P. The use of nanocrystals in biological detection. *Nat Biotechnol* 2004;22(1): 47–52.
12. Murphy CJ, Optical sensing with quantum dots. *Anal Chem* 2002;74(19):520a–526a.
13. Kaji N, Tokeshi M, Baba Y. Quantum dots for single bio-molecule imaging. *Anal Sci* 2007;23(1):21–24.
14. Pathak S, Choi SK, Arnheim N, Thompson ME. Hydroxylated quantum dots as luminescent probes for in situ hybridization. *J Am Chem Soc* 2001;123(17):4103–4104.

15. Mattoussi H, Mauro JM, Goldman ER, Anderson GP, Sundar VC, Mikulec FV, Bawendi MG. Self-assembly of CdSe-ZnS quantum dot bioconjugates using an engineered recombinant protein. *J Am Chem Soc* 2000;122(49):12142–12150.
16. Kim S and Bawendi MG, Oligomeric ligands for luminescent and stable nanocrystal quantum dots, *J Am Chem Soc* 2003;125(48):14652–14653.
17. Guo WZ, Li JJ, Wang YA, Peng XG. Conjugation chemistry and bioapplications of semiconductor box nanocrystals prepared via dendrimer bridging. *Chem Mater* 2003;15(16): 3125–3133.
18. Pinaud F, King D, Moore HP, Weiss S. Bioactivation and cell targeting of semiconductor CdSe/ZnS nanocrystals with phytochelatin-related peptides. *J Am Chem Soc.* 2004; 126(19):6115–6123.
19. Wu XY, Liu HJ, Liu JQ, Haley KN, Treadway JA, Larson JP, Ge NF, Peale F, Bruchez MP. Immunofluorescent labeling of cancer marker Her2 and other cellular targets with semiconductor quantum dots. *Nat Biotechnol* 2003; 21(1):41–46.
20. Gao XH, Cui YY, Levenson RM, Chung LWK, Nie SM. In vivo cancer targeting and imaging with semiconductor quantum dots. *Nat Biotechnol* 2004; 22(8):969–976.
21. Dubertret B, Skourides P, Norris DJ, Noireaux V, Brivanlou AH, Libchaber A. In vivo imaging of quantum dots encapsulated in phospholipid micelles. *Science* 2002;298(5599):1759–1762.
22. Osaki F, Kanamori T, Sando S, Sera T, Aoyama Y. A quantum dot conjugated sugar ball and its cellular uptake on the size effects of endocytosis in the subviral region. *J Am Chem Soc* 2004;126(21):6520–6521.
23. Gerion D, Parak WJ, Williams SC, Zanchet D, Micheel CM, Alivisatos AP. Sorting fluorescent nanocrystals with DNA. *J Am Chem Soc* 2002;124(24):7070–7074.
24. Gao XH, Chan WCW, Nie SM. Quantum-dot nanocrystals for ultrasensitive biological labeling and multicolor optical encoding. *J Biomed Opt* 2002;7(4):532–537.
25. Jaiswal JK, Mattoussi H, Mauro JM, Simon SM. Long-term multiple color imaging of live cells using quantum dot bioconjugates. *Nat Biotechnol* 2003;21(1):47–51.
26. Pierobon P, Cappello G. Quantum dots to tail single bio-molecules inside living cells. *Adv Drug Deliver Rev* 2012;64(2):167–178.
27. Ybo J, Kambara T, Gonda K, Higuchi H. Intracellular imaging of targeted proteins labeled with quantum dots. *Exp Cell Res* 2008;314(19):3563–3569.
28. Rozenzhak SM, Kadakia MP, Caserta TM, Westbrook TR, Stone MO Naik RR. Cellular internalization and targeting of semiconductor quantum dots. *Chem Commun* 2005;7(17):2217–2219.
29. Delehanty JB, Bradburne CE, Boeneman K, Susumu K, Farrell D, Mei BC, Blanco-Canosa JB, Dawson G, Dawson PE, Mattoussi H, Medintz IL. Delivering quantum dot-peptide bioconjugates to the cellular cytosol: escaping from the endolysosomal system. *Integr Biol* 2010;2(5–6):265–277.
30. Okada CY, Rechsteiner M. Introduction of macromolecules into cultured mammalian cells by osmotic lysis of pinocytic vesicles. *Cell* 1982;29(1):33–41.
31. Courty S, Luccardini C, Bellaiche Y, Cappello G, Dahan M. Tracking individual kinesin motors in living cells using single quantum-dot imaging. *Nano Lett* 2006;6(7):1491–1495.
32. Pierobon P, Achouri S, Courty S, Dunn AR, Spudich JA, Dahan M, Cappello G. Velocity, processivity, and individual steps of single myosin V molecules in live cells, *Biophys J* 9 2009;6(10):4268–4275.

33. Nelson SR, Ali MY, Trybus KM, Warshaw DM. Random walk of processive, quantum dot-labeled myosin Va molecules within the actin cortex of COS-7 cells. *Biophys J* 2009; 97(2):509–518.
34. Dubertret B, Skourides P, Norris DJ, Noireaux V, Brivanlou AH, Libchaber A. In vivo imaging of quantum dots encapsulated in phospholipid micelles, *Science* 2002;298(5599): 1759–1762.
35. Derfus AM, Chan WCW, Bhatia SN. Intracellular delivery of quantum dots for live cell labeling and organelle tracking. *Adv Mater* 2004;16(12):961–966.
36. Keren K, Yam PT, Kinkhabwala A, Mogilner A, Theriot JA. Intracellular fluid flow in rapidly moving cells. *Nat Cell Biol* 2009;11(10):1219–1224.
37. Ishihama Y, Funatsu T. Single molecule tracking of quantum dot-labeled mRNAs in a cell nucleus. *Biochem Biophys Res Commun* 2009;381(1):33–38.
38. Kubitscheck U, Kuckmann O, Kues T, Peters R. Imaging and tracking of single GFP molecules in solution. *Biophys J* 2000;78(4):2170–2179.
39. Pertsinidis A, Zhang Y, Chu S. Subnanometre single-molecule localization, registration and distance measurements. *Nature* 2010;466(7306):647–651.
40. Holtzer L, Meckel T, Schmidt T. Nanometric three-dimensional tracking of individual quantum dots in cells. *Appl Phys Lett* 2007;90(5):053902–053903.
41. Thompson MA, Lew MD, Badieirostami M, Moerner WE. Localizing and tracking single nanoscale emitters in three dimensions with high spatiotemporal resolution using a double-helix point spread function. *Nano Lett* 2010;10(1):211–218.
42. Toprak E, Balci H, Blehm BH, Selvin PR. Three-dimensional particle tracking via bifocal imaging. *Nano Lett* 2007;7(7):2043–2045.
43. Ram S, Prabhat P, Chao J, Ward ES, Ober RJ. High accuracy 3D quantum dot tracking with multifocal plane microscopy for the study of fast intracellular dynamics in live cells. *Biophys J* 2008;95(12):6025–6043.
44. McHale K, Berglund AJ, Mabuchi H. Quantum dot photon statistics measured by three-dimensional particle tracking. *Nano Lett* 2007;7(11):3535–3539.
45. Cang H, Xu CS, Montiel D, Yang H. Guiding a confocal microscope by single fluorescent nanoparticles. *Opt Lett* 2007;32(18):2729–2731.
46. Lessard GA, Goodwin PM, Werner JH. Three-dimensional tracking of individual quantum dots. *Appl Phys Lett* 2007;91(22):2224106–224108.
47. Wells NP, Lessard GA, Werner JH. Confocal, three-dimensional tracking of individual quantum dots in high-background environments. *Anal Chem* 2008;80(24): 9830–9834.
48. Levi V, Ruan Q, Gratton E. 3-D particle tracking in a two-photon microscope: application to the study of molecular dynamics in cells. *Biophys J* 2005;88(4):2919–2928.
49. Wells NP, Lessard GA, Goodwin PM, Phipps ME, Cutler PJ, Lidke DS, Wilson BS Werner JH. Time-resolved three-dimensional molecular tracking in live cells, *Nano Lett* 2010;10(11):4732–4737.
50. Dahan M, Levi S, Luccardini C, Rostaing P, Riveau B, Triller A. Diffusion dynamics of glycine receptors revealed by single-quantum dot tracking. *Science* 2003;302(5644): 442–445.
51. Chen H, Titushkin I, Stroscio M, Cho M. Altered membrane dynamics of quantum dot-conjugated integrins during osteogenic differentiation of human bone marrow derived progenitor cells. *Biophys J* 2007;92(4):1399–1408.
52. Haggie PM, Kim JK, Lukacs GL, Verkman AS. Tracking of quantum dot-labeled CFTR shows near immobilization by C-terminal PDZ interactions. *Mol Biol Cell* 2006;17(12): 4937–4945.

53. Choquet D, Triller A. The role of receptor diffusion in the organization of the postsynaptic membrane. *Nat Rev Neurosci* 2003;4(4):251–265.
54. Lowe AR, Siegel JJ, Kalab P, Siu M, Weis K, Liphardt JT. Selectivity mechanism of the nuclear pore complex characterized by single cargo tracking. *Nature* 2010;467(7315):600–603.
55. Tada H, Higuchi H, Wanatabe TM, Ohuchi N. In vivo real-time tracking of single quantum dots conjugated with monoclonal anti-HER2 antibody in tumors of mice. *Cancer Res* 2007;67(3):1138–1144.
56. Saxton MJ. Lateral diffusion in an archipelago. Single-particle diffusion. *Biophys J* 1993; 64(6):1766–1780.
57. Kusumi A, Sako Y, Yamamoto M. Confined lateral diffusion of membrane receptors as studied by single particle tracking (nanovid microscopy). Effects of calcium-induced differentiation in cultured epithelial cells. *Biophys J* 1993;65(5):2021–2040.
58. Bharali DJ, Lucey DW, Jayakumar H, Pudavar HE, Prasad PN. Folate-receptor-mediated delivery of InP quantum dots for bioimaging using confocal and two-photon microscopy. *J Am Chem Soc* 2005;127(32):11364–11371.
59. Tamang S, Beaune G, Texier I, Reiss P. Aqueous phase transfer of InP/ZnS nanocrystals conserving fluorescence and high colloidal stability. *ACS Nano* 2011;5(12):9392–9402.
60. Yong KT, Ding H, Roy I, Law WC, Bergey EJ, Maitra A, Prasad PN. Imaging pancreatic cancer using bioconjugated InP quantum dots *ACS Nano* 2009;3(3):502–510.
61. Yong KT, Roy I, Hu R, Ding H, Cai H, Zhu J, Zhang X, Bergey EJ Prasad PN. Synthesis of ternary CuInS(2)/ZnS quantum dot bioconjugates and their applications for targeted cancer bioimaging. *Integr Biol (Camb)*2010;2(2–3):121–129.
62. Choi J, Wang NS, Reipa V. Conjugation of the photoluminescent silicon nanoparticles to streptavidin. *Bioconjug Chem* 2008;19(3):680–685.
63. Erogbogbo F, Tien CA, Chang CW, Yong KT, Law WC, Ding H, Roy I, Swihart MT, Prasad PN. Bioconjugation of luminescent silicon quantum dots for selective uptake by cancer cells. *Bioconjug Chem* 2011;22(6):1081–1088.
64. Erogbogbo F, Yong KT, Roy I, Xu G, Prasad PN Swihart MT. Biocompatible luminescent silicon quantum dots for imaging of cancer cells. *ACS Nano* 2008;2(5):873–878.
65. Shiohara A, Prabakar S, Faramus A, Hsu CY, Lai PS, Northcote PT, Tilley RD. Sized controlled synthesis, purification, and cell studies with silicon quantum dots. *Nanoscale* 2011;3(8):3364–3370.
66. He GS, Zheng Q, Yong KT, Erogbogbo F, Swihart MT, Prasad PN. Two- and three-photon absorption and frequency upconverted emission of silicon quantum dots. *Nano Lett* 2008;8(9):2688–2692.
67. Banin U, Bruchez M, Alivisatos AP, Ha T, Weiss S, Chemla DS. Evidence for a thermal contribution to emission intermittency in single CdSe/CdS core/shell nanocrystals. *J Chem Phys* 1999;110(2):1195–1201.
68. Basche T. Fluorescence intensity fluctuations of single atoms, molecules and nanoparticles. *J Lumin* 1998;76(7):263–269.
69. Neuhauser RG, Shimizu KT, Woo WK, Empedocles SA, Bawendi MG. Correlation between fluorescence intermittency and spectral diffusion in single semiconductor quantum dots. *Phys Rev Lett* 2000;85(15):3301–3304.
70 Sugisaki M, Ren HW, Nishi K, Masumoto Y. Fluorescence intermittency in self-assembled InP quantum dots. *Phys Rev Lett* 2001;86(21):4883–4886.
71. Nirmal M, Dabbousi BO, Bawendi MG, Macklin JJ, Trautman JK, Harris TD, Brus LE. Fluorescence intermittency in single cadmium selenide nanocrystals. *Nature* 1996;383(6603):802–804.

72. Fomenko V, Nesbitt DJ. Solution control of radiative and nonradiative lifetimes: a novel contribution to quantum dot blinking suppression. *Nano Lett* 2008;8(1):287–293.
73. Chen Y, Vela J, Htoon H, Casson JL, Werder DJ, Bussian DA, Klimov VI, Hollingsworth JA. "Giant" multishell CdSe nanocrystal quantum dots with suppressed blinking. *J Am Chem Soc* 2008;130(15):5026–5027.
74. Mahler B, Spinicelli P, Buil S, Quelin X, Hermier JP, Dubertret B. Towards non-blinking colloidal quantum dots. *Nat Mater* 2008;7(8):659–664.
75. Vela J, Htoon H, Chen YF, Park YS, Ghosh Y, Goodwin PM, Werner JH, Wells NP, Casson JL, Hollingsworth JA. Effect of shell thickness and composition on blinking suppression and the blinking mechanism in 'giant' CdSe/CdS nanocrystal quantum dots. *J Biophotonics* 2010;3(10–11):706–717.
76. Wang XY, Ren XF, Kahen K, Hahn MA, Rajeswaran M, Maccagnano-Zacher S, Silcox J, Cragg GE, Efros AL, Krauss TD. Non-blinking semiconductor nanocrystals, *Nature* 2009; 459(7247):686–689.
77. Aharonovich I, Castelletto S, Simpson DA, Su CH, Greentree AD, Prawer S. Diamond-based single-photon emitters. *Rep Prog in Phys* 2011;74(7).
78. Faklaris O, Joshi V, Irinopoulou T, Tauc P, Sennour M, Girard H, Gesset C, Arnault JC, Thorel A, Boudou JP, Curmi PA, Treussart F. Photoluminescent diamond nanoparticles for cell labeling: study of the uptake mechanism in mammalian cells *ACS Nano* 2009;3(12):3955–3962.
79. Fleischmann M, Hendra PJ, McQuillan AJ. Raman spectra of pyridine adsorbed at a silver electrode. *Chem Phys Lett* 1974;26(2)163–166.
80. Nie S, Emory SR. Probing single molecules and single nanoparticles by surface-enhanced Raman scattering. *Science* 1997;275(5303):1102–1106.
81. Kneipp K, Wang Y, Kneipp H, Perelman LT, Itzkan I, Dasari R, Feld MS. Single molecule detection using surface-enhanced Raman scattering (SERS). *Phys Rev Lett* 1997;78(9):1667–1670.
82. Xu HX, Aizpurua J, Kall M, Apell P. Electromagnetic contributions to single-molecule sensitivity in surface-enhanced Raman scattering. *Phys Rev E* 2000;62(3):4318–4324.
83. Otto A, Mrozek I, Grabhorn H, Akemann W. Surface-enhanced Raman scattering. *J Phys Condens Matter* 1992;4:1143–1212.
84. Park WH, Kim ZH. Charge transfer enhancement in the SERS of a single molecule. *Nano Lett* 2010;10(10):4040–4048.
85 Michaels AM, Jiang J, Brus L. Ag nanocrystal junctions as the site for surface-enhanced Raman scattering of single rhodamine 6G molecules. *J Phys Chem B* 2000;104(50): 11965–11971.
86. Bosnick KA, Jiang J, Brus LE. Fluctuations and local symmetry in single-molecule rhodamine 6G Raman scattering on silver nanocrystal aggregates. *J Phys Chem B* 2002;106(33):8096–8099.
87. Zuloaga J, Prodan E, Nordlander P. Quantum description of the plasmon resonances of a nanoparticle dimer. *Nano Lett* 2009;9(2):887–891.
88. Lombardi JR, Birke RL. A unified view of surface-enhanced Raman scattering. *Acc Chem Res* 2009;42(6):734–742.
89. Lust A, Pucci A, Akemann W, Otto A. SERS of $CO_2$ on cold-deposited Cu: an electronic effect at a minority of surface sites. *J Phys Chem C* 2008;112(30):11075–11077.

90. Otto A, Lust M, Pucci A, Meyer G. "SERS active sites", facts, and open questions. *Can J Anal Sci Spectrosc* 2007;52(3):150–171.
91. Otto A. The 'chemical' (electronic) contribution to surface-enhanced Raman scattering. *J Raman Spectrosc* 2005;36(6–7):497–509.
92. Zhao LL, Jensen L, Schatz GC. Surface-enhanced Raman scattering of pyrazine at the junction between two $Ag_2O$ nanoclusters. *Nano Lett* 2006;6(6):1229–1234.
93. Zhou Q, Li XW, Fan Q, Zhang XX, Zheng JW. Charge transfer between metal nanoparticles interconnected with a functionalized molecule probed by surface-enhanced Raman spectroscopy. *Angew Chem-Int Ed* 2006;45(24):3970–3973.
94. Sun MT, Xu HX. Direct visualization of the chemical mechanism in SERRS of 4-aminothiophenol/metal complexes and metal/4-aminothiophenol/metal junctions. *Chem Phys Chem* 2009;10(2):392–399.
95. Li ZY, Xia Y. Metal nanoparticles with gain toward single-molecule detection by surface-enhanced Raman scattering. *Nano Lett* 2010;10(1):243–249.
96. Lassiter JB, Aizpurua J, Hernandez LI, Brandl DW, Romero I, Lal S, Hafner JH, Nordlander P, Halas NJ. Close encounters between two nanoshells. *Nano Lett* 2008;8(4):1212–1218.
97. Li L, Hutter T, Steiner U, Mahajan S. Single molecule SERS and detection of biomolecules with a single gold nanoparticle on a mirror junction. *Analyst* 2013;138(16):4574–4578.
98. Wang H, Han XM, Ou XM, Lee CS, Zhang XH, Lee ST. Silicon nanowire based single-molecule SERS sensor. *Nanoscale* 2013;5(17):8172–8176.
99. Fang C, Brodoceanu D, Kraus T, Voelcker NH. Templated silver nanocube arrays for single-molecule SERS detection. *Rsc Advances* 2013;3(13):4288–4293.
100. Ameer FS, Pittman CU, Zhang DM. Quantification of resonance raman enhancement factors for rhodamine 6G (R6G) in water and on gold and silver nanoparticles: implications for single-molecule R6G SERS. *J Phys Chem C* 2013;117(51):27096–27104.
101. Potara M, Baia M, Farcau C, Astilean S. Chitosan-coated anisotropic silver nanoparticles as a SERS substrate for single-molecule detection. *Nanotechnology* 2012;23(5):10.
102. Liu HW, Zhang L, Lang XY, Yamaguchi Y, Iwasaki HS, Inouye YS, Xue QK, Chen MW. Single molecule detection from a large-scale SERS-active Au79Ag21 substrate. *Sci Rep* 2011;1(112):5.
103. Doering WE, Nie SM. Spectroscopic tags using dye-embedded nanoparticles and surface-enhanced Raman scattering. *Anal Chem* 2003;75(22):6171–6176.
104. Qian X, Peng XH, Ansari DO, Yin-Goen Q, Chen GZ, Shin DM, Yang L, Young AN, Wang MD, Nie S. *In vivo* tumor targeting and spectroscopic detection with surface-enhanced Raman nanoparticle tags. *Nat Biotechnol* 2008;26(1):83–90.
105. Shenoy D, Fu W, Li J, Crasto C, Jones G, DiMarzio C, Sridhar S, Amiji M. Surface functionalization of gold nanoparticles using hetero-bifunctional poly(ethylene glycol) spacer for intracellular tracking and delivery. *Int J Nanomed* 2006;1(1):51–57.
106. Gentile F, Coluccio ML, Accardo A, Marinaro G, Rondanina E, Santoriello S, Marras S, Das G, Tirinato L, Perozziello G, De Angelis F, Dorigoni C, Candeloro P, Di Fabrizio E. Tailored Ag nanoparticles/nanoporous superhydrophobic surfaces hybrid devices for the detection of single molecule. *Microelectron Eng* 2012;97:349–352.
107. Stranahan SM, Willets KA. Super-resolution optical imaging of single-molecule SERS hot spots. *Nano Lett* 2010;10(9):3777–3784.

108. Saito T, Takahashi S, Obara T, Itabashi N, Imai K. Platinum plasmonic nanostructure arrays for massively parallel single-molecule detection based on enhanced fluorescence measurements. *Nanotechnology* 2011;22(44).
109. Chowdhury MH, Lakowicz JR, Ray K. Ensemble and single molecule studies on the use of metallic nanostructures to enhance the intrinsic emission of enzyme cofactors. *J Phys Chem C* 2011;115(15):7298–7308.
110. Sugawa K, Tamura T, Tahara H, Yamaguchi D, Akiyama T, Otsuki J, Kusaka Y, Fukuda N, Ushijima H. Metal-enhanced fluorescence platforms based on plasmonic ordered copper arrays: wavelength dependence of quenching and enhancement effects. *ACS Nano* 2013;7(11):9997–10010.

# Chapter 8

# Surface Functionalized Carbon Nanotubes for Biomedical Applications

*Bingdi Chen*

*Research Center for Translational Medicine, East Hospital, Tongji University School of Medicine, No. 150 Jimo Road, Shanghai 200120; The Institute for Biomedical Engineering & Nano Science Tongji University, No. 1239 Siping Road, Shanghai, 200092, P. R.China*
*inanochen@tongji.edu.cn*

Carbon nanotubes (CNTs) have many advantages including unique intrinsic physical and chemical properties, have the ability of readily being decorated with various functional materials and biological species and also possess the ability to penetrate into cells, which justifies their worth in being intensively exploited for potential nanobiomedical applications. On the other hand, design and fabrication of multi-functional nanohybrids can significantly improve the early detection and diagnosis of diseases. Thus, it will make great sense in the nanobiomedical area to combine the merits of CNTs and multi-functional nanohybrids together to synthesize CNTs based multi-functional magnetic-fluorescent nanohybrids. In this chapter, we will give an overview on the surface functionalization of CNTs for biomedical applications. Firstly, we will review the water-soluble and biocompatible functionalization of CNTs. Then, we will mainly discuss the combination of CNTs with multi-functional nanohybrids and their applications in multi-modal bioimaging and drug delivery.

## 1. Introduction

The first reported observation of carbon nanotubes (CNTs) was by Iijima in 1991 for multi-walled nanotubes (MWNTs) as shown in **Figure 1**.[1] It took, however, less than two years before single-walled carbon nanotubes (SWCNTs) were discovered experimentally by Iijima at the NEC Research Laboratory in Japan and by Bethune at the IBM Almaden Laboratory in California.[2,3]

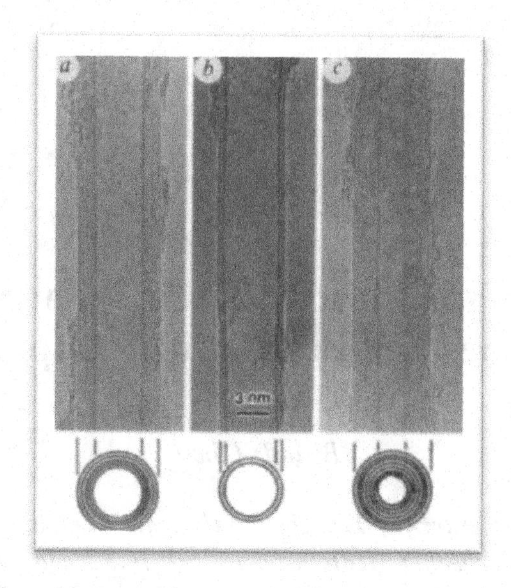

**Figure 1.** The observation by transmission electron microscopy (TEM) of MWNTs with various inner and outer diameters and numbers of cylindrical shells reported by Iijima in 1991.[1]

An ideal nanotube can be considered as a hexagonal network of carbon atoms that has been rolled up to make a seamless hollow cylinder. These hollow cylinders can be tens of micrometers long, but with diameters as small as 0.7 nm, and with each end of the long cylinder "capped with half a fullerene molecule, i.e., 6 pentagons". Single-walled nanotubes (SWNTs), having a cylindrical shell with only one atom in thickness, can be considered as the fundamental structural unit. Such structural units form the building blocks of both MWNTs, containing multiple coaxial cylinders of ever-increasing diameter about a common axis, and nanotube ropes, consisting of ordered arrays of SWNTs arranged on a triangular lattice.

CNTs have been intensively exploited for biomedical applications due to their unique intrinsic physical and chemical properties during the past decades.[4–7] Moreover, CNTs have been considered as an ideal matrix for the synthesis of CNT-based nanohybrids for biomedical applications because there are some advantages in using CNTs as the matrix to synthesize nanohybrids.[8] Firstly, CNTs have a large surface area to load nanoparticles in a one-dimensional direction, preventing the aggregation of nanoparticles in solution. Secondly, it has been reported that CNTs can readily penetrate into cells either by endocytosis or non-endocytic mechanism,[9,10] which assist the functional nanoparticles on the surface of the CNTs to enter into cells and increase their sensitivity and efficiency for detection as a diagnostic agent. Thirdly, CNTs can be further decorated with biological species and drugs to potentially serve as an all-in-one diagnostic and therapeutic tool.[11] As a result, many efforts have been focused on the CNT-functional nanoparticles nanohybrids so far.

CNTs have unique intrinsic physical and chemical properties, some of which can be exploited for biomedical applications. However, most of the intrinsic properties are not suitable for direct use. For example, CNTs have been observed to exhibit infrared emissions,[12,13] but the emissions are so weak and therefore do not have sufficient fluorescence for *in vivo* imaging. So considerable efforts have been focused on the synthesis of luminescent CNTs decorated with fluorescent materials.[14–16] Another example reports that CNTs can lower the transverse relaxation of water proton thus acting as $T_2$ weighted magnetic resonance imaging (MRI) contrast agent.[17] However the MRI signal is too weak to provide sufficient information. So, it is necessary to functionalize CNTs with magnetic materials, such as superparamagnetic iron oxide (SPIO) nanoparticles.[14] The orderly attachment of multiple SPIO nanoparticles on the surface of the CNTs in a one-dimensional direction can enhance the magnetic effect due to the magnetic coupling between SPIO nanoparticles.[18]

## 2. Synthesis and Biocompatible Functionalization of CNTs

### 2.1. *Synthesis of CNTs*

At present, CNTs are produced using one of three main methods: arc discharge,[1,19] laser-ablation,[20–22] and chemical vapor deposition (CVD).[23–25]

The arc discharge technique utilizes two carbon electrodes to generate an arc by direct current (DC). The arc discharge technique produces high quality MWNTs and SWNTs. MWNTs do not need a catalyst for growth, while SWNTs can only be grown in the presence of a catalyst.

In the laser-ablation technique used by Thess *et al.*[20] for producing CNTs, intense laser pulses were utilized to ablate a carbon target. The pulsed laser-ablation of graphite in the presence of an inert gas and catalyst formed SWNTs at 1200°C.

In the CVD method, CNTs are synthesized by taking hydrocarbons (the commonly used sources are methane, ethylene, and acetylene) and using an energy source, such as electron beam or resistive heating, to impart energy to them. The energy source breaks the molecule into reactive radical species in the temperature range of 550–750°C. These reactive species then diffuse down to the substrate, which is heated and coated in a catalyst (usually a first-row transition metal such as Ni, Fe, or, Co), where it remains bonded. As a result, the CNTs are formed.

## 2.2. *Biocompatible functionalization of CNTs*

Surface functionalization of CNTs is the first step to make CNT-based integrated multifunctional nanosystems. The primary CNTs synthesized by the above-mentioned preparation methods come with a number of impurities whose type and amount depend on the technique used. The most common impurities are fullerenes, graphitic polyhedrons with enclosed metal particles, and amorphous carbon. Normally, the purification of CNTs was carried out by wet or dry oxidation. In the arc discharge method, the impurities can be purified by oxidation as the carbonaceous impurities have high oxidation rates. However, in this case, 95% of the starting materials are destroyed and the remaining samples require annealing at high temperature (about 2800°C). For CNTs grown by CVD of carbon monoxide, a typical purification process included sonication, oxidation, and acid washing steps. Firstly, the as-received MWNTs were refluxed in 2.6 M $HNO_3$ for 38 hours. After cooling, the solution was diluted and washed with deionized water until pH 5–6, and dried under vacuum at room temperature. Secondly, the purified MWNTs were suspended in a 3:1 mixture of concentrated. $H_2SO_4$:$HNO_3$ and sonicated for 4 hours at 35–40° C for cutting MWNTs into short pieces. The resultant solution was diluted and washed with deionized water sufficiently, and dried under vacuum at room temperature. Thirdly, the as-prepared short MWNTs were further polished by a 4:1 mixture of concentrated $H_2SO_4$:30% $H_2O_2$ at 70°C for 1 hour under stirring. After washing with deionized water and drying under vacuum at room temperature, the resultant powder was collected. In this process, the CNTs not only can be purified, but also can be cut short, which can be shown in **Figure 2** by TEM. The as-received CNTs synthesized *via* CVD method contain impurities of catalyst, amorphous carbon and fullerenes etc., and the length can be up to micrometers (**Figure 2(a)**). When purified and cut short by sonicating in concentrated $H_2SO_4$:$HNO_3$ mixture and further polished by concentrated $H_2SO_4$:30% $H_2O_2$ mixture, the CNTs can be cut short to about 100–500 nm with a high purity (**Figure 2(b)**).

Due to van der Waals forces between adjacent tubes, CNTs have an inherent tendency to exist as bundles (agglomerates), resulting in poor dispersion and surface area contact, which consequently limits their practical applications in aqueous solvents. After the above-mentioned purification, the surface of CNTs possesses some carboxyl groups. However, the total solubility of the purified CNTs is still too poor for biomedical application. So the further surface modification is significant to make CNTs water-soluble and biocompatible enough for biomedical application. There are several approaches for functionalizing CNTs, which include defect, noncovalent, and covalent functionalization.[26]

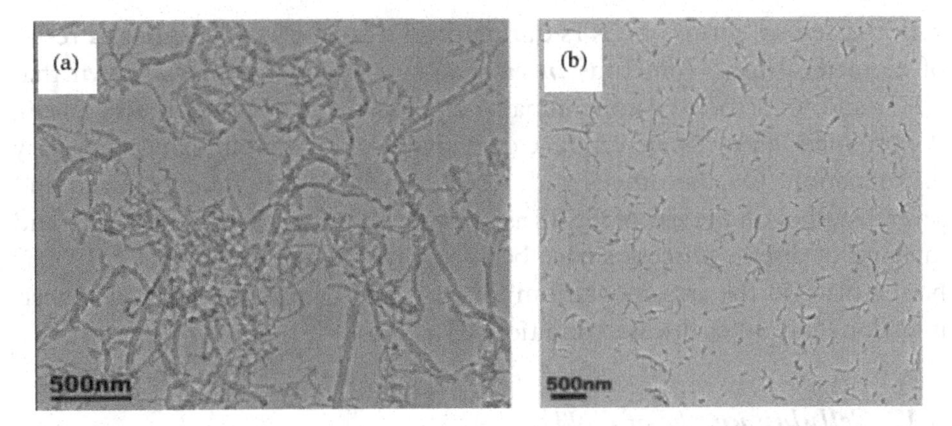


**Figure 2.** TEM images of **(a)** CNTs before purification, which contain catalyst, amorphous carbon and with a length up to micrometers; **(b)** CNTs purified and cut short by sonicating in concentrated $H_2SO_4$:$HNO_3$ mixture and further polished by concentrated $H_2SO_4$:30% $H_2O_2$ mixture.

In defect functionalization, CNTs are purified by oxidative methods to remove metal particles or amorphous carbon from the raw materials. In these methods, defects are preferentially observed at the open ends of CNTs. The noncovalent functionalization of nanotubes improves the solubility without spoiling the physical properties of CNTs. This type of functionalization mainly involves surfactants or wrapping with polymers.[27–29] The main potential disadvantage of noncovalent attachment is that the forces between the wrapping molecule and nanotube might be weak, thus as a filler in a composite, the efficiency of the load transfer might be low. In the case of covalent functionalization, the translational symmetry of CNTs is disrupted by changing $sp^2$ C atoms to $sp^3$ C atoms, and the properties of CNTs, such as electronic and transport properties, are varied from the original CNTs.[30]

## 3. Cytotoxicity of CNTs

There are several exciting prospects for the application of CNTs in medicine. However, concerns over adverse and unanticipated effects on human health have also been raised. In fact, CNTs can serve as an instructive example of the Janus-like properties of nanomaterials. On one hand, studies show that CNTs are toxic and that the extent of that toxicity depends on properties of the CNTs, such as their structure (single-walled or multiple-walled), length and aspects ratios, surface area, degree of aggregation, extent of oxidation, bound functional group(s), method of manufacturing, concentration, and dose. People could be exposed to CNTs either accidentally by coming in contact with the aerosol forms of CNTs during production or by exposure as a result of biomedical use. Numerous *in vitro* and *in vivo* studies have shown that CNTs and/or associated contaminants or catalytic materials that arise during the production process may induce oxidative stress and prominent pulmonary inflammation. On the other hand, CNTs can be readily functionalized and several studies on the use of CNTs as versatile excipients for drug delivery and imaging of disease processes have been reported, suggesting that CNTs may have a place in the armamentarium for treatment and monitoring of cancer, infection, and other disease conditions.

## 3.1. *Cellular uptake of CNTs*

Until now, there is a lack of studies on the mechanism of the behavior of CNTs in the biological interphase. When introducing xenobiotic materials into cells, the first barrier is the plasma membrane around the cell, which is composed of different phospholipids noncovalently linked together by hydrophobic interactions to form the bilayer around the cell, in addition to several embedded proteins. The membrane controls the transport of xenobiotic substances into the cell by various pathways.[31] Following the binding of CNTs to the cell, the possible pathways of their entry into cells are subdivided into energy-dependent endocytosis[32–34] and passive diffusion across the lipid bilayer.[10] In our previous study the two pathways have been observed by TEM, as shown in **Figure 3**. **Figure 3(a)** showed the CNTs interact with cells *via* endocytosis, while **Figure 3(b)** showed the CNTs can directly penetrate the plasma membrane.

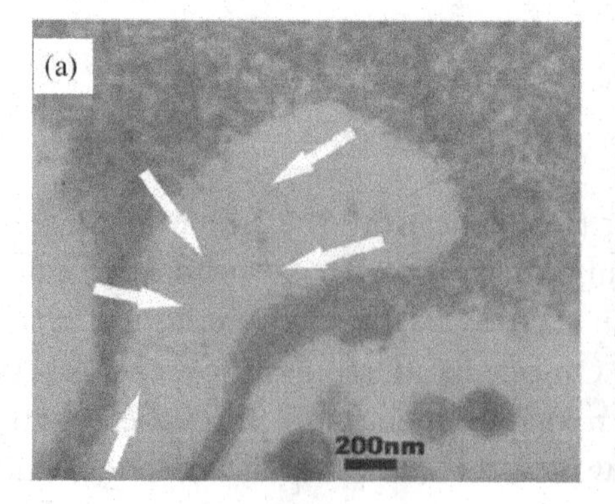


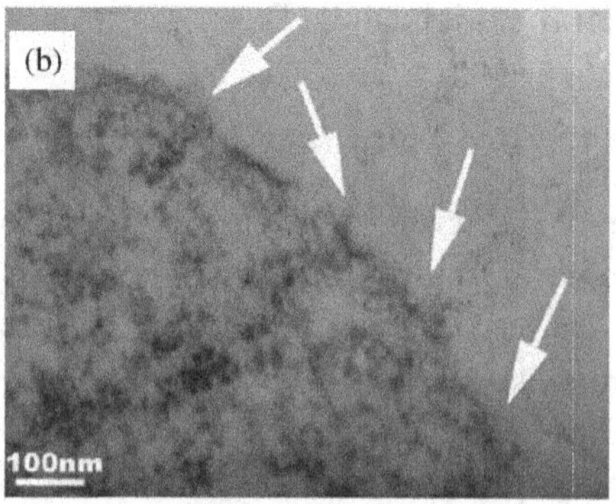


**Figure 3.** Cellular uptake of CNTs *via* endocytosis **(a)** or direct penetration **(b)** observed by TEM.

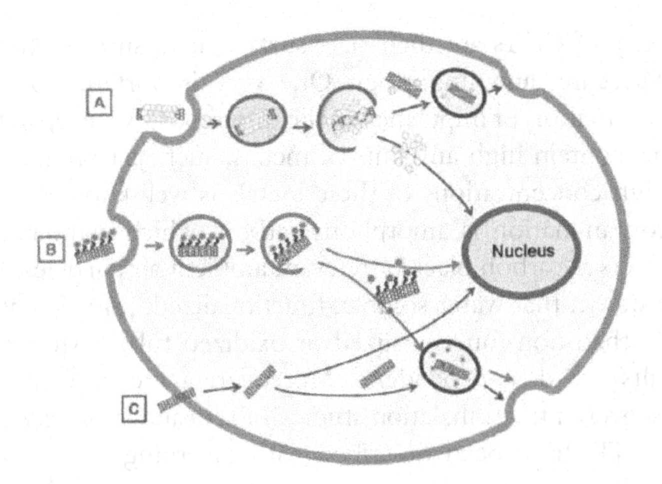


**Figure 4.** Scheme of the possible pathways of cellular uptake of CNTs *via* endocytosis or direct penetration.[38]

There are many factors to affect the cellular uptake of CNTs. Firstly, the size-dependent cellular uptake of CNTs must be considered. Short CNTs (diameter: 10–160 nm, with a maximum length of ~2–3 $\mu$m) were observed to be inside the cells while long CNTs (diameter: 10–160 nm, length: 0.1–12 $\mu$m) were not.[35] Additionally, individual MWNTs may enter cells through direct penetration, whereas bundled MWNTs may enter *via* endocytosis.[36] Secondly, the surface chemistry of CNTs plays an important role for their incorporation into biological systems because of their hydrophobicity and high surface area, which increases interactions with cellular compartments. However, the size effect is more critical to cellular internalization than are the surface properties of CNTs. Thirdly, the functional groups of CNTs can have an effect on the cellular interactions of CNTs with different cell types, as reported in a recent study.[37]

In conclusion, the possible cellular pathways of functionalized CNTs of certain sizes are schematically described in **Figure 4**. Depending on their size and functional groups, the CNTs interact with cells *via* endocytosis or non-endocytosis (direct penetration). Specifically functionalized individual CNTs can directly penetrate the plasma membrane as well as the nuclear membrane.

## 3.2. *Cellular toxicity of CNTs*

Cytotoxicity is of major interest when introducing xenobiotic substances, such as CNTs, into living systems. The important parameters for determining the cytotoxicity of CNTs are their size, surface area, surface chemistry, tendency to aggregate, and impurities. One very important point regarding toxicity is the variation of impurities of the different CNT preparations. Most used fractions contain high amounts of metals, such as iron and nickel. The toxicity of high concentrations of these metals is well-known. Another very important contamination is amorphous carbon, which exhibits comparable biological effects as carbon black or relevant ambient air particles. Cell culture studies have shown that water-soluble, functionalized, pristine nanotubes are less cytotoxic than non-functionalized or oxidized tubes. Most of the cytotoxicity results are dose-dependent. Moreover, a recent study by Wörle-Knirsch[39] discovered that inhalation studies *in vivo* and submerse applications *in vitro* of CNTs have been described with diverging results. One of the important reasons is that the assessment of cytotoxicity by different methods can lead to very different results. Data from A549 cells (Human lung adenocarcinoma epithelial cell line) incubated with CNTs show a strong cytotoxic effect within the MTT [(3-(4,5-Dimethylthiazol-2-yl)-2,5-Diphenyltetrazolium Bromide] assay after 24 hours that reaches roughly 50%, whereas the same treatment with SWCNTs, but detection with water-soluble tetrazolium salt (WST-1), reveals no cytotoxicity. Lactate dehydrogenase (LDH), Fluorescence-activated cell sorting (FACS)-assisted mitochondrial membrane potential determination, and Annexin V/Propidium iodide (PI) staining also reveal no cytotoxicity. SWCNTs do interact with MTT-formazan crystals, formed after reduction of MTT and not with water-soluble tetrazolium salts such as WST-1, XTT, or INT. SWCNTs bind MTT-formazan crystals and stabilize their chemical structure, and, as a consequence, these crystals cannot be solubilized in 2-propanol/HCl, Sodium dodecyl sulfate (SDS), or acetone. The other reason for the controversial results of the cytotoxicity of CNTs is that there is still lack of a standardized CNT reference material to be used. To the best of our knowledge, there is no comprehensive comparison studies on the cytotoxicity of CNTs based on comparable experimental conditions. It is very difficult to determine the standard values by which to evaluate their cytotoxicity, due to the difficulty of controlling the properties of CNTs. Therefore, continuous and extensive studies in this field are required.

## 4. Magnetic-Fluorescent Functionalized CNTs for Multimodal Cellular Imaging

There is considerable interest in using CNTs for various biomedical applications. The physical properties of CNTs, such as mechanical strength, electrical conductivity, and optical properties, could be of great value for creating advanced biomaterials. CNTs can also be chemically modified to present specific moieties (e.g., functional groups, molecules, and polymers) for cell and tissue imaging, drug delivery.

CNTs have many unique properties suitable for biomedical applications. For example, CNTs have optical transitions in the near-infrared (NIR) region, which has been shown to be useful in biological tissue imaging because NIR has greater penetration depth and lower excitation scattering. In addition, fluorescence in the NIR region displays much lower auto-fluorescence than do the ultraviolet or visible ranges. These properties make CNTs potent imaging agents with higher resolution and greater tissue depth for NIR fluorescence microscopy and optical coherence tomography.[40] Another example is the unique Raman signatures of CNTs, which, because of their extensive symmetric carbon bonds make them suitable to be utilized as cellular probes.[41] One more example is that CNTs can lower the transverse relaxation of water proton thus acting as $T_2$-weighted MRI contrast agent.[17,42] However, most of the intrinsic properties are not suitable for direct use in biomedical applications, either the properties are not stable or too weak. So it is necessary to functionalized CNTs with functional moieties.

Recently, considerable efforts have been devoted in designing CNTs based multifunctional magnetic and fluorescent nanohybrids. There are many imaging modalities, including MRI, positron emission tomography (PET), optical, and ultrasound methods, etc. However, each imaging technique has its own merits and drawbacks in terms of sensitivity, resolution, complexity and data acquisition time. For example, the fluorescence optical imaging has high sensitivity and multiplexed detection capabilities for *in vitro* detection and cell labeling, but the low penetration capability of light through biological tissue hampers the use of optical-based labels for non-invasive diagnostic imaging. In contrast, MRI possesses an advantage of its high spatial resolution through deep tissues. As a result, the dual-modal imaging probes to integrate the high sensitivity of optical imaging with the high spatial resolution of MRI are becoming increasingly important as molecular imaging contrast agents. With the magnetic-fluorescent nanohybrids, one can imagine a surgeon identifying the specific location of a tumor in the body before surgery with a MRI scan, then using fluorescence imaging to find and remove all parts of the tumor during the operation. Most of the magnetic-fluorescent nanohybrids are core-shell nanostructures consisting of a magnetic core coated with a silica shell containing fluorescent components or nanocomposites consisting of a magnetic entity and a fluorescent moiety encapsulated within a silica matrix,[43–45] which often compromise their sensitivity and efficiency as a diagnostic tool. Moreover, obtaining magnetic-fluorescent nanohybrids by a simple and effective approach still remains a tremendous challenge. We have done some effort on the development of CNT-based magnetic-fluorescent nanohybrids. Herein, we will summarize our previous works briefly.

## 4.1. *Magnetic-fluorescent nanohybrids of CNTs functionalized with SPIO and QDs*

We developed a simple and novel layer-by-layer (LBL) assembly in combination with covalent connection strategy for the synthesis of multifunctional CNTs-based magnetic-fluorescent nanohybrids as multimodal cellular imaging agents for detecting human embryonic kidney (HEK) 293T cells *via* MRI and confocal fluorescence imaging. The LBL assembly based on the electrostatic attraction between the charged species has been widely used to synthesize nanohybrids or hollow nanostructures.[46–48] Recently, our group has developed the LBL assembly for synthesizing CNT-metal oxide nanoparticles nanohybrids, metal oxide nanotubes and ZnO nanorods-based nanohybrids.[49–52]

The overall synthetic strategy for the CNTs-based magnetic-fluorescent nanohybrids is shown in **Figure 5**. Firstly, the as-received MWNTs synthesized by CVD methods were purified and cut short by a typical purification process included sonication, oxidation, and acid washing steps. As shown in the **Figure 2**, the CNTs were cut short to about 100–500 nm. Secondly, CNTs were noncovalently functionalized with poly (allylamine) hydrochloride (PAH) with amine groups being terminated on their surface. Thirdly, the PEGylated carboxyl modified SPIO were attached on the surface of the functionalized CNTs by a dehydration reaction between the amine and carboxyl using ethyl (dimethylaminopropyl) carbodiimide (EDC)/N-Hydroxysuccinimide (NHS) as agents. Lastly, the CNT–SPIO–CdTe (Cadmium telluride) nanohybrids were achieved by subsequent conjugation of PAH and carboxyl modified CdTe quantum dots (QDs) *via* the same approach used to attach SPIO.

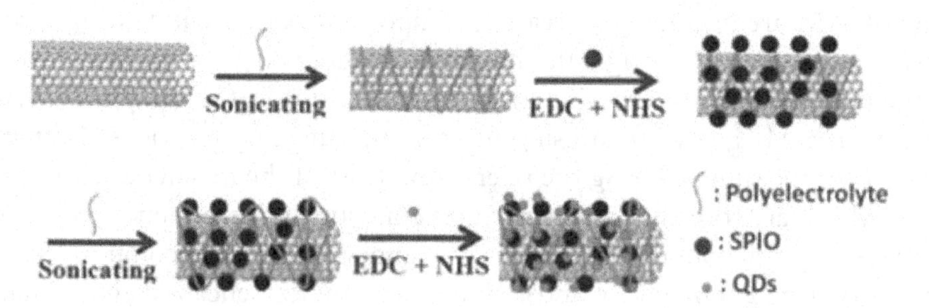


**Figure 5.** Schematic illustration for synthesis of CNTs-based magnetic-fluorescent nanohybrids.

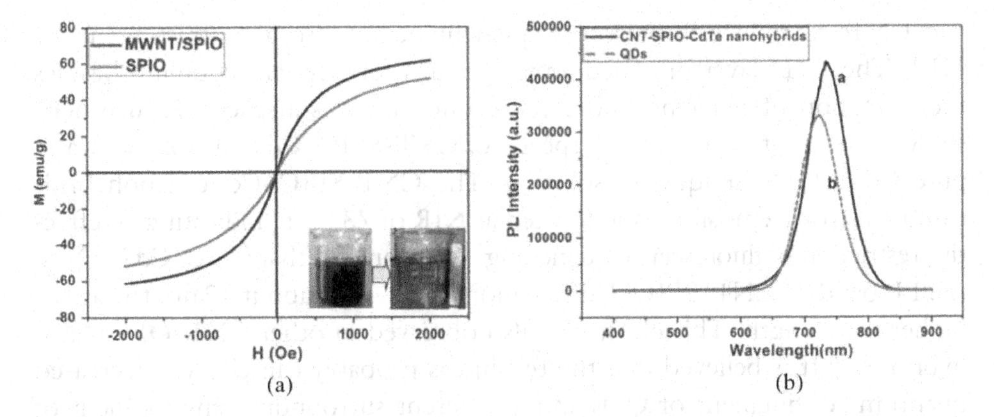


**Figure 6.** **(a)** M–H curve for the CNT–SPIO–CdTe nanohybrids and pure SPIO at room temperature. Inset: photograph of the CNT–SPIO–CdTe nanohybrids dispersed in water before (left) and after (right) application of an external magnet; **(b)** room temperature PL spectra for the CNT–SPIO–CdTe nanohybrids, line a, and pure CdTe QDs, line b.

SPIO and NIR fluorescent CdTe QDs were covalently coupled on the surface of CNTs in sequence *via* LBL assembly. It was indicated that the SPIO layer acted not only as a contrast agent for MRI, but also as a spacer between CdTe QDs and CNTs for prohibiting fluorescence quenching of QDs on the surface of the CNTs.

The magnetic properties of the CNT–SPIO–CdTe nanohybrids are characterized by SQUID measurement, as showed in **Figure 6(a)**. As can be seen, the CNT–SPIO–CdTe nanohybrids exhibit superparamagnetic behavior with a saturation magnetization of about 65 emug$^{-1}$ at room temperature, which was higher than that of pure SPIO (55 emug$^{-1}$). The enhancement of the magnetic saturation is mainly because there are multiple SPIO nanoparticles surrounding the surface of the CNTs, resulting in an enhanced magnetic effect due to the magnetic coupling between SPIO nanoparticles, which has also been reported by others.[53] Moreover, the saturation magnetization of the CNT–SPIO–CdTe nanohybrids can be readily tailored *via* adjusting the SPIO density on the surface of the CNTs.

As we know, photoluminescence (PL) quenching of the QDs directly assembled onto the surface of CNTs is one of the main challenges in the CNT–QDs nanohybrids due to an energy transfer process.[54] Generally, the problem of PL quenching was avoided by providing an appropriate spacer between QDs and CNTs, such as a silica-shell spacer or multiple layers of polyelectrolyte.[55] Herein, the PAH and SPIO layers have been coated on the surface of CNTs as spacer prior to the decoration of CdTe QDs *via* LBL assembly technique. The SPIO layer acted as not only a spacer between QDs and CNTs to reduce fluorescence quenching, but also a contrast agent for MRI. The PAH layer provided amine groups for covalent coupling besides their function of reducing fluorescence quenching as a spacer. **Figure 6(b)** shows the room-temperature PL spectra of CNT–SPIO–CdTe nanohybrids and pure CdTe QDs in aqueous solution. The CNT–SPIO–CdTe nanohybrids exhibit a strong emission band located at NIR of 734 nm, indicating an effective restriction of fluorescence quenching. Compared to that of free CdTe QDs, the PL band of CNT–SPIO–CdTe nanohybrids shifts about 12 nm towards a higher wavelength. This effect was also observed in other CNT–QDs nanohybrids.[55,56] It is believed that the redshift is probably caused by a decreased quantum confinement of QDs and a different surrounding environment of QDs in the CNT–QDs nanohybrids.

The multi-functional CNT–SPIO–CdTe nanohybrids can be used as multimodal cellular imaging agents for detecting HEK 293T cells *via* MRI (**Figure** 7) and confocal fluorescence imaging (**Figure 8**). The MRI images using CNT–SPIO–CdTe nanohybrids and pure SPIO with an equivalent Fe concentration of 50 mg ml$^{-1}$ as contrast agent in the complete medium were performed on a clinical 3.0 Tesla MRI system, as shown in **Figure** 7**(a)**. The CNT–SPIO–CdTe nanohybrids (**Figure 7(a1)**) and pure SPIO (**Figure 7(a2)**) show almost the same MRI contrast. However, when incubated with HEK 293T cells, CNT–SPIO–CdTe nanohybrids exhibit a remarkably stronger MRI contrast than pure SPIO (**Figure 7(b)**). Under the same Fe concentration, the $T_2$-weighted MRI contrast of CNT–SPIO–CdTe nanohybrids is more than 2.5 times stronger than that of pure SPIO for labeling of HEK

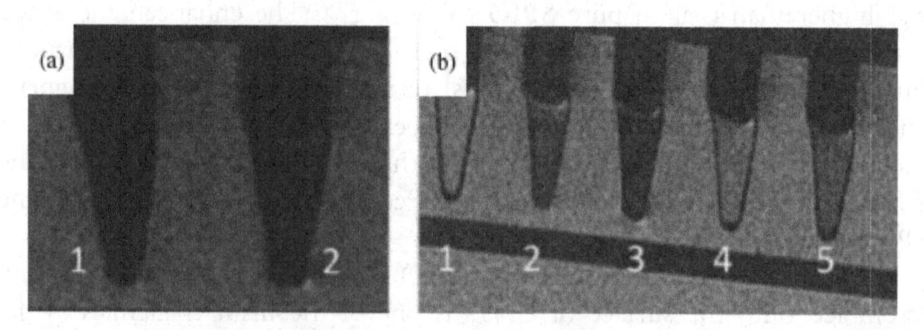


**Figure 7.** $T_2$-weighted MRI images of **(a)** free SPIO **(a1)** and CNT–SPIO–CdTe nanohybrids **(a2)** with an equivalent Fe concentration of 50 mg mL$^{-1}$ in complete medium; **(b)** $T_2$-weighted MRI images of **b1)** HEK 293T cells; **b2, b3)** HEK 293T cells incubated by CNT–SPIO–CdTe nanohybrids with an Fe concentration of 20 mg mL$^{-1}$ and 50 mg mL$^{-1}$ respectively; **b4, b5)** HEK 293T cells incubated by pure SPIO with an Fe concentration of 20 mg mL$^{-1}$ and 50 mg mL$^{-1}$ respectively.

293T cells, which exhibits promising application. The significantly improved MRI signal can be attributed to some advantages of the CNT–SPIO–CdTe nanohybrids. Firstly, there are multiple SPIO nanoparticles surrounding the surface of the CNTs, resulting in an enhanced magnetic effect due to the magnetic coupling between SPIO nanoparticles.[18] Secondly, CNTs can lower the transverse relaxation times, $T_2$, and thus resulting in a stronger MRI contrast.[57] Last and most importantly, CNTs are able to penetrate into cells without assistance of any external transportation system, which strengthens the MRI contrast by increasing the Fe concentration in the cell.

**Figure 8** shows the confocal fluorescence image of CNT–SPIO–CdTe nanohybrids labeled cells. As can be seen, CNT–SPIO–CdTe nanohybrids labeled (**Figure 8**) cells exhibit the red fluorescence, which is consistent with

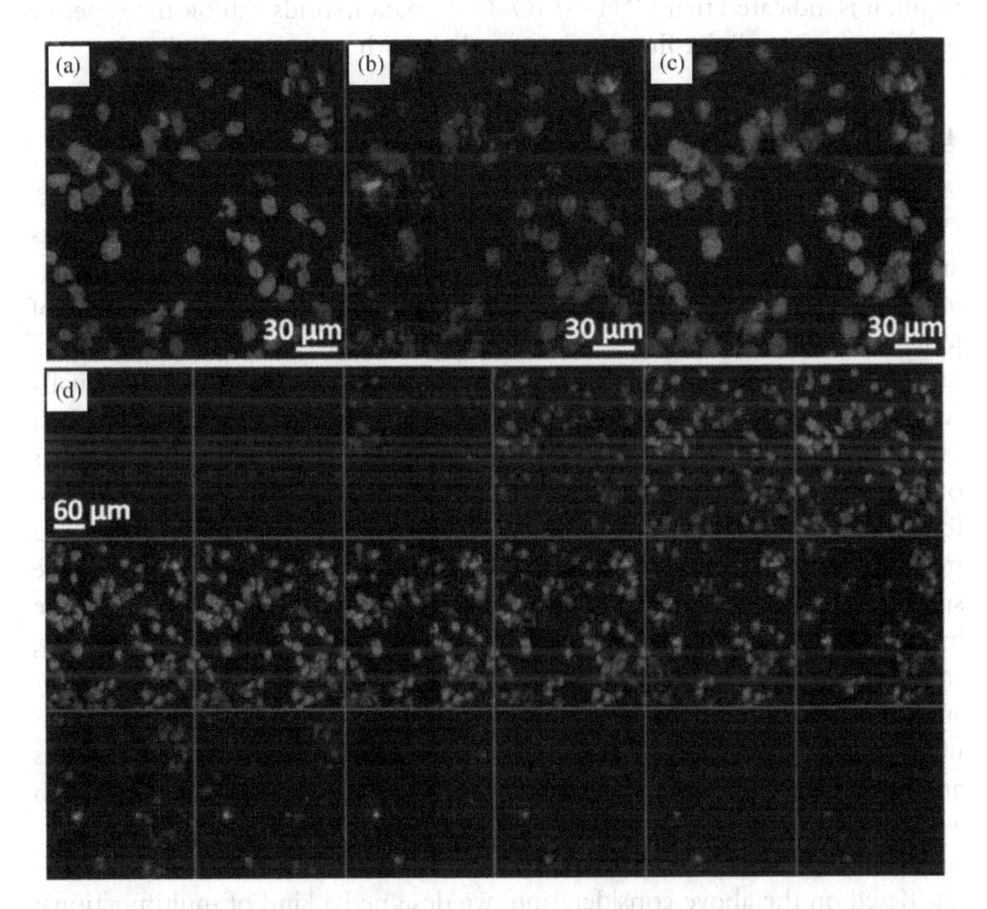


**Figure 8.** Confocal fluorescent images of HEK 293T cells incubated with CNT–SPIO–CdTe nanohybrids with a Cd concentration of 5 mg ml$^{-1}$: **(a)** fluorescent image of nuclei illuminated with blue; **(b)** fluorescent image showing CNT–SPIO–CdTe color only; **(c)** overlaid image of HEK 293T cells **(d)** Z-stack image of interiors of cells.

PL, and no obviously morphologically damaged cells exist. Moreover, most CNT–SPIO–CdTe nanohybrids accumulate on the cell surface or inner *via* penetrating into cells, indicating their superior performance as cellular fluorescence markers. In order to confirm the uptake of the nanohybrids to the cells, Z-stack images of the cells were obtained, as shown in **Figure 8(d)**. Here, nuclei are shown in blue, and the red fluorescence observed in close proximity to the nuclei confirms intracellular accumulation and translocation through the cell membrane of the CNT–SPIO–CdTe nanohybrids. It can be concluded that CNT–SPIO–CdTe nanohybrids are more readily able to accumulate and penetrate into cells compared to SPIO–CdTe, taking advantage of the CNTs. In addition, the CNT–SPIO–CdTe nanohybrids with NIR fluorescence can effectively avoid the influence of cell auto-fluorescence. As a result, it is indicated that CNT–SPIO–CdTe nanohybrids exhibit the superior performance as cellular fluorescence markers with promising application.

## 4.2. *Magnetic-fluorescent nanohybrids of CNTs functionalized with Eu, Gd co-doped* $LaF_3$

CNTs have been observed to exhibit infrared emissions,[12,13] but the emissions are weak and therefore do not have sufficient fluorescence for *in vivo* imaging. As a result, considerable efforts have been focused on the synthesis of luminescent CNTs decorated with fluorescent materials.[16,58,59] For synthesis of luminescent CNTs, semiconductor QDs or organic fluorophores are usually selected as the decoration. Recently, considerable attention has been focused on the lanthanide-based luminescent probes because these kinds of probes offer several advantages over semiconductor QDs and organic fluorophores for optical imaging. Firstly, their inherent long luminescence lifetimes ranging from microseconds (Yb, Nd) to milliseconds (Eu, Tb) enable quantitative spatial determination through time-gating measurements and prevent the interference from any spontaneous background emission sources.[60] Secondly, their emission bands are very narrow, nonoverlapped, independent of their sizes and easily tuned by doping different lanthanide ions.[61] Thirdly, lanthanide doped nanocrystals show superior photostability, high quantum yields and low toxicity.[62–64] Significantly, the co-doped different lanthanide ions such as Gd, Eu in a matrix can readily achieve the magnetic-fluorescent nanohybrids as a multifunctional probe.

Based on the above consideration, we designed a kind of multifunctional magnetic-fluorescent nanohybrids based on the multi-walled carbon nanotubes (MWCNTs) coated with lanthanide co-doped nanocrystals. The nanohybrids were synthesized *in situ*, depositing Eu, Gd co-doped $LaF_3$

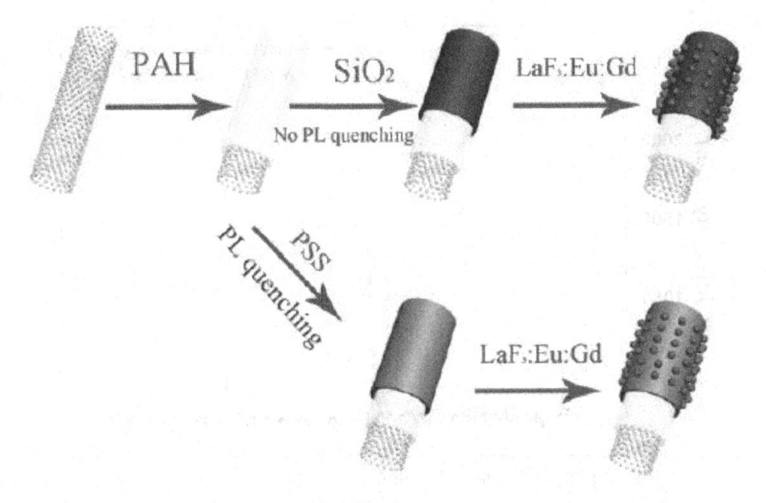


**Figure 9.** Schematic illustration for synthesis of magnetic-fluorescent MWNT/$SiO_2$/$LaF_3$:Eu:Gd nanohybrids.

nanocrystals onto the surface of MWNTs *via* a facile LBL assembly technique. The synthesis procedure is illustrated in **Figure 9**. Briefly, MWNTs were firstly noncovalently functionalized with polyelectrolyte of positively charged PAH and negatively charged poly (sodium 4-styrene sulfonate) (PSS) in sequence *via* a LBL assembly technique. By mixing the polyelectrolyte functionalized MWNTs with $La(NO_3)_3{\cdot}6H_2O$, $EuCl_3{\cdot}6H_2O$ and $GdCl_3{\cdot}6H_2O$ in aqueous solution, the lanthanide ($La^{3+}$, $Eu^{3+}$ and $Gd^{3+}$) ions was strongly absorbed onto the surface of MWNTs because of the electrostatic interaction between the charged species. After dropwise adding $NH_4F$ aqueous solution into the above-mentioned solution, the $LaF_3$:Eu:Gd nanocrystals were deposited *in situ* on the surface of MWNTs, and thus the MWNT/$LaF_3$:Eu:Gd nanohybrids were obtained. As previously reported, PL quenching took place when fluorophores directly assembled on the surface of MWNTs because of the background absorption of the MWNTs and the photo-induced electron transfer (ELT) process.[65,66] No obvious PL emission of Eu ions can be observed from the MWNT/$LaF_3$:Eu:Gd nanohybrids when $LaF_3$:Eu:Gd nanocrystals are directly coated on the surface of the MWNTs, as can be seen from **Figure 10 (line a)**. For circumventing the PL quenching, a $SiO_2$ shell spacer was grown between the MWNTs and $LaF_3$:Eu:Gd nanocrystals to screen the background absorption of the MWNTs and prevent the electronic coupling *via* the hydrolytic precipitation of tetraethyl orthosilicate (TEOS), which was a general approach to circumvent PL quenching in other multifunctional magnetic-fluorescent nanohybrids.[67–69] After coating a $SiO_2$ shell spacer between the $LaF_3$:Eu:Gd nanocrystals and MWNTs, the nanohybrids

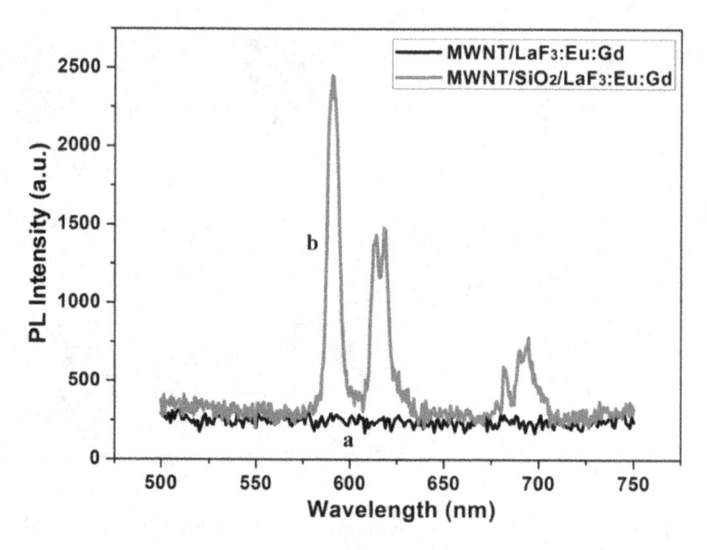


**Figure 10.** Room-temperature PL spectra of the MWNT/$LaF_3$:Eu:Gd nanohybrids **(line a)** and the MWNT/$SiO_2$/$LaF_3$:Eu:Gd nanohybrids **(line b)**.

show a strong PL emission at room temperature, as shown in **Figure 10 (line b)**. The characteristic emission peaks of the $Eu^{3+}$ within the wavelength range from 550 to 720 nm were corresponding to the transitions from the excited $^5D_0$ to $^7F_J$ levels.

The morphological and structural characterizations of the as-synthesized MWNT/$LaF_3$:Eu:Gd nanohybrids were shown in **Figures 11(a) and 11(b)**. As can be seen, most of the MWNTs are compactly coated by numerous nanocrystals with size of 5–10 nm. No isolated nanocrystals can be observed on the surface of MWNTs due to the efficient electrostatic attraction between the polyelectrolyte modified MWNTs and the lanthanide ($La^{3+}$, $Eu^{3+}$ and $Gd^{3+}$). After coating with a homogeneous $SiO_2$ shell along the whole nanotube, the thickness of the $SiO_2$ shell is about 5 nm, as can be clearly seen in **Figure 11(c)** and can be readily tuned by varying the concentration of TEOS and ageing time. Moreover, the strongly and negatively charged $SiO_2$ shell derived from hydroxy of TEOS is favorable to the subsequent LBL assembly of polyelectrolyte and lanthanum fluoride nanocrystals on the MWNTs, as shown in **Figure 11(d)**.

In order to assess the potential application of the MWNT/$SiO_2$/$LaF_3$:Eu:Gd nanohybrids as $T_1$ and/or $T_2$-weighted MRI contrast agent, firstly the relaxivity value of the nanohybrids, longitudinal ($T_1$) and transverse proton relaxation times ($T_2$) were measured as a function of gadolinium ion concentration at 1.5 T. As shown in **Figure 12**, the MWNT/$SiO_2$/$LaF_3$:Eu:Gd nanohybrids exhibit $R_1$ and $R_2$ values of 1.4 and 15.7 $S^{-1}$ per mM of $Gd^{3+}$, respectively. The $R_2$ value of 15.7 is about four times as high as that

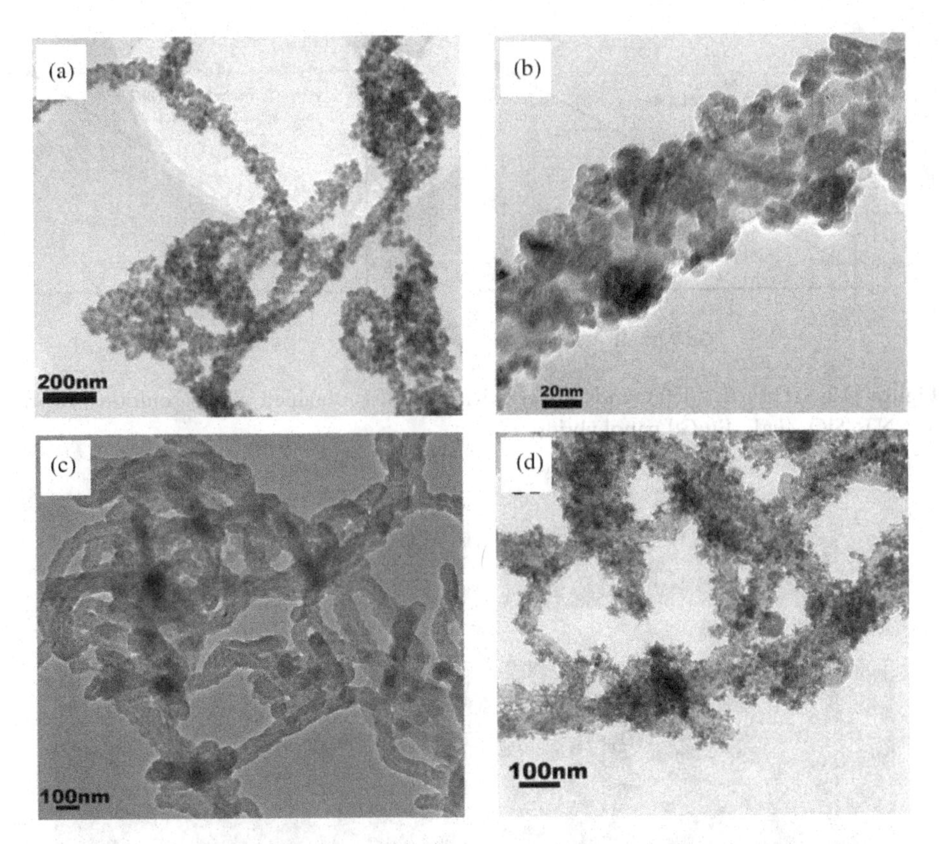


**Figure 11.** Morphological and structural characterizations of the magnetic-fluorescent nanohybrids prepared by LBL assembly technique: **(a)**, **(b)** TEM image of MWNT/$LaF_3$:Eu:Gd nanohybrids, **(c)** TEM image of $SiO_2$ coated MWNTs, **(d)** TEM image of MWNT/$SiO_2$/$LaF_3$:Eu:Gd nanohybrids.

of commercial used Magnevist ($R_2$ = 4.61 $S^{-1}$ per mM of $Gd^{3+}$). The high $R_2/R_1$ value of 11.2 indicates that the nanohybrids can be used as a $T_2$-weighted MRI contrast agent.

*In vitro* MRI analysis showed that the nanohybrids containing higher Gd concentrations are in hypersignal compared to the water reference sample (**Figures 13(a) and 13(b)**). The positive contrast derives from the Gd to reduce the longitudinal water proton relaxivity time. More interestingly, in the $T_2$-weighted MRI images (**Figures 13(c) and 13(d)**); the nanohybrids containing higher Gd concentrations appear in hyposignal. The important $T_2$ effect observed in the nanohybrids can be likely related to the presence of MWNTs, which is consistent with the previous report.[57] It is proposed that the motion of electrons around the circumference of the CNTs causes the magnetic susceptibility,[70,71] resulting in the decrease of transverse relaxation of water proton. It is important to note that the MWNT/$SiO_2$/$LaF_3$:Eu:Gd

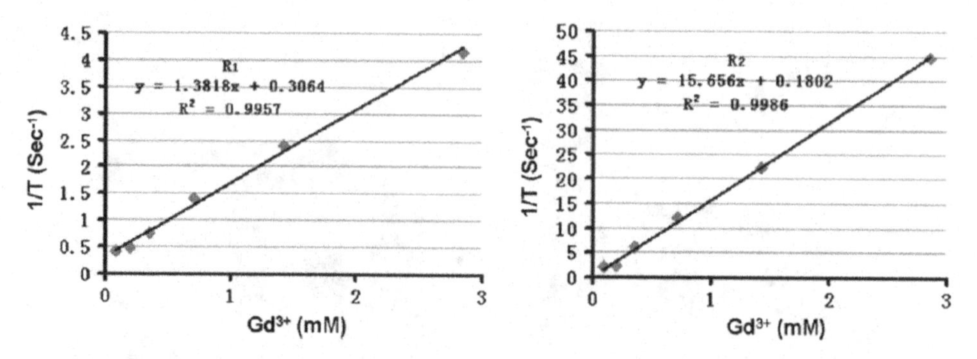


**Figure 12.** The $R_1$ (left) and $R_2$ relaxivity (right) obtained from solutions of the MWNT/$SiO_2$/$LaF_3$:Eu:Gd nanohybrids.

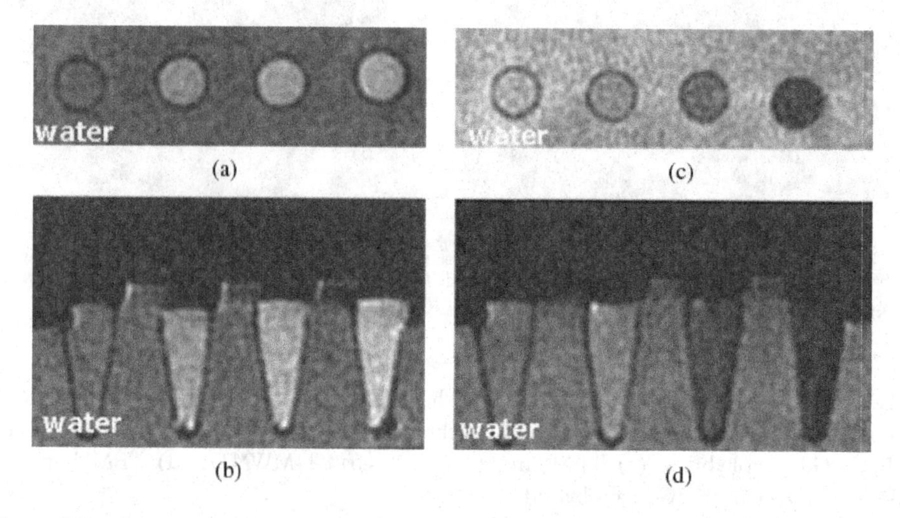


**Figure 13.** MRI images of the MWNT/$SiO_2$/$LaF_3$:Eu:Gd nanohybrids in water: **(a)**, **(b)** $T_1$ mode, **(c)**, **(d)** $T_2$ mode. From left to right in each image, the concentration of $Gd^{3+}$ in the nanohybrids increases.

nanohybrids can be soluble in aqueous and stable in physiological solutions (such as PBS, serum, plasma) by proper surface functionalization. So, the nanohybrids are expected to act as a promising multifunctional probe by MRI/optical imaging.

## 5. Surface Functionalized CNTs for Drug Delivery

Drug delivery has benefited greatly from the advances in nanotechnology by using a variety of nanomaterials (i.e., liposomes, polymersomes, microspheres, and polymer conjugates) as vehicles to deliver therapeutic agents.[72] CNTs have also been investigated extensively as drug delivery systems since CNTs have unique structural characteristics and have been shown to interact with various biomacromolecules (i.e., proteins and DNA) by physical adsorption.[73] In addition, several chemical modification schemes have been developed to conjugate therapeutic molecules or targeting moieties covalently to CNTs.[4] In one of the previous reports by Guo Y *et al.*,[16] CNTs incorporated with poly (lactic-co-glycolic acid) (PLGA) were developed for paclitaxel delivery by plasma polymerization and emulsion technique. The drug-loading efficiency of PLGA-coated CNTs and their cytotoxicity on human prostate cells have been investigated. In another interesting study, Zheng *et al.*[74] provided important insight into the interactions between CNTs and DNA molecules. Molecular dynamics modeling showed that the base of ssDNA interacted with the surface of CNT *via* π–π stacking, resulting in helical wrapping of ssDNA chains around the CNT. This research highlights the potential use of CNTs for gene delivery.

## 6. Future Outlook

Extensive research efforts over the last two decades have elevated the status of CNTs as one of the most widely used classes of nanomaterials. Despite safety concerns over CNTs, many studies have reported the successful use of CNTs in biological applications. Future research needs to be devoted to the development of intelligent and integrated multifunctional CNTs based nanosystems. The multifunctionality is based on the structural integration of several nanospecies with unique properties including QDs for *in vivo* imaging, magnetic nanocomposites for hyperthermia and separation, nanotube capsules for drug storage, and surface functional groups for biological selectivity. The intelligent nanosystems utilize integrated functionalities and perform controlled actions such as switching on/off the drug release, signaling to the sensing device, recognizing a specific cell, and monitoring the concentration level in a biological and living system. The development of these multifunctional systems require both engineering design of the nanostructure and biomedical tuning of the intelligent responses, therefore demanding a true interdisciplinary research in synthesis, nanostructure architecture, surface engineering, and biomedical system integration.

## References

1. Iijima S. Helical microtubules of graphitic carbon. *Nature* 1991:354:56–58.
2. Iijima S, Ichihashi T. Single-shell carbon nanotubes of 1-nm diameter. *Nature* 1993;363: 603–605.
3. Bethune D, Klang C, De Vries M, Gorman G, Savoy R, Vazquez J, Beyers R. Cobalt-catalysed growth of carbon nanotubes with single-atomic-layer walls. *Nature* 1993; 363:605–607.

4. Bianco A, Kostarelos K, Partidos CD, Prato M. Biomedical applications of functionalised carbon nanotubes. *Chem Commun* 2005;5:571–577.
5. Yang W, Thordarson P, Gooding JJ, Ringer SP, Braet F. Carbon nanotubes for biological and biomedical applications. *Nanotechnol* 2007;18:412001–412012.
6. Liu Z, Tabakman S, Welsher K, Dai H. Carbon nanotubes in biology and medicine: *in vitro* and *in vivo* detection, imaging and drug delivery. *Nano Res* 2009;2:85–120.
7. Ji S-R, Liu C, Zhang B, Yang F, Xu J, Long J, Jin C, Fu D-L, Ni Q-X, Yu X-J. Carbon nanotubes in cancer diagnosis and therapy. *Biochim Et Biophys Acta-Rev Cancer* 2010;1806:29–35.
8. Cha C, Shin SR, Annabi N, Dokmeci MR, Khademhosseini A. Carbon-based nanomaterials: multifunctional materials for biomedical engineering. *ACS nano* 2013;7:2891–2897.
9. Kam NWS, Jessop TC, Wender PA, Dai H. Nanotube molecular transporters: internalization of carbon nanotube-protein conjugates into mammalian cells. *J Am Chem Soc* 2004;126:6850–6851.
10. Pantarotto D, Briand J-P, Prato M, Bianco A. Translocation of bioactive peptides across cell membranes by carbon nanotubes. *Chem Commun* 2004;1:16–17.
11. Dhar S, Liu Z, Thomale J, Dai H, Lippard SJ. Targeted single-wall carbon nanotube-mediated Pt (IV) prodrug delivery using folate as a homing device. *J Am Chem* Soc 2008;130:11467–11476.
12. Cherukuri P, Gannon CJ, Leeuw TK, Schmidt HK, Smalley RE, Curley SA, Weisman RB. Mammalian pharmacokinetics of carbon nanotubes using intrinsic near-infrared fluorescence. *Proceedings of the National Academy of Sciences of the United States of America* 2006;103:18882–18886.
13. Welsher K, Liu Z, Sherlock SP, Robinson JT, Chen Z, Daranciang D, Dai H. A route to brightly fluorescent carbon nanotubes for near-infrared imaging in mice. *Nature Nanotechnol* 2009;4:773–780.
14. Chen B, Zhang H, Zhai C, Du N, Sun C, Xue J, Yang D, Huang H, Zhang B, Xie Q. Carbon nanotube-based magnetic-fluorescent nanohybrids as highly efficient contrast agents for multimodal cellular imaging. *J Mater Chem* 2010;20:9895–9902.
15. Chen B, Zhang H, Du N, Zhang B, Wu Y, Shi D, Yang D. Magnetic-fluorescent nanohybrids of carbon nanotubes coated with Eu, Gd co-doped LaF3 as a multimodal imaging probe. *J Colloid Interface Sci* 2012;367:61–66.
16. Guo Y, Shi D, Cho H, Dong Z, Kulkarni A, Pauletti GM, Wang W, Lian J, Liu W, Ren L. *In vivo* imaging and drug storage by quantum–dot–conjugated carbon nanotubes. *Adv Funct Mater* 2008;18:2489–2497.
17. Richard C, Doan B-T, Beloeil J-C, Bessodes M, Tóth É, Scherman D. Noncovalent functionalization of carbon nanotubes with amphiphilic Gd3+ chelates: toward powerful T1 and T2 MRI contrast agents. *Nano Lett* 2008;8:232–236.
18. Wang C, Chen J, Talavage T, Irudayaraj J. Gold nanorod/$Fe_3O_4$ nanoparticle "nano-pearl-necklaces" for simultaneous targeting, dual-mode imaging, and photothermal ablation of cancer cells. *Angew Chem* 2009;121:2797–2801.
19. Journet C, Maser W, Bernier P, Loiseau A, De La Chapelle ML, Lefrant DLS, Deniard P, Lee R, Fischer J. Large-scale production of single-walled carbon nanotubes by the electric-arc technique. *Nature* 1997;388:756–758.
20. Thess A, Lee R, Nikolaev P, Dai H, Petit P, Robert J, Xu C, Lee YH, Kim SG, Rinzler AG. Crystalline ropes of metallic carbon nanotubes. *Science* 1996;273:483–487.

21. Braidy N, El Khakani M, Botton G. Carbon nanotubular structures synthesis by means of ultraviolet laser ablation. *J Mater Res* 2002;17:2189–2192.
22. Vander Wal R, Berger G, Ticich T. Carbon nanotube synthesis in a flame using laser ablation for *in situ* catalyst generation. *Appl Phys A* 2003;77:885–889.
23. Li W, Xie S, Qian LX, Chang B, Zou B, Zhou W, Zhao R, Wang G. Large-scale synthesis of aligned carbon nanotubes. *Science* 1996;274:1701–1703.
24. Jose-Yacaman M, Miki-Yoshida M, Rendon L, Santiesteban J. Catalytic growth of carbon microtubules with fullerene structure. *Appl Phys Lett* 1993;62:202–204.
25. Qin L. CVD synthesis of carbon nanotubes. *J Mater Sci Lett* 1997;16:457–459.
26. Hirsch A. Functionalization of single-walled carbon nanotubes. *Angew Chem Int Ed* 2002;41:1853–1859.
27. Guldi DM, Rahman GA, Jux N, Balbinot D, Hartnagel U, Tagmatarchis N, Prato M. Functional single-wall carbon nanotube nanohybrids associating SWNTs with water-soluble enzyme model systems. *J Am Chem Soc* 2005;127:9830–9838.
28. Léger B, Menuel S, Landy D, Blach J-F, Monflier E, Ponchel A. Noncovalent functionalization of multiwall carbon nanotubes by methylated-$\beta$-cyclodextrins modified by a triazole group. *Chem Commun* 2010;46:7382–7384.
29. Shim M, Kam NWS, Chen RJ, Li Y, Dai H. Functionalization of carbon nanotubes for biocompatibility and biomolecular recognition. *Nano Lett* 2002;2:285–288.
30. Park H, Zhao J, Lu JP. Effects of sidewall functionalization on conducting properties of single wall carbon nanotubes. *Nano Lett* 2006;6:916–919.
31. Besterman JM, Low RB. Endocytosis: a review of mechanisms and plasma membrane dynamics. *Biochem J* 1983;210:1–13.
32. Kam NWS, Liu Z, Dai H. Functionalization of carbon nanotubes *via* cleavable disulfide bonds for efficient intracellular delivery of siRNA and potent gene silencing. *J Am Chem Soc* 2005;127:12492–12493.
33. Liu Y, Wu DC, Zhang WD, Jiang X, He CB, Chung TS, Goh SH, Leong KW. Polyethylenimine-grafted multiwalled carbon nanotubes for secure noncovalent immobilization and efficient delivery of DNA. *Angew Chem* 2005;117:4860–4863.
34. Porter AE, Gass M, Muller K, Skepper JN, Midgley PA, Welland M. Direct imaging of single-walled carbon nanotubes in cells. *Nature Nanotechnol* 2007;2:713–717.
35. Simon-Deckers A, Gouget B, Mayne-L'Hermite M, Herlin-Boime N, Reynaud C, Carriere M. *In vitro* investigation of oxide nanoparticle and carbon nanotube toxicity and intracellular accumulation in A549 human pneumocytes. *Toxicology* 2008;253:137–146.
36. Mu Q, Broughton DL, Yan B. Endosomal leakage and nuclear translocation of multiwalled carbon nanotubes: developing a model for cell uptake. *Nano Lett* 2009;9:4370–4375.
37. Kostarelos K, Lacerda L, Pastorin G, Wu W, Wieckowski S, Luangsivilay J, Godefroy S, Pantarotto D, Briand J-P, Muller S. Cellular uptake of functionalized carbon nanotubes is independent of functional group and cell type. *Nature Nanotechnol* 2007;2:108–113.
38. Lee Y, Geckeler KE. Carbon nanotubes in the biological interphase: the relevance of noncovalence. *Adv Mater* 2010;22:4076–4083.
39. Wörle-Knirsch J, Pulskamp K, Krug H. Oops they did it again! Carbon nanotubes hoax scientists in viability assays. *Nano Lett* 2006;6:1261–1268.
40. Cherukuri P, Bachilo SM, Litovsky SH, Weisman RB. Near-infrared fluorescence microscopy of single-walled carbon nanotubes in phagocytic cells. *J Am Chem Soc* 2004;126: 15638–15639.

41. Liu Z, Davis C, Cai W, He L, Chen X, Dai H. Circulation and long-term fate of functionalized, biocompatible single-walled carbon nanotubes in mice probed by Raman spectroscopy. *Proceedings of the National Academy of Sciences of the United States of America* 2008;105:1410–1415.
42. Chen BD, Zhou T, Zhang B, Yao AH. Novel single walled carbon nanotube based magnetic-fluorescent nanohybrids as dual-modal MRI/optical imaging probes. *Adv Mater Res* 2012;476:1134–1137.
43. Shi D, Cho HS, Chen Y, Xu H, Gu H, Lian J, Wang W, Liu G, Huth C, Wang L. Fluorescent polystyrene–$Fe_3O_4$ composite nanospheres for *in vivo* imaging and hyperthermia. *Adv Mater* 2009;21:2170–2173.
44. Kim J, Kim HS, Lee N, Kim T, Kim H, Yu T, Song IC, Moon WK, Hyeon T. Multifunctional uniform nanoparticles composed of a magnetite nanocrystal core and a mesoporous silica shell for magnetic resonance and fluorescence imaging and for drug delivery. *Angew Chem Int Ed* 2008;47:8438–8441.
45. Lu C-W, Hung Y, Hsiao J-K, Yao M, Chung T-H, Lin Y-S, Wu S-H, Hsu S-C, Liu H-M, Mou C-Y. Bifunctional magnetic silica nanoparticles for highly efficient human stem cell labeling. *Nano Lett* 2007;7:149–154.
46. Decher G. Fuzzy nanoassemblies: toward layered polymeric multicomposites. *Science* 1997;277:1232–1237.
47. Caruso F, Caruso RA, Möhwald H. Nanoengineering of inorganic and hybrid hollow spheres by colloidal templating. *Science* 1998;282:1111–1114.
48. Caruso F. Nanoengineering of particle surfaces. *Adv Mater* 2001;13:11–22.
49. Du N, Zhang H, Chen B, Ma X, Liu Z, Wu J, Yang D. Porous indium oxide nanotubes: layer-by-layer assembly on carbon-nanotube templates and application for room-temperature $NH_3$ gas sensors. *Adv Mater* 2007;19:1641–1645.
50. Chen B, Zhang H, Du N, Li D, Ma X, Yang D. Hybrid nanostructures of Au nanocrystals and ZnO nanorods: layer-by-layer assembly and tunable blue-shift band gap emission. *Mater Res Bull* 2009;44:889–892.
51. Du N, Zhang H, Chen J, Sun J, Chen B, Yang D. Metal oxide and sulfide hollow spheres: layer-by-layer synthesis and their application in lithium-ion battery. *J Phys Chem B* 2008;112:14836–14842.
52. Du N, Zhang H, Yang D. One-dimensional hybrid nanostructures: synthesis *via* layer-by-layer assembly and applications. *Nanoscale* 2012;4:5517–5526.
53. Correa-Duarte MA, Grzelczak M, Salgueiriño-Maceira V, Giersig M, Liz-Marzan LM, Farle M, Sierazdki K, Diaz R. Alignment of carbon nanotubes under low magnetic fields through attachment of magnetic nanoparticles. *J Phys Chem B* 2005;109:19060–19063.
54. Guldi DM, Rahman GA, Sgobba V, Kotov NA, Bonifazi D, Prato M. CNT–CdTe versatile donor-acceptor nanohybrids. *J Am Chem Soc* 2006;128:2315–2323.
55. Grzelczak M, Correa-Duarte MA, Salgueiriño-Maceira V, Giersig M, Diaz R, Liz-Marzán LM. Photoluminescence quenching control in quantum dot–carbon nanotube composite colloids using a silica-shell spacer. *Adv Mater* 2006;18:415–420.
56. Shi D, Guo Y, Dong Z, Lian J, Wang W, Liu G, Wang L, Ewing RC. Quantum dot-activated luminescent carbon nanotubes *via* a nano scale surface functionalization for *in vivo* imaging. *Adv Mater* 2007;19:4033–4037.
57. Richard C, Doan BT, Beloeil JC, Bessodes M, Toth E, Scherman D. Noncovalent functionalization of carbon nanotubes with amphiphilic Gd3+ chelates: toward powerful T-1 and T-2 MRI contrast agents. *Nano Lett* 2008;8:232–236.

58. Kang H, Clarke ML, Tang J, Woodward JT, Chou SG, Zhou Z, Simpson JR, Walker ARH, Nguyen T, Hwang J. Multimodal, nanoscale, hyperspectral imaging demonstrated on heterostructures of quantum dots and DNA-wrapped single-wall carbon nanotubes. *ACS Nano* 2009;3:3769–3775.
59. Bhirde AA, Patel V, Gavard J, Zhang G, Sousa AA, Masedunskas A, Leapman RD, Weigert R, Gutkind JS, Rusling JF. Targeted killing of cancer cells *in vivo* and *in vitro* with EGF-directed carbon nanotube-based drug delivery. *ACS Nano* 2009;3:307–316.
60. Picot A, D'Aléo A, Baldeck PL, Grichine A, Duperray A, Andraud C, Maury O. Long-lived two-photon excited luminescence of water-soluble europium complex: applications in biological imaging using two-photon scanning microscopy. *J Am Chem Soc* 2008;130: 1532–1533.
61. Thibon A, Pierre VC. Principles of responsive lanthanide-based luminescent probes for cellular imaging. *Anal Bioanal Chem* 2009;394:107–120.
62. Yi G, Lu H, Zhao S, Ge Y, Yang W, Chen D, Guo L-H. Synthesis, characterization, and biological application of size-controlled nanocrystalline $NaYF_4$: Yb, Er infrared-to-visible up-conversion phosphors. *Nano Lett* 2004;4:2191–2196.
63. Wu S, Han G, Milliron DJ, Aloni S, Altoe V, Talapin DV, Cohen BE, Schuck PJ. Non-blinking and photostable upconverted luminescence from single lanthanide-doped nanocrystals. *Proceedings of the National Academy of Sciences of the United States of America* 2009;106:10917–10921.
64. Kuningas K, Ukonaho T, Päkkilä H, Rantanen T, Rosenberg J, Lövgren T, Soukka T. Upconversion fluorescence resonance energy transfer in a homogeneous immunoassay for estradiol. *Anal Chem* 2006;78:4690–4696.
65. Guldi DM, Rahman G, Jux N, Tagmatarchis N, Prato M. Integrating single-wall carbon nanotubes into donor–acceptor nanohybrids. *Angew Chem* 2004;116: 5642–5646.
66. Cao L, Chen HZ, Zhou HB, Zhu L, Sun JZ, Zhang XB, Xu JM, Wang M. Carbon-nanotube-templated assembly of rare-earth phthalocyanine nanowires. *Adv Mater* 2003;15:909–913.
67. Gai S, Yang P, Li C, Wang W, Dai Y, Niu N, Lin J. Synthesis of magnetic, up-conversion luminescent, and mesoporous core–shell-structured nanocomposites as drug carriers. *Adv Funct Mater* 2010;20:1166–1172.
68. Li C, Lin J. Rare-earth fluoride nano-/microcrystals: synthesis, surface modification and application. *J Mater Chem* 2010;20:6831–6847.
69. Yang P, Quan Z, Hou Z, Li C, Kang X, Cheng Z, Lin J. A magnetic, luminescent and mesoporous core–shell structured composite material as drug carrier. *Biomaterials* 2009;30:4786–4795.
70. Minot E, Yaish Y, Sazonova V, McEuen PL. Determination of electron orbital magnetic moments in carbon nanotubes. *Nature* 2004;428:536–539.
71. Ramirez A, Haddon R, Zhou O, Fleming R, Zhang J, McClure S, Smalley R. Magnetic susceptibility of molecular carbon: nanotubes and fullerite. *Science* 1994;265:84–86.
72. Soppimath KS, Aminabhavi TM, Kulkarni AR, Rudzinski WE. Biodegradable polymeric nanoparticles as drug delivery devices. *J Control Release* 2001;70:1–20.
73. Wang J, Liu G, Jan MR. Ultrasensitive electrical biosensing of proteins and DNA: carbon-nanotube derived amplification of the recognition and transduction events. *J Am Chem Soc* 2004;126:3010–3011.
74. Zheng M, Jagota A, Semke ED, Diner BA, McLean RS, Lustig SR, Richardson RE, Tassi NG. DNA-assisted dispersion and separation of carbon nanotubes. *Nat Mater* 2003;2:338–342.

# Chapter 9

# Intracellular siRNA Delivery by Multifunctional Nanoparticles

*Kun Lu*[*,†], *Lei Wang*[*,‡], *Xiaohui Tan*[*,§], *Yao Chen*[*,‡], *Changhui Wang*[†] *and Lifeng Qi*[*,¶]

[*] *Institute for Biomedical Engineering and Nano Science, East Hospital, Tongji University School of Medicine, No. 150 Jimo Road, Shanghai 200120, P. R. China*

[†] *Shanghai Tenth People's Hospital, Tongji University School of Medicine, Shanghai 200072, P. R. China*

[‡] *Shanghai Tongji Hospital, Tongji University School of Medicine, Shanghai 200065, P. R. China*

[§] *Shanghai Shuguang Hospital, Shanghai University of Traditional Chinese Medicine, Shanghai 201203, P. R. China*

[¶] *Corresponding author.*

*Email: leonqi168@163.com*

RNAi is a sequence-specific post-transcriptional gene silencing (PTGS), which is triggered by double-stranded RNA (dsRNA). Currently, there has been a great interest in the use of small interfering RNA (siRNA) as a research tool to study gene function and drug target validation, and the therapeutic application of siRNAs already draws worldwide attention and appears to show much promise for novel drug discovery. However, RNAi efficiency is largely dependent on the development of many delivery vehicles that can deliver the siRNAs to target cells. This chapter mainly describes intracellular siRNA delivery by various nanoparticle systems, including liposomes, cationic polymers and inorganic nanoparticles. What is more, we introduced our interests on siRNA delivery for cancer therapy which are currently ongoing. At last, the chemical modification of siRNA duplex was also discussed to achieve increased stability and sensitive detection of siRNA.

## Introduction

Novel tools for evaluating gene function *in vivo* such as ribozymes and RNA interference (RNAi) are emerging as the most highly effective strategies.[1,2,3] RNAi is sequence-specific post-transcriptional gene silencing (PTGS), which is triggered by double-stranded RNA (dsRNA). This evolutionally conserved gene-silencing pathway was first described in the nematode worm *Caenorhabditis elegans.*[4] This process has been linked to many previously described phenomena such as PTGS in plants.[5] The difficulty of using RNAi in somatic mammalian cells was overcome when Tuschl and his colleagues discovered that small interfering RNAs (siRNAs) (21 nt), normally generated from long dsRNA during RNAi, could be used to inhibit specific gene targets.[6] Currently, there has been a great interest in the use of siRNA as a research tool to study gene function and drug target validation.[7,8] Recently, the therapeutic application of siRNAs, however, is largely dependent on the development of many delivery vehicles that can efficiently deliver the siRNAs to target cells. In addition, such delivery vehicles should be administered efficiently, safely and repeatedly.

## 1. Cationic Liposome-Mediated Nucleic Acid Delivery

Cationic liposomes represent one of the few multifunctional nanoparticles that can meet these requirements of intracellular nucleic acid delivery.[9] These agents are composed of positively charged lipid bilayers, and can be complexed to negatively charged siRNA duplexes. The routes of delivery include direct intratumoral injection, intravenous, intraperitoneal, intraarterial, intracranial, and others. Although the first generation of liposomal delivery reagents were hampered by low levels of *in vivo* delivery and subsequent low gene expression, optimization strategies directed at the lipid formulation have indicated that liposomal delivery remains an appropriate approach for somatic gene therapy.[10] In this respect, cationic liposome-mediated intravenous gene delivery has been shown to produce significant levels of reporter gene expression in all tissues examined.[11] Furthermore, Thierry and colleagues (1995) demonstrated that the expression of a transgene through systemic gene transfer via liposomal delivery was efficient and depends on both liposome preparation and the design of the plasmid expression vector. The demonstration that siRNAs can be used to inhibit gene expression in mammals has opened another door to analysis of gene function and drug target validation. The success of siRNAs as therapeutics, however, is largely dependent on the development of a delivery vehicle that can efficiently deliver them *in vivo*. For drug delivery, liposomes have been widely investigated as versatile carriers and shown to facilitate gene transfer *in vivo*.

## 1.1. *Intravenous (iv) siRNA/miRNA delivery by liposomes*

Similar to antisense oligonucleotides and ribozymes, synthetic siRNAs are generally delivered to cells via liposome-based transfection reagents.[12] These reagents offer the possibility of developing pharmaceutical siRNAs for local or systemic delivery. Notably, most patients who die from cancer have subclinical metastatic disease present at the time of diagnosis.[13] Thus, it is essential that molecular medicine-based therapies treat cancer bead by systemic administration. In addition, when catalytic RNAs such as siRNAs are made *in vitro*, a variety of 2 and backbone chemical modifications can be introduced into the molecules to increase their half-life in biological fluids.[14] In the past, several groups have shown that lipid-mediated intravenous delivery of gene can produce high level, systemic expression of biologically and therapeutically relevant genes.[15] In addition, siRNA has been successfully complexed with cationic, anionic, and neutral liposomes, as well as with various mixtures thereof.

Being unstable in biological fluids, the *in vivo* stability and uptake of antisense oligonucleotides and ribozymes have been improved by the use of cationic and anionic liposomes after intravenous administration. In this respect, intravenous pretreatment of mice with cationic lipids and antisense against intercellular adhesion molecule-1 (ICAM) significantly decreased its expression in the lung following LPS challenge.[16] Systemic administration of cationic liposomes and a plasmid encoding for a ribozyme against the transcription factor NF-κB suppressed NF-κB expression in metastatic melanoma cells and significantly reduced metastatic spread.[17] Srensen *et al.* have found that intravenous injection of liposome-formulated siRNAs against GFP inhibited GFP gene expression in various organs such as the liver and spleen.[18] In the liver and spleen, most of the uptake was in endothelial, Kupffer cells and macrophages. Taking together, these studies demonstrated the utility of *in vivo* gene targeting via intravenous delivery of nucleic acids such as antisense oligonucleotides, ribozymes, and siRNAs. However, still lipid-based systems have important drawbacks, including the lack of specific targeting and variation arising during fabrication. Setting the optimal liposome formulations for intravenous delivery of siRNAs would be an important step to identify gene function in whole animals and to develop therapeutic siRNAs.

## 1.2. *Intraperitoneal (i.p.) siRNA delivery by liposomes*

The appropriate delivery route to assess the efficacy of siRNAs can facilitate the development of therapeutic siRNAs. Notably, malignant ascites is a major cause of morbidity in patients with intra-abdominal dissemination of various neoplasms such as ovarian, colorectal and breast cancer.[19] Treatment of patients with this troubling clinical condition would involve systemic or intra-abdominal chemotherapy, generally not successful due to drug resistance of such tumors. Notably, the peritoneal cavity represents the largest one of human body and its anatomic structures, along with residual leukocyte populations, play an important role in the defense against invading microorganisms, in particular those breaching the gut integrity. Thus, the development of efficient i.p. delivery agents would mark a new potential to explore the efficacy of new therapeutics *in vivo.* In this respect, Srensen *et al.* have found that cationic liposomes could enhance the uptake of anti-TNF-α ribozymes and siRNAs by peritoneal macrophages and were able to down-regulate the expression of TNF-α *in vivo.*[20] Similarly, Kisick *et al.* demonstrated that cationic liposomes could deliver ribozymes to peritoneal cells.[21]

Adherent peritoneal cells are more susceptible to *in vivo* transfection than their non-adherent counterparts. The *in vivo* uptake of siRNAs can differ dramatically in the cell types as well as in the status of cell differentiation. Different cell types may require significantly different transfection conditions to yield optimal uptake of siRNAs. For biological activity, liposomes must deliver their contents into the cytoplasm, which can be cell-dependent. Srensen *et al.* have analysed the *in vivo* transfection efficiency of adherent and non-adherent peritoneal cells. A large proportion of adherent peritoneal cells were transfected *in vivo* with an FITC-labeled siRNA. In contrast, most of the non-adherent peritoneal cells were not transfected. Thus, there might be some mechanism in adherent macrophages for effectively taking up the liposome-formulated siRNA molecules. These macrophages could be the major source of TNF-α production, since i.p. delivery of siRNAs against mouse TNF-α reduced the expression of TNF-α.[8] Mice pretreated with anti-TNF-α siRNA prior to LPS challenge showed less severe clinical symptoms than those treated with an inactive siRNA. These data would suggest that liposome-mediated delivery of anti-TNF-α siRNAs might be valuable in a number of conditions such as rheumatoid arthritis in which TNF-α is a pathogenic mediator.[22]

### 1.3. *Concluding remarks of nucleic acid liposomal delivery*

The study of nucleic acids has revealed remarkable properties of RNA molecules that could make them attractive therapeutic agents, independent of their well-known ability to encode biologically active proteins. This application is now mainly driven by siRNAs that serve as guide sequences to induce target-specific mRNA cleavage via the activation of a highly conserved regulatory mechanism, called RNA interference or posttranscriptional gene silencing.[6] However, crucial for the success of siRNAs as a pharmaceutical or a basic research tool is *in vivo* transfection efficiency. A significant degree of *in vivo* cell uptake of siRNAs via liposomal delivery was observed in our studies. Similarly, hydrodynamic force (rapid injection via tail vein) has been applied to deliver naked DNA and siRNAs to hepatocytes, with significant delivery efficiencies.[15, 23, 24] Although, delivery of siRNAs via such method can induce gene silencing *in vivo*, the technique is not feasible for clinical application in humans. From the perspective of human cancer therapy, however, cationic liposomes have been shown to be safe and effective for *in vivo* gene delivery, and are currently being used in many approved clinical trials.[25] Notably, chemically synthesized siRNAs are expected to be useful in pathological situations where immediate and/or short-term effects are required. In addition to delivery, the effectiveness of siRNAs as therapeutic agents may depend on their stability and bioavailability into the targeted tissues or organs. Chemical modifications can also contribute to changes in the tissue distribution and pharmacokinetics of siRNAs. The analysis of these parameters (e.g., stability in blood circulation, extravasations into tissues) in whole animals should provide important information for liposome optimization. Despite being the most commonly used methods of nucleic acid delivery, however, liposome-based delivery still has important drawbacks, including the lack of specific targeting and potential side effects.

## 2. Cationic Polymers-Mediated Nucleic Acid Delivery

Compared with the nonviral vectors, cationic polymers are more stable, easier to change molecular weight and shape. A variety of functional devices can be added to further optimize the systems. Cationic polymers siRNA delivery vectors can be divided into synthetic and natural vectors. The former include PEI, PLGA, PAMAM dendrimers, etc. The latter include chitosan and its derivatives and cyclodextrin inclusion compounds, etc.

## 2.1. *Synthetic cationic polymers*

### 2.1.1. *polyethyleneimine (PEI)*

PEI (polyethyleneimine) is the most widely studied cationic polymer during recent years. It has 2 types: linear and branched chain type. Branched PEIis a highly branching polymer and can release endosome with proton sponge ability to deliver siRNA.[26] In Schiffelers'[27] research, self-assembling nanoparticles with siRNA were constructed with PEI that is PEGylated with an Arg–Gly–Asp (RGD) peptide ligand attached at the distal end of the polyethylene glycol (PEG), as a means to target tumor neovasculature expressing integrins and used to deliver siRNA-inhibiting vascular endothelial growth factor receptor-2 (VEGF R2) expression and thereby tumor angiogenesis. In laroui's[28] study, polyethyleneimine/TNFα siRNA nanocomplexes were constructed with polyethyleneimine (PEI) that is PEGylated with an RGD peptide ligand attached at the distal end of the PEG, as a means to target tumor neovasculature expressing integrins and used to deliver siRNA inhibiting vascular endothelial growth factor receptor-2 (VEGF R2) expression and thereby tumor angiogenesis. In Abbasi's[29] article, polyethylenimine (PEI)-coated human serum albumin (HSA) nanoparticles exhibited efficient siRNA delivery to the MCF-7 breast cancer cell line. Liu[30] uses a carrier, which was composed of a cationic oligomer (PEI1200), a hydrophilic polymer (PEG) and a biodegradable lipid-based crosslinking moiety, to deliver siRNA. And the result shows that this novel siRNA delivery carrier system with an MDR1-targeting siRNA (siMDR1) effectively reduces the expression of MDR1 in human colon CSCs (CD133+-enriched cell population), resulting in significantly increasing the chemosensitivity to paclitaxel.

### 2.1.2. *PLGA*

PLGA (poly (lactic-co-glycolic) acid) is a copolymer which is used in a host of FDA-approved therapeutic devices, owing to its biodegradability and biocompatibility. It has been widely studied on siRNA delivery. PLGA has a weak siRNA entrapment ability and a low transfection efficiency in cells. So researchers use cationic polymers to modify PLGA to enhance the efficiency of gene silencing. In Nguyen's[31] study, a potential siRNA carrier for pulmonary gene delivery was assessed by encapsulating siRNA into biodegradable polyester nanoparticles consisting of tertiary-amine-modified polyvinyl alcohol (PVA) backbones grafted to PLGA. The result shows that amine-modified-PVA–PLGA/siRNA nanoparticles could be a promising siRNA carrier for pulmonary gene delivery due to their fast degradation and potent gene knockdown profile. Sureban[32] has utilized PLGA Nanoparticle (NP) technology to deliver DCAMKL-1-specific siRNA to knock down potential key cancer regulators. Lastly, DAPT-mediated inhibition of Notch-1 resulted in HCT116 tumor growth arrest and down-regulation of Notch-1 via an *miR-144* dependent mechanism. Patil's research [33] in 2008 used PEI incorporate in the PLGA matrix to improve siRNA encapsulation. This PLGA-PEI nanoparticle matrix increased siRNA encapsulation by about 2-fold and also improved the siRNA release profile. PLGA-PEI nanoparticles carrying luciferase-targeted siRNA enabled effective silencing of the gene in cells stably expressing luciferase as well as in cells that could be induced to overexpress the gene. Yuan's[34] developed the biodegradable chitosan-modified poly (D,L-Lactide-co-glycolide)(CHT-PLGA) nanoparticles with positive surface charge, high transfection efficiency, and low toxicity. And this CHT-PLGA NPs showed siRNA's excellent binding ability and effective protection of oligos from RNase degradation.

### 2.1.3. *Poly(amido amine) (PAMAM) dendrimers*

PAMAM dendrimers represent an exciting new class of macromolecular architecture called "dense star" polymers. Unlike classical polymers, dendrimers have a high degree of molecular uniformity, narrow molecular weight distribution, specific size and shape characteristics, and a highly-functionalized terminal surface. The manufacturing process is a series of repetitive steps starting with a central initiator core. Each subsequent growth step represents a new "generation" of polymer with a larger molecular diameter, twice the number of reactive surface sites, and approximately double the molecular weight of the preceding generation. In Weber's research,[35] their group developed amino-terminated carbosilane denderimers (CBS) to protect and transport siRNA. CBS/siRNA dendriplexes were shown to silence GAPDH expression and reduce HIV replication in SupT1 and PBMC. These results point to the possibility of utilizing dendrimers such as CBS to deliver and transfect siRNA into lymphocytes thus allowing the use of RNA interference as a potential alternative therapy for HIV infection. Agrawal[36] reported the development of dendrimer-conjugated magnetofluorescent nanoworms that we call "dendriworms" as a modular platform for siRNA delivery *in vivo*. This platform maximizes endosomal escape to robustly produce protein target knockdown *in vivo*, and is tolerated well in mouse brain. They show that dendriworms carrying siRNA against the epidermal growth factor receptor (EGFR) reduce protein levels of EGFR in human glioblastoma cells by 70–80%.

#### 2.1.4. *Others*

In Beh's report,[37] cationic nanoparticles self-assembled from the amphiphilic copolymer poly(*N*-methyldietheneamine sebacate)-co-[(cholesteryl oxocarbonylamido ethyl) methyl bis(ethylene) ammonium bromide] sebacate) (P(MDS-co-CES) were synthesized and used to deliver Bcl-2-targeted siRNA into HepG2, HeLa and MDA-MB-231 cell lines, and to down-regulate Bcl-2 mRNA expression levels. Sun[38–39] designed a novel amphiphilic and cationic triblock copolymer consisting of monomethoxy poly(ethylene glycol), poly(ε-caprolactone) (PCL) and poly(2-aminoethyl ethylene phosphate) denoted as $mPEG_{45}$-b-$PCL_{100}$-*b*-$PPEEA_{12}$ and synthesized for siRNA delivery. They show clear evidence that the micelleplex is capable of delivering siRNA and paclitaxel simultaneously to the same tumoral cells both *in vitro* and *in vivo*. We further demonstrate that systemic administration of the micelleplex carrying polo-like kinase 1 (Plk1)-specific siRNA, and paclitaxel can induce a synergistic tumor suppression effect in the MDA-MB-435s xenograft murine model, requiring a thousand-fold less paclitaxel than needed for paclitaxel monotherapy delivered by the micelleplex and without activation of the innate immune response or generation of carrier-associated toxicity. Wilson[40] and Lee[41] present a delivery vehicle for siRNA, termed thioketal nanoparticles (TKNs), through localized oral administration and injection from jugular vein in mouse, can protect mice from ulcerative colitis and acute hepatic damage.

### 2.2. *Natural cationic polymer*

#### 2.2.1. *Chitosan and derivative*

Chitosan is a linear polysaccharide composed of randomly distributed β-(1-4)-linked D-glucosamine (deacetylated unit) and N-acetyl-D-glucosamine (acetylated unit). It is made by treating shrimp and other crustacean shells with the alkali sodium hydroxide. Chitosan enhances the transport of polar drugs across epithelial surfaces, and is biocompatible and biodegradable. Chitosan has a number of commercial and possible biomedical uses, such as nucleic acid drug delivery. Howard[42] demonstrated that knockdown of TNF-α expression in systemic macrophages by i.p. administration of chitosan/siRNA nanoparticles in mice down-regulates systemic and local inflammation. Chitosan nanoparticles containing an unmodified anti-TNF-α Dicer-substrate siRNA (DsiRNA) mediated TNF-α knockdown (~66%) in primary peritoneal macrophages *in vitro*. Down-regulation of TNF-α-induced inflammatory responses arrested joint swelling in collagen-induced arthritic (CIA) mice dosed i.p. with anti-TNF-α DsiRNA nanoparticles. Their work showed nanoparticle-mediated TNF-α knockdown in peritoneal macrophages as a method to reduce both local and systemic inflammation, thereby presenting a novel strategy for arthritis treatment. In Ji's research,[43] nanoparticles which were formulated with chitosan/siRNA exhibited irregular, lamellar and dendritic structures with a hydrodynamic radius size of about 148 nm and net positive charges with zeta-potential value of 58.5 mV. Their result showed that FHL2 siRNA formulated within chitosan nanoparticles could knock down about 69.6% FHL2 gene expression, and blocking FHL2 expression by siRNA could also inhibit the growth and proliferation of human colorectal cancer Lovo cells. Nielsen[44] formed aerosolized chitosan/siRNA nanoparticles using a nebulizing catheter. *In vitro* silencing effects of aerosolized and non-aerosolized formulations were evaluated in an enhanced green fluorescent protein (EGFP) endogenous-expressing H1299 cell line by flow cytometry. The results showed minimal alteration in gene silencing efficiency before (68%) and after (62%) aerosolization in EGFP-expressing H1299 cells. Howard's research [45] in 2006 introduced a novel chitosan-based siRNA nanoparticle delivery system for RNA interference *in vitro* and *in vivo*. The formation of interpolyelectrolyte complexes between siRNA duplexes (21-mers) and chitosan polymer into nanoparticles. Rapid uptake (1 h) of Cy5-labeled nanoparticles into NIH 3T3 cells, followed by accumulation over a 24 h period. Nanoparticle-mediated knockdown of endogenous EGFP was demonstrated in both H1299 human lung carcinoma cells and murine peritoneal macrophages (77.9% and 89.3% reduction in EGFP fluorescence, respectively). Western analysis showed 90% reduced expression of BCR/ABL-1 leukemia fusion protein while BCR expression was unaffected in K562 (Ph+) cells after transfection using nanoparticles containing siRNA specific to the BCR/ABL-1 junction sequence. Lee's study,[46] chitosan nanoparticles encapsulating siRNA were prepared using a coacervation method in the presence of polyguluronate (PG), which showed low cytotoxicity and were useful in delivering siRNA to HEK 293FT and HeLa cells. Chitosan/PG nanoparticles were considered promising for siRNA delivery due to their low cytotoxicity and ability to transport siRNA into cells, which can effectively inhibit induction of targeting mRNA. Han's group[47] developed an Arg–Gly–Asp (RGD) peptide-labeled chitosan nanoparticle (RGD-CH-NP) as a novel tumor targeted delivery system for short interfering RNA (siRNA). And the results showed that RGD-CH-NP is a novel and highly selective delivery system for siRNA with the potential for broad applications in human disease.

#### 2.2.2. *Cyclodextrin polymers*

Cyclodextrins (sometimes called cycloamyloses) is a family of compounds made up of sugar molecules bound together in a ring (cyclic oligosaccharides). They are composed of 5 or more α-D-glucopyranoside units linked 1->4, as in amylose (a fragment of starch). Cyclodextrins are able to form host–guest complexes with hydrophobic molecules given the unique nature imparted by their structure. Because cyclodextrins are hydrophobic inside and hydrophilic outside, they can form complexes with hydrophobic compounds. Thus they can enhance the solubility and bioavailability of such compounds, such as siRNA. In Heidel's report,[48] their nanoparticles consist of a synthetic delivery system that uses a linear, cyclodextrin-containing polycation, transferrin (Tf) protein targeting ligand, and siRNA. And the nanoparticles are well tolerated when administered to cynomolgus monkeys at 3–9 mg siRNA/kg. Bartlett[49] used positron emission tomography (PET) and bioluminescent imaging to quantify the *in vivo* biodistribution and function of nanoparticles formed with cyclodextrin-containing polycations and siRNA. Conjugation of 1,4,7,10-tetraazacyclododecane-1,4,7,10-tetraacetic acid to the 5′ end of the siRNA molecules allows labeling with 64Cu for PET imaging.

## 3. Inorganic Nanomaterials-Mediated Nucleic Acid Delivery

### 3.1. *Quantum dots*

RNA interference (RNAi) is a key technology for sequence-specific suppression of genes. Its potential applications of synthetic siRNA have been considered for treatment of various diseases, including cancer, in recent years. However, therapeutic applications of siRNA are still limited due to poor cellular uptake and accelerated degradation in biological uids. Most current transfection reagents, including nanoparticle-based carriers, have been used with RNAi in a serum-free medium for their protection of siRNA from nuclease degradation. Lack of efficient delivery system with stable siRNA *in vivo*, and high specificity to the desired tissue site is still a challenge to RNAi therapy. The current approaches in siRNA delivery include liposomes, polymers, peptides, virus-based vectors, etc. However, these carriers are not fluorescent, and therefore are unable to monitor the siRNA delivery process. Typical strategies to track siRNA delivery include monitoring fluorescently end-modified siRNA or co-transfecting reporter plasmids. But these methods experienced rapid photobleaching, are incapable of simultaneous monitoring of multiple siRNA molecules, and are insensitive to different heterogeneous siRNA delivery. Recently, quantum dots (QDs) have been used as co-transfection reagents or delivery systems for siRNA delivery tracking.[50] On the other hand, these QDs such as L-arginine or peptide-modified QDs suffer from limited transfection efficiency. High transfection efficiency is particularly required for high siRNA concentration at 100 nM.

QDs are nanocrystals formed by semiconductor materials (such as CdSe, ZnSe, GaAs, InAs), displaying attractive photophysical properties, including high quantum yield, resistance to photobleaching, and tunable photoluminescence, making them potentially powerful tools in a range of biomedical applications. But the toxicity displayed by many QDs hampered its application in this area. Luckily, the toxicity displayed by unmodified QDs can be reduced by various surface coatings such as BSA and cross-linked lysine residues. S. Klein *et al.*[51] presented a class of transfection tools that are based on biocompatible, water-soluble luminescent SiQDs to serve for *ABCB1* siRNA to silence the *ABCB1* gene in Caco-2 cells. The functionalized SiQDs achieve high gene-transfection efficiency for *ABCB1* siRNA by real-time PCR and transcellular transport studies. Adela Bonoiu *et al.*[52] evaluated the specificity and efficiency of quantum dot (QD) complexed with MMP-9-siRNA (nanoplex) in down-regulating the expression of *MMP-9* gene in brain microvascular endothelial cells. Almost 80% of the QD-mediated gene silencing efficiency was obtained in BMVECs. Li *et al.*[53] synthesized two L-arginine (L-Arg)-modified CdSe/ZnSe QDs as siRNA carriers to silence HPV18 E6 gene in HeLa cells. These QDs showed significantly low cellular cytotoxicity and good siRNA protection. Furthermore, the properties and capabilities of these QDs showed that amino acid-modified QDs could be used as useful siRNA carriers to effectively silence a target gene as well as fluorescence probes to analyze intracellular imaging *in vivo.*

In our previous work,[54] we developed a proton sponge polymer and amphipol PMAL-coated mono-dispersed QDs for siRNA delivery. The proton-sponge-coated QDs are prepared based on ligand exchange reactions and the endosome-disrupting polymers as amphipol PMAL, grafted with hyperbranched PEI. A new generation of nanoparticle carrier that allows efficient delivery and real-time imaging of siRNA in live cells has been developed by combining two distinct types of nanomaterials, semiconductor quantum dots, and amphipols as shown in **Figure 1**. An important finding is that although amphipols are broadly used for solubilizing and delivering hydrophobic proteins into the lipid bilayers of cell membrane, when combined with nanoparticles, they offer previously undiscovered functionalities including cytoplasm delivery, siRNA protection, and endosome escape. Compared with the classic siRNA carriers such as Lipofectamine™ and polyetheleneimine. This new class of nanocarrier works in both serum-free and complete cell culture media,

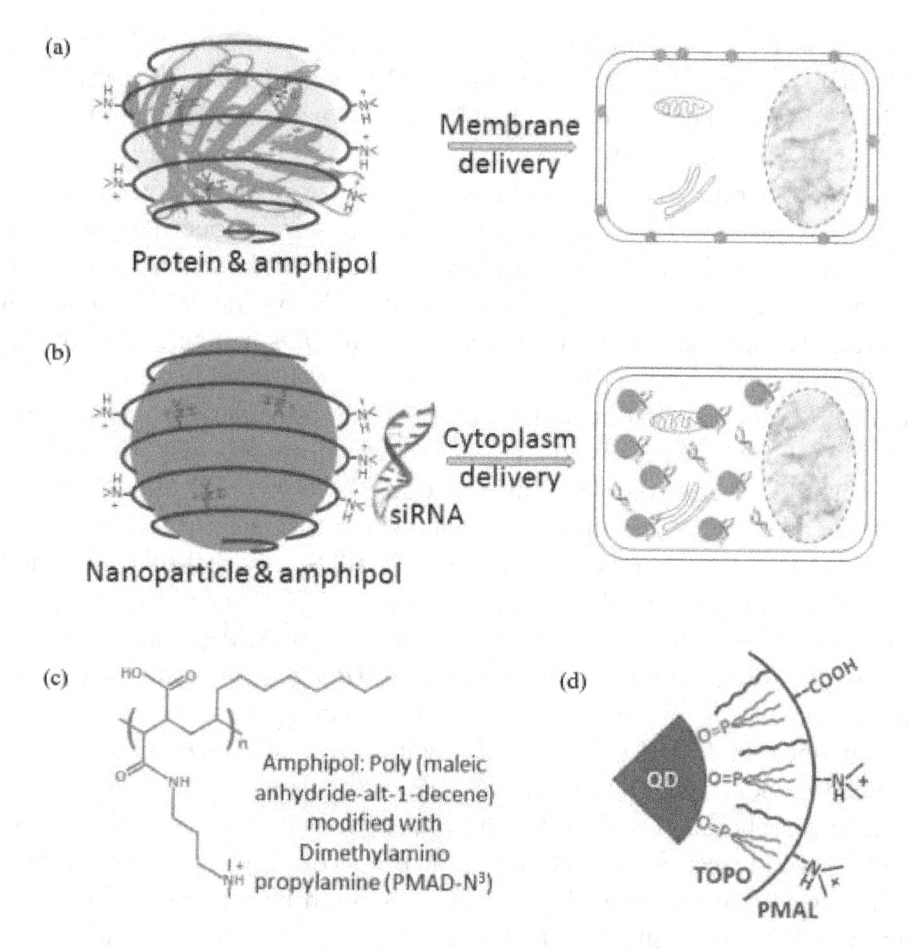


**Figure 1.** Schematic drawing of the hybrid structure of QD and amphipol for siRNA delivery and real-time imaging in live cells. (**a**) Solublization of hydrophobic proteins and delivery into cell membrane lipid bilayers. (**b**) Hydrophobic QDs encapsulated by amphipol for siRNA intracellular delivery. The siRNA molecules are attached to the QD surface via electrostatic interaction. (**c**) Molecular structure of the amphipol polymer used in the current study. The polymer has both a hydrophobic domain (hydrocarbons) and a hydrophilic domain (carboxylic acids and tertiary amines). (**d**) Schematic drawing of the hydrophobic interaction between TOPO-coated QDs and the amphipol. The amphipol and QDs are bound to each other via multivalent hydrophobic interaction.

which is advantageous over Lipofectamine. It also outperforms polyethyleneimine in gene silencing under both conditions with significantly reduced toxicity. Furthermore, the intrinsic fluorescence of QDs provides a mechanism for real-time imaging of siRNA delivery in live cells. This new multifunctional, compact, and traceable nanocarrier is expected to yield important information on rational design of siRNA carriers and to have widespread applications of siRNA delivery and screening *in vitro* and *in vivo*.

### 3.2. *Gold nanoparticles*

Gold nanoparticles (AuNPs), the particle size of which usually lies between 1–150 nm, exhibit good chemical and physical properties. They are optimal nontoxic carriers for gene therapy. AuNPs can be used in siRNA gene delivery, forming complexes with positively charged polymer carrier, in order to improve the load capacity and stability of siRNA. Xiao[55] *et al.* found a multifunctional gold nanorod (NR)-based nanocarrier. They made Au NR conjugate with DOX, polyarginine, a cationic polymer for complexing siRNA, and octreotide (OCT), a tumor-targeting ligand, to specifically target NE cancer cells with overexpressed somatostatin receptors. Au-DOX-OCT-ASCL1 siRNA (Au-DOX-OCT complexed with ASCL1 siRNA) resulted in significantly higher gene silencing in NE cancer cells than Au-DOX-ASCL1 siRNA (non-targeted Au-DOX complexed with ASCL1 siRNA) as measured by an immunoblot analysis. Additionally, Au-DOX-OCT-ASCL1 siRNA was the most efficient nanocarrier at altering the NE phenotype of NE cancer cells and showed the strongest anti-proliferative effect. Suresh Acharya[56] *et al.* conjugated gold nanoparticles (AuNP) with cysteine-terminated KDEL (Lys-Asp-Glu-Leu) peptide and siRNA directed against NADPH oxidase 4 (Nox4), found that the cellular uptake of AuNP nanoconstructs was more efficient than Lipofectamine-mediated transfection in differentiated myotubes ($P < 0.05$) compared to undifferentiated myoblasts, suggesting that AuNP nanoconstructs provide an efficient platform for siRNA delivery to differentiated myotubes. In order to identify the most adequate nanoparticles to efficiently transport siRNAs, João Conde[57] *et al.* use a hierarchical approach including three biological systems of increasing complexity: *in vitro* cultured human cells, *in vivo* freshwater polyp (*Hydra vulgaris*), and *in vivo* healthy nude mice. The ionic linkage of siRNA on the AuNPs showed efficiency in cells and in Hydra. Their research suggest that RGD gold nanoparticles-mediated delivery of siRNA by intratracheal instillation in mice leads to successful suppression of tumor cell proliferation and respective tumor size reduction.

### 3.3. *Magnetic nanoparticles*

Magnetic nanoparticles with $Fe_3O_4$ as main ingredients are solid phase carrier. As $Fe_3O_4$ nanoparticles (NPs) offer a large surface area, good superparamagnetic property, and other excellent physical and biological properties, it could be applied to targeted drug and siRNA delivery and contrast agent for MR imaging due to the superparamagnetism and biodegradability. Their biocompatibilities could be significantly improved through modification of the surface properties, such as grafting peptides, nucleic acid (DNA, RNA), folic acid, polymers, and some bimolecules, which could increase their stability and biocompatibility and allow them to be directed to the desired target. Peng[59] *et al.* have found that PLGA-modified $Fe_3O_4$ nanoclusters loaded with siRNA, synthesized via one-pot method and characterized by SEM, TEM, and XRD, got the abilities of protecting and releasing siRNA by gel electrophoresis and Quant-iT Picogreen assay *in vitro*. PLGA-modified magnetic nanoclusters were considered as a novel siRNA delivery vehicle and exhibited excellent abilities of protecting and releasing siRNA *in vitro*. Wei[58] *et al.* found that $Fe_3O_4$ nanoparticles coated with WSG-peptide, prepared *via* a facile biomineralization technique at room temperature, particles were 35.92 emu/g, slightly higher than that of $Fe_3O_4$ without WSG peptides and had a good cellular compatibility. In addition, compared with $Fe_3O_4$ NPs, the mineralized $Fe_3O_4$ NPs coated with WSG peptides could more easily assemble into the cancer cell, indicating that the WSG–$Fe_3O_4$ nanoparticles possess cancer-targeting property.

Our group recently developed cell-penetrating magnetic nanoparticles (CPMNs) for highly efficient intracellular siRNA delivery to breast cancer cells.[60] In this study, amphipol polymer and protamine peptide were employed to modify magnetic nanoparticles to form CPMNs. The unique CPMN could efficiently deliver the eGFP siRNA intracellularly and silence the eGFP expression in cancer cells, which was verified by fluorescent imaging of cancer cells. Compared with lipofectamine and polyethyleneimine (PEI), CPMNs showed superior silencing efficiency and biocompatibility with minimum siRNA concentration as 5 nm in serum-containing medium as shown in **Figure 2**. CPMN was proved to be an efficient siRNA delivery system, which will have great potential in applications as a universal transmembrane carrier for intracellular gene delivery and simultaneous MRI imaging. This novel system could be applied for MRI imaging and fluorescent imaging of siRNA delivery simultaneously.

## 3.4. *Mesoporous silica nanoparticles (MSN)*

MSNs are emerging as one of the most appealing candidates for drug delivery. Such materials offer advantages such as good physical and chemical stability, high loading capacity, and opportunities for controlled drug release.

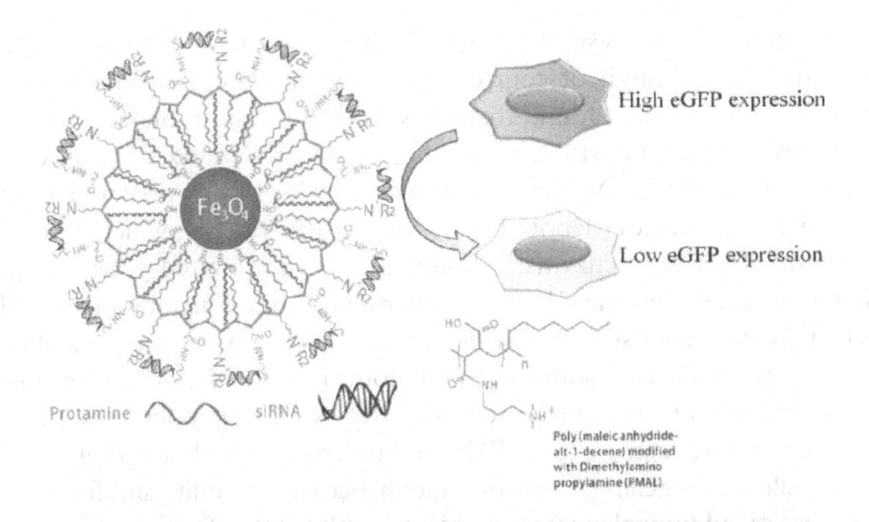


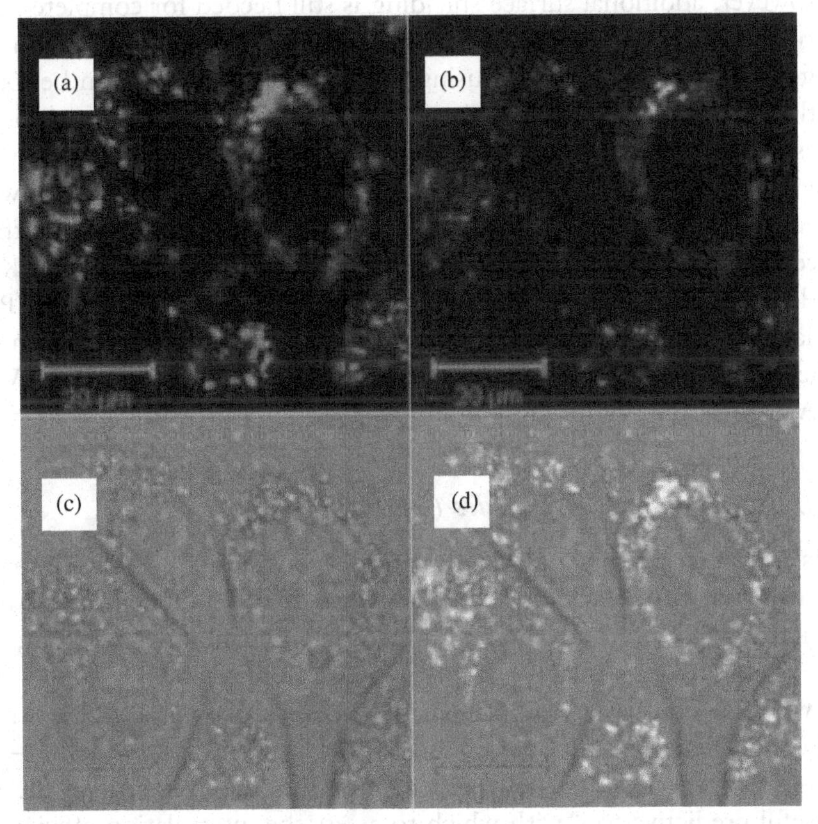


**Figure 2.** Schematic Strategy of the CPMN Hybrid Nanoparticles for eGFP siRNA Delivery, and intracellular distribution of Cy3-labeled siRNA (red color) in U251 cells transfected by CPMN. **(a)** siRNA evenly distributed in cytoplasm; **(b)** distribution of lysotracker (green color); **(c)** phase contrast image control; **(d)** siRNA partially colocalized with lysotracker, shown by the merged picture of A, B, and C.

In addition, mesoporous silica has been found to be relatively "nontoxic" and biocompatible, although of course depending on dose and administration route. Wang[61] *et al.* have found that siRNA could be encapsulated within the mesopores of magnetic MSNs under a strongly dehydrated solution condition, the siRNA-loaded M-MSNs were capped with PEI to prepare a type of siRNA delivery vehicles (denoted as M-MSN_siRNA@PEI), which efficiently protected siRNA from enzymatic degradation and exhibited great potential in initiating gene silencing. Via their experiment, they concluded that highly efficient MSNs-based siRNA delivery vectors must possess the capability of initiating effectively endosomal escape before the degradation of their packaged siRNA in endolysosomes. Noha Gouda[62] *et al.* wrap poly (PEG)-block-polycation/siRNA complexes (PEGylated polyplexes) with a hydrated silica, termed "silica nanogelling," in order to enhance their stability and functionality. However, additional surface shielding is still needed for complete inhibition of those undesired interactions. In this regard, they focused on silica nanogel shielding of polyplexes using soluble silicates, which can be formed directly on the polyplex surface through condensation of anionic silicic acid species in the vicinity of polycations to form a gel layer sufficiently covering the net positive charge of polyplexes. Xu Li[63] *et al.* encapsulated siRNA within the mesopores of M-MSNs, followed by the coating of PEI on the external surface of siRNA-loaded M-MSNs and the chemical conjugation of KALA peptides. Via their experimental observation, the knockdown of EGFP and vascular endothelial growth factor (VEGF) in tumor cells were both with excellent RNAi efficiencies. The intratumoral injection of M-MSN_VEGF siRNA@PEI-KALA significantly inhibited the tumor growth.

## 3.5. *Layered double hydroxides (LDHs)*

LDHs constitute a family of layered materials consisting of positively charged layers with charge-balancing anions between them. LDH attracted much attention because of its advantages of high biocompatibility, high drug loading, and low toxicity, shape and particle size controllability. Wong[64] *et al.* observed that efficacy of LDH-mediated delivery of four different siRNAs into cortical neurons and NIH 3T3 cells was found to vary widely (6–80%, and 2–11%, respectively). They further found that the mass ratio at the PZC is a useful predictive tool with which to assess the intercalation efficiency of selected siRNA sequences into the LDH interlayer and subsequent internalization into the cell cytoplasm. Chen[65] *et al.* found that LDHs with the *Z*-average particle size of approximately 110 nm can mediate siRNA delivery in mammalian cells, resulting in gene silencing. However, short double-stranded nucleic acids are mostly adsorbed onto the external surface and not well protected by LDHs. They made a further research in order to enhance the intercalation of siRNA into the LDH interlayer and the efficiency of subsequent siRNA delivery, and found dsDNA/siRNA is more effectively intercalated into smaller LDHs with the *Z*-average particle size of approximately 45 nm. More dsDNA/siRNA is transfected into HEK 293T cells, and more efficient silencing of the target gene is achieved using smaller LDHs. Li *et al.*[66] took advantage of the LDH anion exchange capacity to intercalate 5-FU into its interlayer spacing and load siRNA on the surface of LDH nanoparticles. The combination significantly enhanced cytotoxicity to three cancer cell lines, e.g., MCF-7, U2OS and HCT-116.

## 4. Intracellular siRNA Delivery as JAM2-siRNA and CAP1-siRNA for Cancer Therapy

RNA interference (RNAi) is emerging as one of the most powerful technologies for sequence-specific suppression of genes and has potential applications ranging from functional gene analysis to therapeutics, especially for antitumor drug target screen.[67] Here we introduce our recent work for novel siRNA candidates for anti-tumoral studies.

### 4.1. *JAM2-siRNA*

JAM-2 is a part of a subfamily of junctional adhesion molecules (JAMs) comprising JAM-A, JAM-2 (JAM-B), and JAM-3(JAM-C). JAM-2 is specifically enriched in cell–cell contacts at the level of the tight junction, and highly expressed in lymphatic and vascular endothelial cells, mostly in high endothelial venules (HEVs). JAM-2 is preferentially expressed in the endothelium of arterioles in and around tumors and sites of inflammation. In our previous study, JAM-2 was found to be over-expressed in glioma by IHC (immunohistochemistry) assay. The intracellular JAM-2 siRNA delivery was achieved in real-time imaging mediated by proton-sponge-coated QD. This delivery system has been found particularly suitable at 5 nm siRNA concentration in serum containing medium. The gene silencing efficiency against JAM-2 expression reached more than 90%. Moreover, JAM-2 knockdown was reported for the first time that improves glioma cell migration inhibition by the Notch pathway.[68]

A novel class of proton-sponge-coated QDs, PMAL grafted with PEI loaded with CdSe/ZnSe QDs have been prepared through direct ligand-exchange reactions and surface modifications as multifunctional complexes

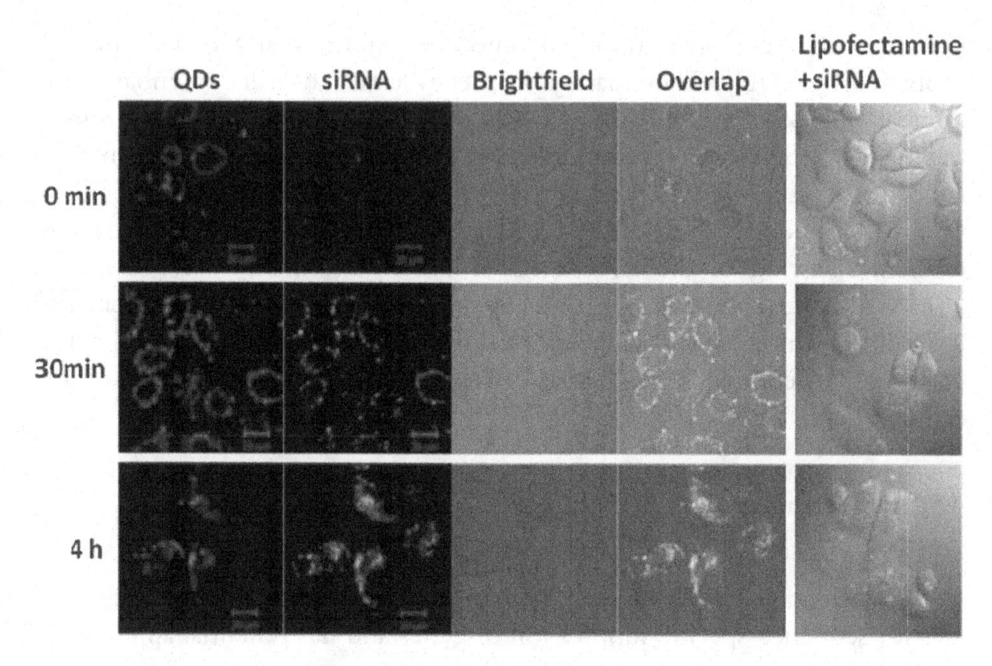


**Figure 3.** Real-time imaging of siRNAAF488 release from QD-PMAL-PEI by confocal microscopy. Confocal microscopy was used to monitor the time-dependent accumulation of QD–siRNAAF488 complexes in U251 cells. The concentration of both QDs and siRNA is 5 nm. The QD–siRNA AF488 complexes were attached to the cell membrane immediately after being added in cell medium, and a bright red ring standing for QDs fluorescence was observed. After internalization, green fluorescence from siRNAAF488 increases significantly at a time dependent mode (**Figure 4**), indicating that the siRNA. Molecules begin to dissociate from QD–siRNAAF488 complexes step by step. Until 6 hours, most siRNA molecules have already diffused inside the cytoplasm. Lipofectamine-mediated siRNA AF488 delivery was used as positive control, and observed in different time intervals comparable with that of the QDs group. Lipofectamine–siRNA complexes showed much lower intracellular siRNA release even after 4 hours' incubation. This indicates that lipofectamine required much longer time for interaction with cell membrane to achieve sufficient intracellular siRNA delivery, thus leading to lower gene silencing efficiency.

for siRNA delivery and real-time intracellular imaging. Systematic biological experiments show that QD-PMAL-PEI exhibits low cytotoxicity. These QD–siRNA complexes can be readily internalized into cells confirmed by flow cytometric and confocal microscopic analyses (**Figure 3**). Importantly, superior gene silencing efficiency (more than 90%) is achieved by using the QD–siRNA complexes as determined by real-time fluorescence quantitative PCR. In addition, the QD–siRNA complexes, which target *JAM-2* oncogene, can inhibit U251 cell migration significantly. It is worth noting that all transfection experiments of QD-PMAL-PEI were conducted in serum-containing medium. The QDs show superior gene silencing properties compared with lipofectamine. The minimum concentration of siRNA for transfection is only 5 nm, much lower than that of others transfection reagents previously reported. The dual modality of QD–siRNA complexes allows real-time tracking of QDs and siRNA release during siRNA delivery. The unique properties of proton-sponge-coated QDs can be further optimized *in vivo* for tumor targeting, drug delivery and simultaneous imaging due to their high stability and superior silencing efficiency in serum-containing environments.

### 4.2. *CAP1-siRNA*

Adenylate cyclase-associated protein 1 (CAP1) is an actin-monomer-binding protein coded by the *CAP1* gene[69] which was originally cloned from budding yeast and is located in the downstream of the *ras* gene.[70] Human homology of *CAP1* was identified in the early 1990s.[71] Both mammal and yeast CAPs interact with actin[72] and play a role in actin turnover.[73] Given the critical role for actin filament reorganization in cell migration and the regulatory role for CAP1 in actin filament reorganization,[74,75] it is logical to hypothesize that *CAP1* may be associated with tumor metastasis. Currently, however, studies on the correlation between *CAP1* expression and tumor metastasis are scarce.[76] The objective of the present study was to evaluate the potential value of *CAP1* in early diagnosis and prognostic prediction of progressive lung cancer.

siRNA-based RNA interference (RNAi) is a post-transcriptional process triggered by the introduction of dsRNA which leads to gene silencing in a sequence-specific manner.[77] As one of the most notable discoveries of the past decade in functional genomics, RNAi has become an important tool for analyzing gene functions in eukaryotes. Using this advanced technology, we successfully silenced the *CAP1* gene in invasive lung cancer 95-D cells, which resulted in a significant reduction in the capacity of 95-D cells to migrate (see **Figure 4**). This observation further supports a functional role for *CAP1* in the process of lung cancer cell metastasis. At present, the molecular mechanisms underlying the effect of *CAP1* on lung cancer cell migration and metastasis are largely unknown. Cofilin is a family of actin-binding proteins which disassembles actin filaments. Cofilin activity is present in the malignant, invasive cancer cells[78] and may function as a factor to regulate cancer cell migration/invasion phenotypes. We previously demonstrated that *CAP1* is a downstream target of cofilin,[79] suggesting that the *CAP1* overexpression-associated lung cancer cell migration and metastasis may involve a cofilin-mediated signaling pathway. Nevertheless, this requires further evaluation.

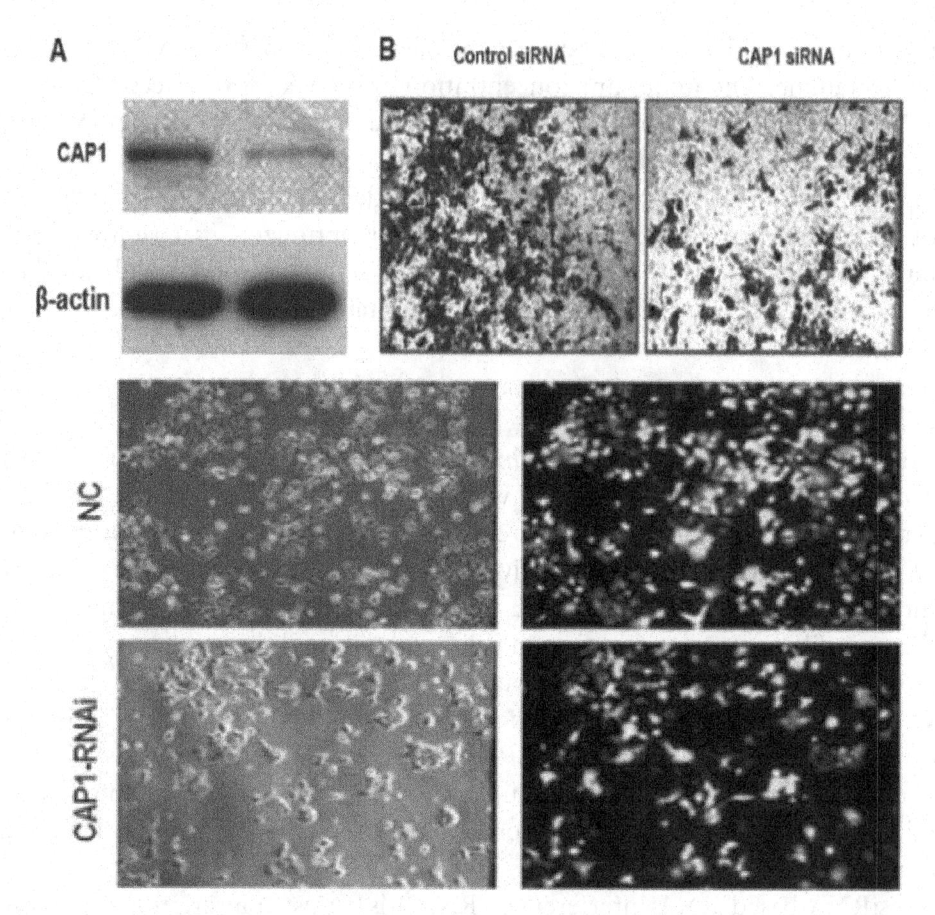


**Figure 4.** A representative image of invasive 95-D lung cancer cells transfected with the control siRNA and the cyclase-associated protein 1 (CAP1)-specific siRNA respectively after migrating over the transwell membrane in the *in vitro* cell migration assay. Migrated cells were stained. Reduced migration of invasive lung cancer cells following CAP1 knockdown. To validate the positive correlation between *CAP1* and metastasis of lung tumors, we attempted to silence the *CAP1* gene in invasive lung cancer 95-D cells with a small interfering RNA (siRNA) molecule specific for human *CAP1* and to determine the migration capacity of the CAP1-siRNA carrying cells *in vitro*. As shown in the figure, the *CAP1* gene was adequately silenced by sequence-specific *CAP1*-siRNA and correspondingly the number of cells that migrated across the transwell membrane was significantly reduced in the *CAP1*-gene-silenced 95-D cells.

## 5. Chemical Modifications to Achieve Increased Stability and Sensitive Detection of siRNA

Now it was demonstrated that exogenously delivered chemically synthesized siRNA can function as trigger for this specific silencing mechanism in mammalian cells.[80] Initial experiments indicated that, in contrast to long dsRNA, siRNA does not stimulate an unspecific inhibition of protein synthesis mediated by activation of protein kinase R (PKR) in mammals.[81] Because of its specificity and high efficiency as well as simple practicability, siRNA-triggered RNAi has become rapidly accepted as the method of choice for studying gene function in cell culture systems. Moreover, recent reports demonstrated the successful siRNA-mediated down-regulation of reporter genes[82,83] as well as endogenous target genes in mice.[84,85,86] Now, researchers in academia and industry are attempting to utilize RNAi as a platform for the development of therapeutics. Whereas there is no need to alter the all-ribo nature of siRNA for *in vitro* knockdown experiments (besides the possibility that modified siRNA might be helpful in elucidating the still-not-fully-understood underlying mechanism of RNAi), this may change when considering the development of drugs based on siRNA. Pre-requisites for a clinical development of siRNA-based drugs are, besides the *in vivo* reconfirmation of target down-regulation found in cell culture, a sufficient stability towards nucleases present in a biological environment and a favorable pharmacokinetic profile. Therefore, in order to be successful in the transformation of siRNA from a powerful and well established tool for manipulating gene expression in mammalian cells into a functional therapeutic molecule, it might be necessary to utilize chemical modifications of the siRNA. Ideally, these modifications should improve in both resistance towards cellular degradation processes as well as cellular uptake.[87] Although it is well known that dsRNA is more stable towards nucleases than its single-stranded counterpart, it is likely that an all-ribo siRNA carrying no further chemical modification would not match the criteria for drug development mentioned above.

Chemical alterations have substantially contributed to the advancement of antisense oligonucleotide (ASON)[88–90] and ribozyme-based therapeutics.[91,92] These modifications were vital as they prolonged the presence of the active compounds in a biological milieu significantly. For example, the half-life of an unmodified all-ribo RNA hammerhead ribozyme in human serum was determined to be less than six seconds.[93] With the aid of suitable chemically modified nucleotides in the ribozyme, the half-life could be extended to as much as 8 hours without significantly compromising the enzymatic activity.[94,95] Among the great variety of different chemical modifications, the two most frequently used are the 2-O-methylnucleotides and the phosphorothioate linkage (PTL), in which a non-bridging oxygen of the phosphodiester between adjacent nucleotides in the oligomer chain is replaced by a sulphur atom.[96] These two modifications were also successfully employed in the gapmer approach of second generation ASONs.[97] Moreover, especially in the ribozyme field, further 2-sugar modifications such as 2-O-allyl, 2-amino-2-deoxy and 2-deoxy-2-fluororibofuranose nucleotides were exploited.[98] Generally, chemical modifications in oligoribonucleotides are synthesized using the well-established phosphoramidite technology.[99,100] This solid-phase chemistry adds suitably protected nucleotides stepwise to the growing ribonucleotide chain immobilized on a solid support. By means of this approach, it is possible to synthesize small quantities of oligoribonucleotides in a high-throughput fashion as well as to manufacture kilogram amounts of material for clinical trials. Up to now, published reports assess the consequences of modifications in siRNA on activity *in vitro* and stability in biological media such as serum. Most reports focus on sugar–phosphate backbone modifications, whereas studies on alterations of the nucleobases in siRNA are rare.[101]

## 5.1. *Probing RNAi's tolerance towards chemical modifications*

In an early study on the function of long dsRNA, chemical modifications were introduced on both RNA strands.[102] Alteration of the sense strand proved to be generally better tolerated than that of the antisense strand. Furthermore, modified nucleotides which promote the A-form helical structure of a natural RNA–RNA duplex, such as 2-deoxyfluoro nucleotides, did not adversely affect RNAi activity, whereas modified nucleotides with a tendency to destabilize the A form type duplex, such as 2-deoxy or 2-aminodeoxy nucleotides, resulted in a reduced inhibitory activity of the dsRNA. Several reports carefully address the influence of sugar–phosphate backbone modifications on siRNA function.

Analysis of siRNA function in a Drosophila embryo lysate revealed that a change from the all-ribo design to all-2-deoxy oligonucleotides in one or both siRNA strands abolished RNAi.[103] It is interesting that up to four consecutive 2-deoxyribonucleotide substituents at both 3-ends produced appreciably efficient siRNA that was marginally less active than the all-ribo siRNA variants in an *in vitro* mRNA cleavage assay. Nevertheless, one should be careful when adding stretches of 2-deoxy residues to siRNA since the resulting hetero-duplex could be prone to RNase-H-mediated cleavage of the RNA-containing strand. This may help explain why substitution of either RNA strand in an siRNA by 2-deoxy oligonucleotides results in the described reduction of RNAi activity in the aforementioned Drosophila embryo lysate,[103] in a Drosophila *in vivo* assay,[104] and in HeLa.[103] Full modification of either one or both siRNA strands with 2-O-methyl nucleotides almost abolished siRNA activity in different systems,[103] and siRNA with only an all-2-O-methyl sense strand retained significant residual activity.[101,105] For 19 and 21 basepair blunt-ended siRNA, the sense and antisense strands were synthesized in an alternating fashion of 2-O-methyl-modified and unmodified nucleotides that furnished an overall duplex structure in which the 2-O-methyl nucleotides were either facing each other or a modified nucleotide was *vis-à-vis* a regular one. Some of these siRNA molecules mediated a protein knockdown in HeLa cells comparable to an all-ribo blunt end siRNA, but positional effects of 2-O-methyl modifications were observed. Similarly, an siRNA design with alternating blocks of five 2-O-methyl-modified nucleotides in a row followed by five regular ribose nucleotides proved to be as efficient as the parent all-ribo/dTdT-overhang siRNA.[105] In another study, six 2-O-methyl nucleotides were introduced into each siRNA strand (four at the 3-ends and two at the 5-ends) without losing significant activity. Increasing the number of 2-O-methyl modification in increments of two up to 28 out of 42 nucleotides in the siRNA duplex resulted in a gradual attenuation of inhibitory activity.[106]

Several groups studied the effects of 2-deoxyfluoro (2-F) modifications on siRNA function. Introducing up to four 2-F-uridines in the flanks of the duplex did not affect siRNA activity.[107] Replacing all pyrimidines in an siRNA with the corresponding 2-F nucleotides in either the sense strand (ten 2-F residues), the antisense strand (nine 2-F residues) or the whole siRNA had no effect on siRNA ability to inhibit gene expression in a HeLa cell system. Combining the 2-F modifications with partially or full replacement of purine residues by the corresponding 2-deoxy nucleotide led to a maximum of 50% reduction of siRNA activity. Restricting the deoxy-residues in the 2-F background to the 3- or the 5-half of the antisense strand demonstrates that the 5-region of the siRNA (defined by the antisense strand) reacts more sensitively towards these modifications.[101] Similarly, replacing all the 19 pyrimidines of a given siRNA with 2-F nucleotides and four 2-deoxythymidines on the overhangs did not affect activity in a HeLa cell system. Additional substitution of three phosphodiesters with PTL at the 3-end of each strand also did not influence the RNAi machinery significantly. Complete replacement of all phosphodiesters with PTLs on either strand or in the entire siRNA reduced silencing activity by roughly 50%.[107] Again, the antisense strand was more sensitive towards modification than the sense strand.[101] When up to 12 PTLs were incorporated at the flanks of the siRNA molecule, activity was marginally lower compared to unmodified siRNA.[106, 108] At the same time it was demonstrated that long stretches of consecutive PTLs caused toxicity in HaCaT cells.[106] Also siRNA with alternating phosphodiesters and phosphorothioates still displayed full silencing efficiency. The observed reduction in cell growth and viability was attributed to cytotoxic effects of the PTL modifications.[108]

### 5.2. *Improving siRNA's resistance towards nucleolytic degradation*

siRNA with sugar phosphate backbone modifications was tested for improved stability towards nucleolytic degradation and for prolonged inhibitory potency. Comparing the results of different research groups is difficult, as diverse assay systems were utilized to address siRNA stability. Complete 2-O-methylation rendered the siRNA serum nuclease-resistant, although gene silencing activity was almost completely abolished. Five nucleotide-blockwise modification of siRNA with 2-O-methyl nucleotides preserved activity and at the same time resulted in an increased stability. siRNA with every other nucleotide in both strands modified by a 2-O-methyl nucleotide were notably stabilized, but only the version with a modified nucleotide in one strand facing a nonmodified nucleotide in the other strand retained activity. Stabilization of siRNA by either blockwise or alternating 2-O-methyl modification results in a drastically prolonged target knock-down, with inhibition of target expression extended from 48 hours to 120 hours. After that time the parent unmodified siRNA had completely lost its ability to reduce protein expression[105]. The enhanced serum stability of the modified siRNA may be paralleled by an improved intracellular stability resulting in prolonged gene silencing. 2-O-methyl modifications could also change association kinetics of siRNA with the RNA-induced silencing complex (RISC) or directly influence the enzymatic activity of a siRNA-loaded RISC. 2-O-methylation of terminal siRNA residues also resulted in a prolonged inhibitory potency of the corresponding siRNA.[106] Although not demonstrated, an increased intracellular stability might be responsible for this observation.

The serum stability of single- and double-stranded phosphodiester RNA and the corresponding PTL containing counterparts was determined.[107] As expected, after a 30-second incubation in 5% fetal bovine serum, single-stranded RNA was no longer detectable. Strikingly, single-stranded phosphorothioate-RNA was also readily degraded in this assay system. Even more surprisingly, an unmodified siRNA was stable for up to 72 hours and this stability could not be further improved by a complete phosphorothiolation of both siRNA strands. On the other hand, upon incubation in a HeLa cell extract, the beneficial effect of phosphorothioates on the half-life of siRNA and single-stranded RNA was demonstrated.[101] Introduction of an inverted deoxy, a basic modification (iB) at the termini of the siRNA, which is known to protect nucleic acids against degradation by exonucleases, did not result in a protection against degradation by serum nucleases. Furthermore, siRNA carrying aminohexyl phosphodiester modifications on all termini was tested for stability in serum. Because of the rapid degradation of this "end-capped," two 2-deoxythymidine 3-overhang carrying siRNA in serum, it was reasoned that siRNA is predominantly degraded by serum endonucleases.[105]

## 5.3. *Fluorescent labeling of siRNA for the sensitive detection*

Despite the change in the overall physicochemical properties of siRNA caused by the attachment of bulky hydrophobic fluorescence dyes, these conjugates are widely used to address the following issues: (1) Analysis of cellular distribution of siRNA; (2) Assessment of transfection efficiency; (3) Expression attenuation; (4) Correlation of siRNA up-take with down-regulation of target protein (in combination with labeled antibodies). The 5-end of the sense strand is ideally suited to attach the fluorescent reporter group as the dye can be linked to the oligoribonucleotide during solid-phase synthesis yielding a full-length product which differs in the chromatographic properties from the ($n$–1) product, thus facilitating purification by means of reversed phase or ion exchange high-performance liquid chromatography. Fluorescent siRNA is commonly used to assess transfection efficiencies and to test for the optimal transfection reagent for a given cell type. Fluorescein and the Cy-dyes are preferentially selected for that purpose.

5-Cy3-labeled siRNA was delivered to Kasumi-1 cells to analyze electroporation efficacy as well as intracellular distribution. As judged by fluorescence-activated cell sorting (FACS) analysis, almost all cells harbored Cy3 fluorescence sixteen hours after transfection. Fluorescence microscopy revealed a cytoplasmic distribution of the fluorescent label, whereas the nuclear regions were only weakly stained.[108] The sub-cellular distribution of siRNA was monitored by confocal microscopy after lipofection of HeLa cells with 3-end Alexa488 labeled siRNA.[109] Fluorescence was localized to discrete foci on the cytoplasmic side of the nuclear membrane. Nevertheless, it was suspected that cellular uptake of fluorescently labeled siRNA correlates with gene silencing, as formation of small foci of concentrated fluorescent dyes in endosomal compartments were observed independent of a knockdown phenotype. It was also pointed out that cell sorting could be misleading when utilizing the transfected fluorophore on the siRNA in order to enrich the silenced-cell population.[109] siRNA fluorescently labeled on the 5-terminus of the sense strand was utilized in studies with non-adherent cell lines.[110] With the aid of fluorescein-tagged siRNA, transfection efficiencies of different commercially available transfection agents were investigated employing FACS analysis. For some cells, successful internalization of siRNA was not sufficient to achieve target silencing. An siRNA introduced into myeloma cells by means of lipofection did not result in gene silencing whereas the same siRNA delivered by electroporation triggered RNAi, indicating that these two different transfection techniques deliver the active agents to different cellular compartments.

Researchers from Ambion, Inc. used fluorescently labeled siRNA to track the molecules inside HeLa S3 cells. They demonstrated that the labeled and unlabeled siRNA performed equivalently in terms of target down-regulation and specificity. These properties were not affected by the label being placed either on the sense and/or the antisense strand. Furthermore, fluorescence resonance energy transfer experiments were employed to confirm strand separation of siRNA as part of the RNAi mechanism in transfected mammalian cells. These data nicely complement results where a correlation between target silencing and siRNA strand separation were observed.[111] Generally, fluorescence reporter groups are preferentially attached to at least one of the termini to follow the fate of siRNA inside cells employing fluorescence microscopy. As for radioactive phosphate end-labeled oligonucleotides, one should keep in mind that the fluorophore might potentially get cleaved by a cellular nuclease. As a consequence, analysis of the reporter may not yield the desired information on the parental molecule.

## Summary

Numerous studies indicate that RNAi could be efficient ways to investigate the gene work mechanisms and screen novel drug targets for therapeutics. The key problems which need further research are the highly efficient system delivery vectors for *in vivo* studies. Nanoparticles exhibited much promise due to their multifunctional properties and easily targeted modifications for intracellular siRNA delivery *in vitro* and *in vivo*. A substantial amount of chemical modification in both siRNA strands is tolerated by the RNAi machinery. These modifications have the potency to improve siRNA stability as well as pharmacokinetic properties of the molecule. Overall, we are optimistic that chemical modification in addition to conjugation strategies with biodegradable nanosystems for various biomolecules will play a pivotal role in the transition of siRNA from a laboratory knock-down tool into a valuable platform for the development of therapeutics towards clinical applications.

## Acknowledgments

This work was financially supported by the grant from National Science Foundation of China (NSFC NO. 81371682, 4342234007 and 43422340010), Shanghai Nano Project (43422360007), Shanghai Natural Science Foundation

(13ZR1443900), and The Foundation for Innovative Research Groups of the National Natural Science Foundation of China (81221001). We are also grateful for the helpful discussions with Professor Kensuke Egashira, MD, PhD from the Graduate School of Medical Sciences of Japanese Kyushu University, and Dr. Jidong Zhang from the Department of Biological Sciences, Graduate School of Science and Technology, Kumamoto University, 2-39-1 Kurokami 8608555, Japan.

## References

1. Sioud M. Nucleic acid enzymes as a novel generation of anti-gene agents. *Curr Mol Med* 2001;1:575–588.
2. Sioud M, Rensen DRS. Cationic liposome-mediated delivery of siRNAs in adult mice. *Biochem Biophys Res Commun* 2003;312:1220–1225.
3. Hannon GJ. RNA interference. *Nature* 2002;418:244–251.
4. Fire A, Xu S, Montgomery MK, Kostas SA, Driver SE, Mello CC. Potent and specific genetic interference by double-stranded RNA in *Caenorhabditis elegans. Nature* 1998;391:806–811.
5. Jorgensen R. Altered gene expression in plants due to trans-interaction between homologous genes. *Trends Biotechnol* 1990;8:340–344.
6. Elbashir SM, Harborth J, Lendeckel W, Yalcin A, Weber K, Tuschl T. Duplexes of 21-nucleotide RNAs mediate RNA interference in cultured mammalian cells. *Nature* 2001;411:494–498.
7. Dykxhoorn DM, Novina CD, Sharp PA. Killing the messenger: short RNAs that silence gene expression. *Nat Rev Mol Cell Biol* 2003;4:457–467.
8. Rensen DRS, Leirdal M, Sioud M. Gene silencing by systemic delivery of synthetic siRNAs in adult mice. *J Mol Biol* 2003;327:761–766.
9. Templeton SN. Liposomal delivery of nucleic acids *in vivo. DNA Cell Biol* 2002;21: 859–867.
10. Ulrich AS. Biophysical aspects of using liposomes as delivery vehicles. *Biosci Rep* 2002;22:129–150.
11. Zhu N, Liggitt D, Liu Y, Debs R. Systemic gene expression after intravenous DNA delivery in adult mice. *Science* 1993;261:209–211.
12. Thierry AR, Lunardi-Iskandar Y, Bryant JL, Rabinovich P, Gallo RC, Mahan LC. Systemic gene therapy: biodistribution and long-term expression of a transgene in mice. *Proc Natl Acad Sci* 1995;92:9742–9746.
13. Timme TL, Satoh T, Tahir SA. Therapeutic targets for metastatic prostate cancer. *Curr Drug Targets* 2003;4:251–261.
14. Harborth J, Elbashir SM, Vandenburgh K, Manninga H, Scaringe SA, Weber K, Tuschl T. Sequence, chemical, and structural variation of small interfering RNAs and short hairpin RNAs and the effect on mammalian gene silencing. *Antisense Nucleic Acid Drug Dev* 2003;13:83–105.
15. Liu Y, Liggitt D, Zhong W, Tu G, Gaensler K, Debs R. Cationic liposome- mediated intravenous gene delivery. *J Biol Chem* 1995;270:24864–24870.

16. Ma Z, Zhang J, Alder S, Dileo J, Negishi Y, Stolz D, Watkins S, Huang L, Pitt B, Li S. Lipid-mediated delivery of oligonucleotide to pulmonary endothelium. *Am J Respir Cell Mol Biol* 2002;27:151–159.
17. Kashani-Sabet M, Liu Y, Fong S, Desprez P-Y, Liu S, Tu G, Nosrati M, Handum-Rongkul C, Liggitt D, Thor AD, Debs RJ. Identification of gene function and functional pathways by systemic plasmid-based ribozyme targeting in adult mice. *Proc Natl Acad Sci* 2002;99:3878–3883.
18. Sledz CA, Holko M, de Veer MJ, Silverman RH, Williams BRG. Activation of the interferon system by short-interfering RNAs. *Nat Cell Biol* 2003;5:834–839.
19. Parsons P, Lang M, Steele R. Malignant ascites: a 2-year review from a technical hospital. *J Surg Oncol* 1996;22:237–239.
20. Sioud M. Ribozyme modulation of lipopolysaccharide-induced tumor necrosis factor-alpha production by peritoneal cells *in vitro* and *in vivo*. *Eur J Immunol* 1996;26: 1026–1031.
21. Kisich KO, Malone RW, Feldstein PA, Erickson KL. Specific inhibition of macrophage TNF-α expression by *in vivo* ribozyme treatment. *J Immunol* 1999;163:2008–2016.
22. Beutler BA. The role of tumour necrosis factor in health and diseases. *J Rheumatol* 1999;57:16–21.
23. Lewis DL, Hagstom G, Haley B, Zamore PD. Efficient delivery of siRNA for inhibition of gene expression in post-natal mice. *Nat Genet* 2002;32:107–108.
24. McCaffrey AP, Meuse L, Pham TT, Conklin DS, Hannon GJ, Kay MA. RNA interference in adult mice. *Nature* 2002;418:38–39.
25. Caplan L, Ceruti M, Dosio F. From conventional to stealth liposomes: a new frontier in cancer chemotherapy. *Tumori* 2003;89:237–249.
26. Khalil IA, Kogure K, Akita H. Uptake pathways and subsequent intracellular trafficking in non-viral gene delivery. *Pharmacol Rev* 2006;58:32–45.
27. Schiffelers RM, Ansari A, Xu J. Cancer siRNA therapy by tumor selective delivery with ligand-targeted sterically stabilized nanoparticle. *Nucleic Acids Res* 2004;32:e149.
28. Laroui H, Theiss AL, Yan Y. Functional TNFα gene silencing mediated by polyethyleneimine/TNFα siRNA nanocomplexes in inflamed colon. *Biomaterials* 2011;32: 1218–1228.
29. Abbasi S, Paul A, Prakash S. Investigation of siRNA-loaded polyethylenimine-coated human serum albumin nanoparticle complexes for the treatment of breast cancer. *Cell Biochem Biophys* 2011;61:277–287.
30. Liu C, Zhao G, Liu J. Novel biodegradable lipid nano complex for siRNA delivery significantly improving the chemosensitivity of human colon cancer stem cells to paclitaxel. *J Control Release* 2009;140:277–283.
31. Nguyen J, Steele TWJ, Merkel O. Fast degrading polyesters as siRNA nano-carriers for pulmonary gene therapy. *J Control Release* 2008;132:243–251.
32. Sureban SM, May R, Mondalek FG. Nanoparticle-based delivery of siDCAMKL-1 increases microRNA-144 and inhibits colorectal cancer tumor growth *via* a Notch-1 dependent mechanism. *J Nanobiotechnol* 2011;9:40.
33. Patil Y, Panyam J. Polymeric nanoparticles for siRNA delivery and gene silencing. *Int J Pharm* 2009;367:195–203.
34. Yuan X, Shah BA, Kotadia NK. The development and mechanism studies of cationic chitosan-modified biodegradable PLGA nanoparticles for efficient siRNA drug delivery. *Pharm Res* 2010;27:1285–1295.

35. Weber N, Ortega P, Clemente MI. Characterization of carbosilane dendrimers as effective carriers of siRNA to HIV-infected lymphocytes. *J Control Release* 2008;132:55–64.
36. Agrawal A, Min DH, Singh N. Functional delivery of siRNA in mice using dendriworms. *ACS Nano* 2009;3:2495–2504.
37. Beh CW, Seow WY, Wang Y. Efficient delivery of Bcl-2-targeted siRNA using cationic polymer nanoparticles: downregulating mRNA expression level and sensitizing cancer cells to anticancer drug. *Biomacromolecules* 2008;10:41–48.
38. Sun TM, Du JZ, Yan LF. Self-assembled biodegradable micellar nanoparticles of amphiphilic and cationic block copolymer for siRNA delivery. *Biomaterials* 2008;29: 4348–4355.
39. Sun TM, Du JZ, Yao YD. Simultaneous delivery of siRNA and paclitaxel *via* a "two-in-one" micelleplex promotes synergistic tumor suppression. *ACS Nano* 2011;5: 1483–1494.
40. Wilson DS, Dalmasso G, Wang L. Orally delivered thioketal nanoparticles loaded with TNF-α–siRNA target inflammation and inhibit gene expression in the intestines. *Nat mater* 2010;9:923–928.
41. Lee S, Yang SC, Kao CY. Solid polymeric microparticles enhance the delivery of siRNA to macrophages *in vivo*. *Nucleic Acids Res* 2009;37:e145.
42. Howard KA, Paludan SR, Behlke MA. Chitosan/siRNA nanoparticle-mediated TNF-α knockdown in peritoneal macrophages for anti-inflammatory treatment in a murine arthritis model. *Mol Ther* 2008;17:162–168.
43. Ji AM, Su D, Che O. Functional gene silencing mediated by chitosan/siRNA nanocomplexes. *Nanotechnol* 2009;20:405103.
44. Nielsen EJB, Nielsen JM, Becker D. Pulmonary gene silencing in transgenic EGFP mice using aerosolised chitosan/siRNA nanoparticles. *Pharm Res* 2010;27:2520–2527.
45. Howard KA, Rahbek UL, Liu X. RNA interference *in vitro* and *in vivo* using a chitosan/siRNA nanoparticle system. *Mol Ther* 2006;14:476–484.
46. Lee DW, Yun KS, Ban HS. Preparation and characterization of chitosan/polyguluronate nanoparticles for siRNA delivery. *J Control Release* 2009;139:146–152.
47. Han HD, Mangala LS, Lee JW. Targeted gene silencing using RGD-labeled chitosan nanoparticles. *Clin Cancer Res* 2010;16:3910–3922.
48. Heidel JD, Yu Z, Liu JYC. Administration in non-human primates of escalating intravenous doses of targeted nanoparticles containing ribonucleotide reductase subunit M2 siRNA. *Proc Natl Acad Sci* 2007;104:5715–5721.
49. Bartlett DW, Su H, Hildebrandt IJ. Impact of tumor-specific targeting on the biodistribution and efficacy of siRNA nanoparticles measured by multimodality *in vivo* imaging. *Proc Natl Acad Sci* 2007;104:15549–15554.
50. Qi L, Gao X. Quantum dot-amphipol nanocomplex for intracellular delivery and real-time imaging of siRNA. *ACS Nano* 2008;2:1403–1410.
51. Klein S, Zolk O, Fromm MF. Functionalized silicon quantum dots tailored for targeted siRNA delivery. *Biochem Biophys Res Commun* 2009;387:164–168.
52. Bonoiu A, Mahajan SD, Ye L. MMP-9 gene silencing by a quantum dot-siRNA nanoplex delivery to maintain the integrity of the blood brain barrier. *Brain Res* 2009;1282: 142–155.
53. Li J-M, Zhao M-X, Su H. Multifunctional quantum dot-based siRNA delivery for HPV18 E6 gene silence and intracellular imaging. *Biomaterials* 2011;32:7978–7987.
54. Qi L, Shao W, Shi D. JAM-2 siRNA intracellular delivery and real-time imaging by proton-sponge coated quantum dots. *J Mater Chem B* 2013;1:654–660.

55. Xiao YL, Jaskula-Sztul R, Javadi A. Co-delivery of doxorubicin and siRNA using octreotide-conjugated gold nanorods for targeted neuroendocrine cancer therapy. *Nanoscale* 2012;4:7185–7193.
56. Acharya S, Hill RA. High efficacy gold-KDEL peptide-siRNA nanoconstruct-mediated transfection in C2C12 myoblasts and myotubes. *Nanomedicine* 2014;10:329–337.
57. Conde J, Tian F, Hernández Y. *In vivo* tumor targeting *via* nanoparticle-mediated therapeutic siRNA coupled to inflammatory response in lung cancer mouse models. *Biomaterials* 2013;34:7744–7753.
58. Wei Y, Yin G, Ma C. Synthesis and cellular compatibility of biomineralized $Fe_3O_4$ nanoparticles in tumor cells targeting peptides. *Colloids Surf B Biointerfaces* 2013;107: 180–188.
59. Peng X, Chen J, Cheng T. PLGA modified $Fe_3O_4$ nanoclusters for siRNA delivery. *Mater Lett* 2012;81:102–104.
60. Qi L, Wu L, Zheng S. Cell-penetrating magnetic nanoparticles for highly efficient delivery and intracellular imaging of siRNA. *Biomacromolecules* 2012;13:2723–30.
61. Wang M, Li X, Ma Y. Endosomal escape kinetics of mesoporous silica-based system for efficient siRNA delivery. *Int J Pharm* 2013;448:51–57.
62. Gouda N, Miyata K, Christie RJ. Silica nanogelling of environment-responsive PEGylated polyplexes for enhanced stability and intracellular delivery of siRNA. *Biomaterials* 2013;34:562–570.
63. Li X, Chen Y, Wang M. A mesoporous silica nanoparticle–PEI–Fusogenic peptide system for siRNA delivery in cancer therapy. *Biomaterials* 2013;34:1391–1401.
64. Wong Y, Cooper HM, Zhang K. Efficiency of layered double hydroxide nanoparticle-mediated delivery of siRNA is determined by nucleotide sequence. *J Colloid Interface Sci* 2012;369:453–459.
65. Chen M, Cooper HM, Zhou JZ. Reduction in the size of layered double hydroxide nanoparticles enhances the efficiency of siRNA delivery. *J Colloid Interface Sci* 2013;390:275–281.
66. Li L,Gu W, Chen J. Co-delivery of siRNAs and anti-cancer drugs using layered double hydroxide nanoparticles. *Biomaterials* 2014;35:3331–3339.
67. Zuber J, Shi J, Wang E. RNAi screen identifies Brd4 as a therapeutic target in acute myeloid leukaemia. *Nature* 2011;478:524–528.
68. Qi L, Shao W, Shi D. JAM-2 siRNA intracellular delivery and real-time imaging by proton-sponge coated quantum dots. *J Mater Chem B* 2013;1:654–660.
69. Rudolph C, Plank C, Lausier J, Schillinger U, Müller RH, Rosenecker J. Oligomers of the arginine-rich motif of the HIV-1 TAT protein are capable of transferring plasmid DNA into cells. *J Biol Chem* 2003;278:11411–11418.
70. Suh J, Wirtz D, Hanes J. Efficient active transport of gene nanocarriers to the cell nucleus. *Proc Natl Acad Sci USA* 2003;100:3878–3882.
71. Field J, Vojtek A, Ballester R. Cloning and characterization of CAP, the *S. cerevisiae* gene encoding the 70 kD adenylyl cyclase-associated protein. *Cell* 1990;61:319–327.
72. Matviw H, Yu G, Young D. Identification of a human cDNA encoding a protein that is structurally and functionally related to the yeast adenylyl cyclase-associated CAP proteins. *Mol Cell Biol* 1992;12:5033–5040.
73. Freeman NL, Chen Z, Horenstein J, Weber A, Field J. An actin monomer binding activity localizes to the carboxyl-terminal half of the *Saccharomyces cerevisiae* cyclase-associated protein. *J Biol Chem* 1995;270:5680–5685.

74. Moriyama K, Yahara I. Human CAP1 is a key factor in the recycling of cofilin and actin for rapid actin turnover. *J Cell Sci* 2002;115:1591–1601.
75. Hubberstey AV, Mottillo EP. Cyclase-associated proteins: CAPacity for linking signal transduction and actin polymerization. *FASEB J* 2002;16:487–499.
76. Loisel TP, Boujemaa R, Pantaloni D, Carlier MF. Reconstitution of actin-based motility of Listeria and Shigella using pure proteins. *Nature* 1999;401:613–616.
77. Yamazaki K, Takamura M, Masugi Y. Adenylate cyclase-associated protein 1 overexpressed in pancreatic cancers is involved in cancer cell motility. *Lab Invest* 2009;89: 425–432.
78. Macrae IJ, Zhou K, Li F. Structural basis for double-stranded RNA processing by dicer. *Science* 2006;311:195–198.
79. Van Troys M, Huyck L, Leyman S, Dhaese S, Vandekerkhove J, Ampe C. Ins and outs of ADF/cofilin activity and regulation. *Eur J Cell Biol* 2008;87:649–667.
80. Elbashir SM, Harborth J, Lendeckel W, Yalcin A, Weber K, Tuschl T. Duplexes of 21-nucleotide RNAs mediate RNA interference in cultured mammalian cells. *Nature* 2001;411:494–498.
81. Caplen NJ, Parrish S, Imani F, Fire A, Morgan RA. Specific inhibition of gene expression by small double-stranded RNAs in invertebrate and vertebrate systems. *Proc Natl Acad Sci* 2001;98:9742–9747.
82. McCaffrey AP, Meuse L, Pham TT, Conklin DS, Hannon GJ, Kay MA. RNA interference in adult mice. *Nature* 2002;418:38–39.
83. Lewis DL, Hagstrom JE, Loomis AG, Wolff JA, Herweijer H. Efficient delivery of siRNA for inhibition of gene expression in postnatal mice. *Nat Genet* 2002;32: 107–108.
84. Xia H, Mao Q, Paulson HL, Davidson BL. siRNA-mediated gene silencing *in vitro* and *in vivo*. *Nat Biotechnol* 2002;20:1006–1010.
85. Song E, Lee SK,Wang J, Ince N, Ouyang N, Min J, Chen J, Shankar P, Lieberman J. RNA interference targeting Fas protects mice from fulminant hepatitis. *Nat Med* 2003;9:347–351.
86. Rubinson DA, Dillon CP, Kwiatkowski AV, Sievers C, Yang L, Kopinja J, Rooney DL, Ihrig MM, McManus MT, Gertler FB, Scott ML, van Parijs LA. Lentivirus-based system to functionally silence genes in primary mammalian cells, stem cells and transgenic mice by RNA interference. *Nat Genet* 2003;33:401–406.
87. Cook PD. Making drugs out of oligonucleotides: a brief review and perspective. *Nucleosides Nucleotides* 1999;18:1141–1162.
88. Manoharan M. 2'-carbohydrate modifications in antisense oligonucleotide therapy: importance of conformation, configuration and conjugation. *Biochim Biophys Acta* 1999;1489:117–130.
89. Crooke ST. *Antisense Drug Technology*. Marcel Dekker Inc., Basel, Switzerland, 2001.
90. Opalinska JB, Gewirtz AM. Nucleic-acid therapeutics: basic principles and recent applications. *Nat Rev Drug Discov* 2002;1:503–514.
91. Wincott F, DiRenzo A, Shaffer C, Grimm S, Tracz D, Workman C, Sweedler D, Gonzalez C, Scaringe S, Usman N. Synthesis, deprotection, analysis and purification of RNA and ribozymes. *Nucleic Acids Res* 1995;23:2677–2684.
92. Sun LQ, Cairns MJ, Saravolac EG, Baker A, Gerlach WL. Catalytic nucleic acids: from lab to applications. *Pharmacol Rev* 2000;52:325–347.

93. Jarvis TC, Wincott FE, Alby LJ, McSwiggen JA, Beigelman L, Gustofson J, DiRenzo A, Levy K, Arthur M, Matulic-Adamic J, Karpeisky A, Gonzalez C, Woolf TM, Usman N, Stinchcomb DT. Optimizing the cell efficacy of synthetic ribozymes. Site selection and chemical modifications of ribozymes targeting the protooncogene c-myb. *J Biol Chem* 1996;271:29107–29112.
94. Beigelman L, McSwiggen JA, Draper KG, Gonzalez C, Jensen K, Karpeisky AM, Modak AS, Matulic-Adamic J, DiRenzo AB, Haeberli P, Sweedler D, Tracz D, Grimm S, Wincott F, Thackray VG, Usman N. Chemical modification of hammer head ribozymes. Catalytic activity and nuclease resistance. *J Biol Chem* 1995;270: 25702–25708.
95. Heidenreich O, Benseler F, Fahrenholz A, Eckstein F. High activity and stability of hammer head ribozymes containing 2'-modified pyrimidine nucleosides and phosphorothioates. *J Biol Chem* 1994;269:2131–2138.
96. Eckstein F. Developments in RNA chemistry, a personal view. *Biochimie* 2002;84: 841–848.
97. Kurreck J. Antisense technologies. Improvement through novel chemical modifications. *Eur J Biochem* 2003;270:1628–1644.
98. Usman N, Blatt LM. Nuclease-resistant synthetic ribozymes: developing a new class of therapeutics. *J Clin Invest* 2000;106:1197–1202.
99. Beaucage SL, Radhakrishnan PI. Advances in the synthesis of oligonucleotides by the phosphoramidite approach. *Tetrahedron* 1992;48:2223–2311.
100. Wincott F, DiRenzo A, Shaffer C, Grimm S, Tracz D, Workman C, Sweedler D,Gonzalez C, Scaringe S, Usman N. Synthesis, deprotection, analysis and purification of RNA and ribozymes. *Nucleic Acids Res* 1995;23:2677–2684.
101. Chiu Y.L, Rana TM. siRNA function in RNAi: a chemical modification analysis. *RNA* 2003;9:1034–1048.
102. Parrish S, Fleenor J, Xu S, Mello C, Fire A. Functional anatomy of a dsRNA trigger: differential requirement for the two trigger strands in RNA interference. *Mol Cell* 2000;6: 1077–1087.
103. Elbashir SM, Martinez J, Patkaniowska A, Lendeckel W, Tuschl T. Functional anatomy of siRNAs for mediating efficient RNAi in *Drosophila melanogaster* embryo lysate. *Eur Mol Biol Organ J* 2001;20:6877–6888.
104. Boutla A, Delidakis C, Livadaras I, Tsagris M, Tabler M. Short 5-phosphorylated double-stranded RNAs induce RNA interference in Drosophila. *Curr Biol* 2001;11:1776–1780.
105. Czauderna F, Fechtner M, Dames S, Aygun H, Klippel A, Pronk GJ, Giese K, Kaufmann J. Structural variations and stabilising modifications of synthetic siRNAs in mammalian cells. *Nucleic Acids Res* 2003;31:2705–2716.
106. Amarzguioui M, Holen T, Babaie E, Prydz H. Tolerance for mutations and chemical modifications in a siRNA. *Nucleic Acids Res* 2003;31:589–595.
107. Braasch DA, Jensen S, Liu Y, Kaur K, Arar K, White MA, Corey DR. RNA interference in mammalian cells by chemically modified RNA. *Biochem* 2003;42:7967–7975.
108. Heidenreich O, Krauter J, Riehle H, Hadwiger P, John M, Heil G, Vornlocher H-P, Nordheim A. AML1/MTG8 oncogene suppression by small interfering RNAs supports myeloid differentation of t(8;21)-positive leukemic cells. *Blood* 2003;101:3157–3163.
109. Harborth J, Elbashir SM, Vandenburgh K, Manninga H, Scaringe SA, Weber K, Tuschl T. Sequence, chemical, and structural variation of small interfering RNAs and short hairpin

RNAs and the effect on mammalian gene silencing. *Antisense Nucleic Acid Drug Dev* 2003;13:83–105.
110. Walters DK, Jelinek DF. The effectiveness of double-stranded short inhibitory RNAs (siRNAs) may depend on the method of transfection. *Antisense Nucleic Acid Drug Dev* 2002;12:411–418.
111. Nykanen A, Haley B, Zamore PD. ATP requirements and small interfering RNA structure in the RNA interference pathway. *Cell* 2001;107:309–321.

# Chapter 10

# Chemical and Biological Sensors based on Nanowire Transistors

*Xiaohan Wu* and Jia Huang*,†*

**School of Materials Science and Engineering, Tongji University, Shanghai 201804, P. R. China*
*†The Institute for Biomedical Engineering and Nano Science, Tongji University School of Medicine, Shanghai 200092, P. R. China*

Semiconducting nanowires have attracted attention as being excellent sensing materials due to their superior physical properties such as high surface-to-volume ratio and tunable electrical properties. Chemical and biological sensors based on nanowire transistors perform extremely well, with ultra-high sensitivity and selectivity. This chapter reviews the application of semiconducting nanowire transistors in chemical and biological sensors, as well as their synthesis and surface functionalization. It also discusses sensor measurements and key parameters for optimum performance, including sensor sensitivity, selectivity and reusability.

## 1. Introduction

Transistors provide an ideal platform for building chemical and biological sensors. The miniaturization of transistors is well advanced in the semiconductor industry, so that transistors can now be integrated easily into portable electronic devices at low cost, without the need for heavy equipment. A field-effect transistor (FET) is an electronic device that consists of a source electrode, drain electrode, gate electrode, semiconductor and dielectric layer (**Figure 1**). When a gate voltage is applied between the gate and source electrodes, an electric field is induced across the dielectric layer and semiconductor material. The conductance of the semiconductor is controlled by the

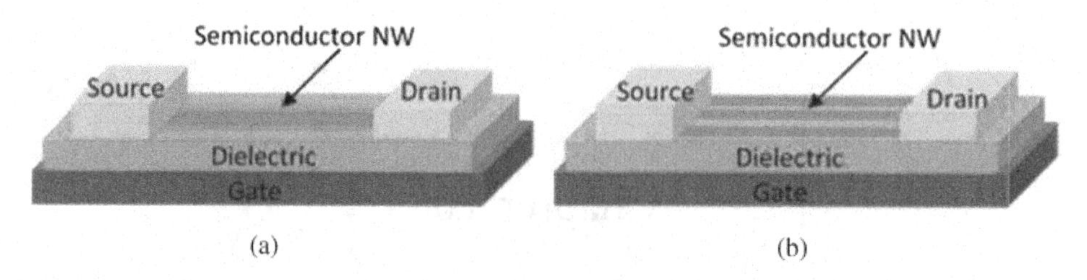


**Figure 1.** Device configuration of (**a**) a single-NW-based FET sensor and (**b**) multiple-NW-based FET sensor.

electric field. FETs operate by modulating the flow of charge carriers in the semiconductor. Using p-type semiconductors as an example, application of a positive gate voltage depletes the positive charge carriers and reduces conductance of the semiconductor. Inversely, a negative gate voltage leads to a higher density of positive carriers and an increase in conductance. FET sensors include nanowires (NWs) that serve as both the semiconductor materials and the sensing materials, in the form of either a single-NW-based FET sensor (**Figure 1(a)**) or a multiple-NW-based sensor (**Figure 1(b)**).[2] In these devices, adsorption of a charged analyte to the surface of the NWs can induce changes in charge carrier density, charge carrier mobility, and conductance of the semiconductor NWs. This interaction between the analyte and semiconductor surface leads to changes in the output current of FETs, which are dependent on the physical and chemical properties and concentrations of the analyte.

Nanostructural semiconductors exhibit properties that make them superior to their bulky counterparts, including electrical, magnetic, optical, thermoelectric and chemical traits. These unique properties can be attributed to many factors, including the quantum effects of electronic conduction, high density of electronic states, diameter-dependent band gap, enhanced surface scattering of electrons and phonons, increased excitation binding energy, high surface to volume ratio, and large aspect ratio.[3,5] Therefore, nanostructural semiconductors have a vast range of potential applications in electronic (transistors, logic devices, diodes), photonic (laser, photodetector), biological (sensors, drug delivery), energy (batteries, solar cells, thermoelectric generators), and magnetic (spintronic, memory) devices.[6]

For sensor applications based on FETs, nanostructural semiconductors have many advantages over conventional materials. Firstly, their high surface-to-volume ratio can enhance the sensitivity of FET sensors. Events occurring at the surface of semiconductor NWs can lead to larger signal changes than those in a bulky structure made of the same material. Due to the large

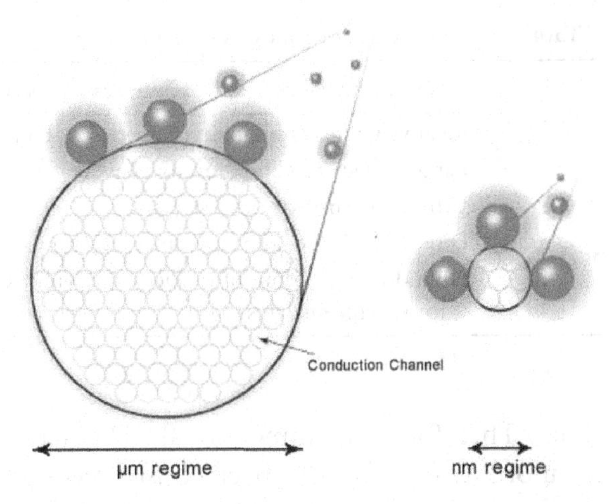


**Figure 2.** The effect of diameter on conductance change in a wire.[7] Copyright 2008 Multidisciplinary Digital Publishing Institute (MDPI).

surface-to-volume ratio, most of the atoms in NWs are capable of sensing the events that occur at the interface or surface. Thus, a higher shift from the output signal baseline is expected from the nanostructure-based devices than from traditional ones.[8] A sensor device based on a thick wire has a relatively small surface-to-volume ratio (**Figure 2**). When analytes approach the wire surface, only the conductance of semiconductors near the surface is affected, while the large interior volume of the wire might not be influenced by analytes.[7] If the wire diameter is reduced to nanometer scale (**b**), most atoms in the wire are close to the wire surface, and the surface-to-volume ratio is much higher. Then, most of the NW's interior area can be influenced by the external electrical field induced by analytes, so that the change in its conductance (which can be measured as the sensor response) is greatly enhanced. Secondly, nanostructures enable the fabrication of sensors at a higher device density on a given chip,[6] which allows for more efficient control of charge carriers in the channel of FET devices, and for a more compact sensor unit. Finally, FET sensors have been fabricated successfully with various nanostructures such as nanoparticles (NPs), nanobelts (NBs), nanorods (NRs), nanotubes (NTs), and NWs, made of a wide range of semiconductor materials including element, compound semiconductors and metal oxide semiconductors. This diversity offers a wide array of options in FET devices, for sensing different chemical or biological analytes.[2]

Among these nanostructures, NPs typically have a relatively higher surface-to-volume ratio, and hence generally exhibit higher sensitivity for analyte detection, but their sensing properties often degrade over time due to the

**Table 1.** Advantages of nanowires for FET sensors

| | |
|---|---|
| Preparation | Simple synthesis methods; Large-scale production |
| Properties | Relatively high crystallinity; Superior stability<br>Large surface to volume ratios; Fast reaction kinetics |
| Performance of NW sensors | Ultrahigh sensitivity and selectivity<br>Direct, label-free, and real-time detection<br>High density loading of surface modifying receptor species for specific analytes |

aggregation of NPs. Therefore, one-dimensional (1D) nanostructures like NWs and carbon nanotubes (CNTs) have been considered as the most favorable candidates for future sensors because of their considerable advantages[2] (**Table 1**). Sensors based on CNTs have also been studied extensively — they have some similar advantages as NWs in terms of morphological and electrical properties, but the difficulty in control of orientation, chirality, and electronic band structure limits their practical application.[8] Furthermore, NWs can be synthesized with high reproducibility, and they offer a rich chemistry and leave enough scope for fine-tuning their band structure through controlled doping and compositional variation.[8] Therefore, this chapter focuses on the application of NW transistors in chemical and biological sensors, including the synthesis of NWs, their surface modification, sensing applications of NW-based FETs, and the relevant parameters.

## 2. Synthesis of Semiconductor Nanowires

The growth of needle-shaped Silicon (Si) whiskers was demonstrated as early as 1964, with a vapor-phase method,[2] but enormous progress in the fabrication of nanostructures did not occur until Morales and Lieber's work on the synthesis of single-crystalline Si NWs with the laser ablation method in 1998.[1] The miscellaneous fabrication processes of 1D nanostructure semiconductors can be sorted into several strategies with different levels of control over the growth parameters (**Table 2**). These strategies include the use of anisotropic growth of crystals, the introduction of a solid–liquid interface for symmetry reduction of a seed, and the use of templates for direct formation of 1D structures within pores.[2,3]

The vapor phase synthesis provides an easy and cost-effective way to produce numerous single-crystalline 1D nanomaterials with a diameter down to the scale of 10 nm, and it is also capable of producing NWs on a large scale.[1] Thermal and

**Table 2.** General synthetic methods of inorganic 1D nanostructure.

| | | |
|---|---|---|
| Vapor phase synthesis | Thermal evaporation CVD/ Laser ablation | Metal seeded/Non-metal seeded |
| Liquid phase synthesis | Crystal structure-governed nucleation in solution | |
| | Metal-assisted growth in liquids | |
| | Template synthesis | Hard template |
| | | Soft template |
| | Electro spinning | |

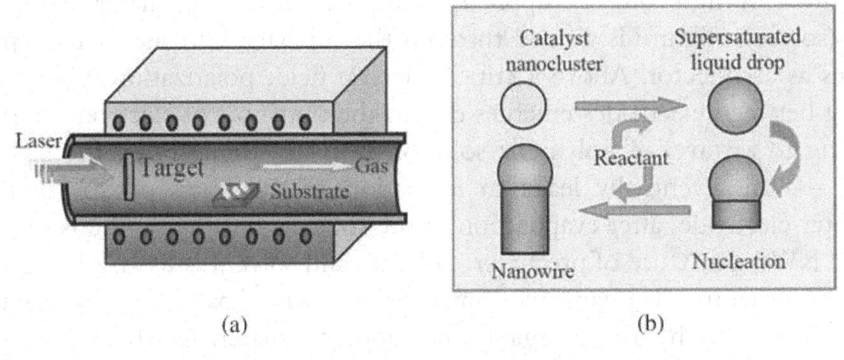


**Figure 3.** (**a**) Schematic diagram of a laser-assisted CVD setup and (**b**) Illustration of the vapor–liquid–solid (VLS) mechanism.[1] Copyright 2008 Institute of Electrical and Electronics Engineers (IEEE).

laser induced evaporation and chemical vapor deposition (CVD) are technologies widely used to grow NWs *via* the vapor phase process (e.g., **Figure 3(a)**). Clean substrates are coated with metal clusters and then placed into a quartz tube at the downstream end of a furnace, where the metal clusters function as catalysts for NW growth. A target is placed at the upstream end of the furnace, and then laser ablated to supply the vapor source. The vapor first diffuses into the catalytic particles and forms liquid drops (**Figure 3(b)**). Continued addition of vapor into the drops brings the alloy beyond super-saturation, leading to nucleation. Depending on the choice of materials, the nucleated material may or may not react with other vapors in the furnace to form semiconducting NWs. Further supply of the target vapor feeds NW growth. The diameter of the NW can be controlled by modifying the catalytic particle size.[1]

Among all liquid phase NW synthesis methods, the anodic aluminum oxide (AAO) membrane-assisted method is one of the most attractive. It can produce NWs with the desirable qualities of regular pore distribution, high pore density and a high aspect ratio of pores. The AAO templates are prepared using controlled electrochemical anodization, cooperating with oxidation of pure aluminum. The desired distribution, size and length of pores in the AAO templates can be obtained by adjusting anodization parameters including voltage, current density, temperature and composition. Metals and semiconductors can then be injected into the pores of an AAO template, and solidified to form NWs. The resulting wires can be collected by chemically dissolving the alumina template. The main advantage of this approach is the ability to synthesize NWs with a large number of various semiconductors. However, its drawback is the predominantly polycrystalline nature of the material.[3]

Electro-spinning has been commonly used as a method to produce polymer fibers. A high voltage is applied to a capillary and a metallic plate. A polymer/solvent mixture is guided through the capillary, and the metallic plate serves as a collector. Above a critical electric field, polarization of the fluid string between electrodes emerges due to the transport of electrode carriers, leading to a travel of polymeric solution from the capillary to the metallic plate — this eventually leads to formation of the fibers on the metallic counter-electrode, after evaporation of the solvent. For the synthesis of inorganic NWs, a mixture of precursor, polymer and solvent is used. The precursors are either metal-organic monomers or pre-formed particles. The resulting spun fibers are a hybrid of organic and inorganic materials, which are generally used after transfer into a purely inorganic medium *via* calcinations.[3] For example; Zinc Oxide (ZnO) NWs are routinely fabricated with this electro-spinning technique (**Figure 4**).

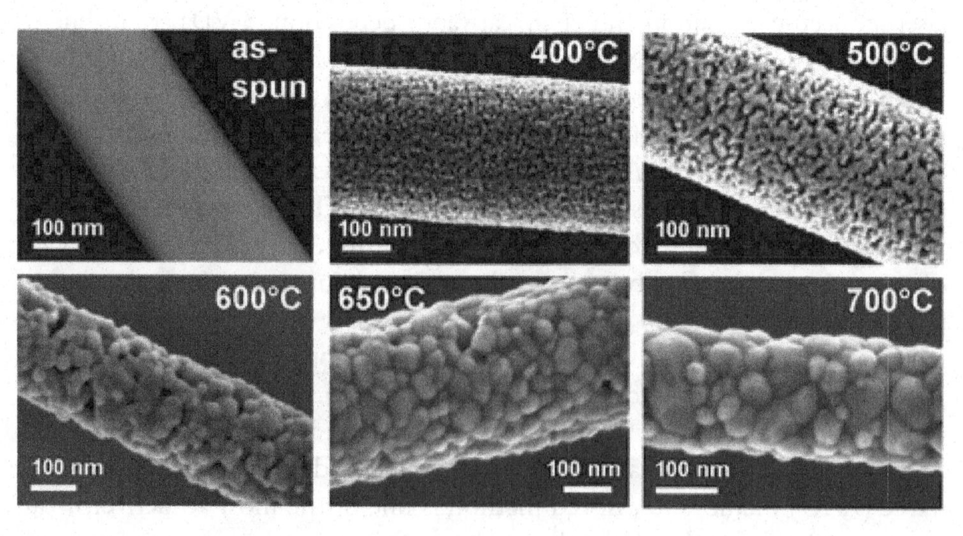


**Figure 4.** SEM images of crystal growth in ZnO NWs calcined at various temperatures for 4 hours in air. The ZnO NWs were initially fabricated by electro-spinning.[9] Copyright 2013 Royal Society of Chemistry (RSC).

Enormous semiconductors have been used as sensors in the form of 1D nanostructure because of their outstanding performance. Some examples of the synthetic strategies of several common semiconductors, and their applications in a broad range of chemical and biological sensors, are illustrated in **Table 3**. It should be noted that the detection limits of these sensors increase markedly with the introduction of nanostructures. For instance, the detection limit of Indium Oxide ($In_2O_3$) NW FET for Nitrogen Dioxide ($NO_2$) is 20 ppb, while bulky $In_2O_3$ is almost insensitive to $NO_2$.[2]

**Table 3.** Synthetic strategies and general sensing applications of semiconducting 1D nanostructures.[1–4]

| Semiconductor | | Synthetic strategy | 1D Nanostructure | Sensing application |
|---|---|---|---|---|
| Element | Si | CVD | NWs/NBs | Chemical sensor: $NH_3$, pH, Sodium, VOCs; Biosensor: ATP, DNA, Glucose, Glutathione, IgG/IgA, PSA |
| | | Laser ablation | NWs | |
| | | MBE | NWs | |
| | | Physical transport | NWs/NBs | |
| | | Template synthesis | NWs/NTs | |
| | | Solvent/SCF | NWs | |
| | Ge | CVD | NWs | |
| | | Laser ablation | NWs | |
| | | Physical transport | NWs | |
| | | Template synthesis | NWs | |
| | | Solvent/SCF | NWs | |
| Metal oxide | ZnO | CVD | NWs | Chemical sensor: $H_2$, CO, $O_3$, $H_2S$, NO, $N_2O$, $NO_2$, $NH_3$, HCHO, Et-OH, LPG |
| | | Laser ablation | NWs | |
| | | MBE | NWs | |
| | | Physical transport | NWs/Rings | |
| | | Carbothermal | NWs/NBs | |
| | | Template synthesis | NWs | |
| | | Solvent | NWs | |
| | | Electrospinning | NWs | |
| | $SiO_2$ | CVD | NWs/NBs | Chemical sensor: $H_2$, CO, $O_3$, $NO_2$, $NH_3$, DMMP, Et-OH, Humidity |
| | | Laser ablation | NWs/NBs | |
| | | Physical transport | NBs | |
| | | Carbothermal | NWs/NBs | |
| | | Template synthesis | NWs | |
| | | Electrospinning | NWs | |
| Compound | InAs | CVD | NWs | Chemical sensor: $H_2$, $CH_4$, Flow rate sensor |
| | GaN | CVD | NWs/NBs | |
| | | Laser ablation | NWs | |
| | | Template synthesis | NWs/NTs | |

## 3. Surface Modification of Nanowires

To further improve sensor selectivity and sensitivity, the NW surface can be modified with a variety of linker molecules or bioaffinitive agents.[8] Such surface modification of the NW can be classified into three categories based on the different materials used: inorganic nanostructures, organic molecules and bio receptors.

Inorganic NPs can be decorated on NWs with simple deposition methods such as sputtering, CVD, and self-assembling. Metal NPs have been affixed to ZnO NWs *via* electrochemical reduction of the desired metal salt.[10]

The covalent attachment of organic molecules and biological receptors onto NWs usually requires the use of linker molecules. Selection of the appropriate linker molecule depends on the surface chemistry of the NWs. In the case of SiNWs coated with a native Silicon Dioxide ($SiO_2$) layer, alkoxysilane derivatives are the preferred linker molecules. For example, 3-(trimethoxysilyl) propyl aldehyde (TMSPA) is a widely used linker that enables the $SiO_2$-coated SiNW to generate an aldehyde-rich surface,[8,11] while a surface coated with amino groups can be established using 3-aminopropyltriethoxysilane (APTES).[8,12] SiNWs that do not have a $SiO_2$ coating can be activated with olefin derivatives *via* UV-mediated hydrosilylation. Creation of an amino group coating on a SiNW surface is normally achieved using 10-N-Boc-amino-dec-1-ene followed by deportation. For metal oxide NWs, phosphonate derivatives are the optimal linker molecules. For instance, 3-phosphonopropanoic acid binds strongly to the metal oxide NW in aqueous solutions and in polar solvents.[8] Activation of these groups enables organic molecules and bioreceptors to attach to the linker molecules that are bound to the NW surface.[8,13]

When NW FET sensors are used in biomedical diagnosis of a particular target such as an antigen, the corresponding biological receptors such as antibodies are usually linked onto the NW surface. When there is a strong and specific binding affinity between the target antigen and antibody, the receptor-functionalized NW FETs can serve as biological sensors with ultra high selectivity (e.g., **Figure 5(a)**). The bioreceptors attached to the semiconductor NWs can facilitate recognition of the target analytes, with their high specificity and strong binding affinity.[14] The interaction between receptors and analytes can change the surface potential of the semiconductor NW and consequently change the FET output current, which can be interpreted as a sensor output signal.[7]

SiNW FETs that have been functionalized with bioreceptors are used widely as biosensor platforms. Due to their large surface-to-volume ratio,

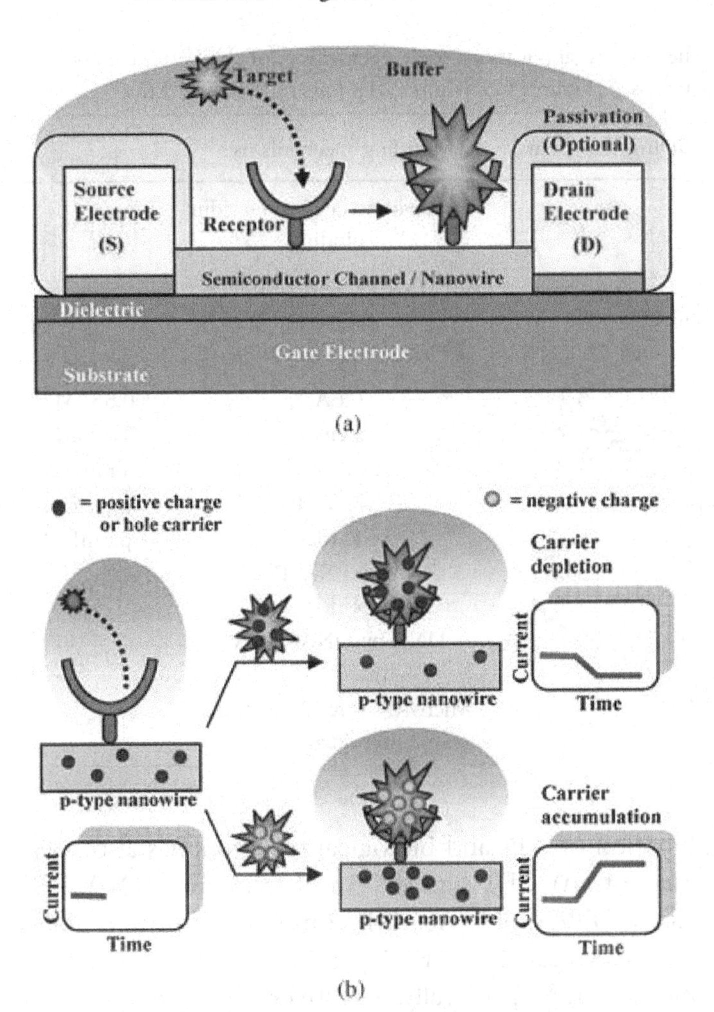


**Figure 5.** (**a**) Illustration of a NW FET biosensor with biological receptors. Target analytes can be recognized by a receptor attached to the NW surface *via* a strong binding affinity. (**b**) A p-type NW FET sensor. When positively charged analytes bind to the receptor, positive carriers (holes) are depleted in the semiconductor NW, decreasing the FET output current. Negatively charged analytes induce an accumulation of hole carriers in semiconductor NWs, increasing the FET output current.[14] Copyright 2008 Institute of Electrical and Electronics Engineers (IEEE).

tunable electrical properties, and biocompatibility, SiNW FET sensors exhibit ultra-high sensitivity and selectivity for the direct detection of a wide range of analytes, such as metal ions, nucleic acids, proteins, DNA, small molecule–protein interactions, cells, and viruses. The performance of a few types of biosensors, based on SiNW FETs that are functionalized by various receptors with bioaffinity, is summarized in **Table 4**.[15]

**Table 4.** The sensing applications and mechanisms of SiNW FETs modified with various bioaffinitive agents.[15] (Figures Copyright 2013 Future Medicine Ltd).[16]

| | Modifying bio agents | Binding mechanisms | Detection Limit | References |
|---|---|---|---|---|
| SiNW | Ligand of protein | Streptavidin via Biotin–avidin binding | 10 pM | 17 |
| | | TnI by Protein–protein interaction | 7 nM | 18 |
| | Antibody | PSA | 30 aM | 19 |
| | | CEA | 0.55 fM | 11 |
| | | Mucin-1 | 0.49 fM | 11 |
| | | cTnT | 1 fg mL–1 | 20 |
| | | CK-MM | 1 pg mL–1 | 21 |
| | | CK-MB | 1 pg mL–1 | 21 |
| | DNA | DNA by PNA–DNA hybridization | 10 fM | 22 |
| | | DNA by DNA–DNA hybridization | 10 pM | 23 |
| | | MicroRNA by PNA–RNA hybridization | 1 fM | 24 |

A combination of NPs and biological receptors has also been utilized in the production of NW FET sensors. An ultrasensitive SiNW FET biosensor modified with a DNA–gold (Au) NP complex has proven ability to detect matrix metalloproteinase-2 (MMP-2), and has attracted particular interest as a protein biomarker.[25] Specifically, negatively charged DNA–Au NP complexes coupled with peptide were attached to the SiNWs as receptors, which were used to enhance the conductance change of the p-type SiNW *via* MMP-2 cleavage reaction of the specific peptide. MMP-2 was measured successfully within a range of 100 fM to 10 nM, and the conductance signal of the p-type SiNW induced by the MMP-2 cleavage reaction was enhanced over 10-fold with use of the DNA–Au NP complexes, compared to control samples. This newly developed SiNW FET sensor possesses several advantages, such as good analytical performance, clear conductance response, high sensitivity, and low detection limits within an appropriate linear range.[25]

## 4. Sensing Applications of Nanowire Transistors

In practice, NW FET sensors have been used to detect physical changes, chemical and biological analytes in a wide range of applications (**Table 5**).

Table 5. A list of target objects that have been sensed using NW transistors.

| | Target object for sensing by NWs | Types of NW Applied | Ref. |
|---|---|---|---|
| Chemical analytes | Gas ($H_2$, $O_2$, $O_3$, CO, $CO_2$, $CH_4$, NO, $N_2O$, $NO_2$, $NH_3$, $H_2S$, $SO_2$) | Si, ZnO, $In_2O_3$, $WO_3$, $TiO_2$, $SnO_2$, $V_2O_5$, GaN, CuPC NWs | 1,26–28 |
| | n-hexane, n-octane, n-decane, ethanol, 1-hexanol, 1-octanol, 1-decanol, buthylether, cyclohexanone, 1,3,5-trimethylbenzene, chlorobenzene, water | SiNWs modified with silane molecules containing various end groups and chain lengths | 29–31 |
| | pH | Coaxial-structured ZnO/ SiNWs | 32 |
| | Cation | Gold-coated SiNW | 33 |
| | Humidity | Single Sb-doped $SnO_2$ NW | 34 |
| Biological analytes and events | Antigen (PSA, CEA, Mucin-1, cTnT, CK-MM, CK-MB) | SiNW modified with corresponding antibody | 15,16,35 |
| | Protein and interaction between proteins | SiNW modified with corresponding protein | 7,36 |
| | DNA hybridization | SiNW modified with corresponding DNA | 7,37 |
| | Carbohydrate–protein interactions | SiNW array modified with carbohydrates | 38 |
| | Uric acid | ZnO NW coated by enzyme | 39 |

Typically, gas analytes can be classified into reductive gases such as Hydrogen ($H_2$), Carbon Monoxide (CO), Carbon Dioxide ($CO_2$), Hydrogen Sulfide ($H_2S$), Ammonia ($NH_3$) and Methane ($CH_4$), and oxidative ones such as $NO_2$, Nitric Oxide (NO), Oxygen ($O_2$) and Ozone ($O_3$). When reductive gas molecules adsorb to the surface of a NW, such as ZnO NWs, they react with adsorbed $O_2^-$, $O^-$ or $O^{2-}$ ions and release electrons to the ZnO NW[40] (**Figure 6(a)**). The large specific surface area and high binding site density enable the NW to absorb a large quantity of gas molecules, generating a measurable change in the FET output current that functions as a sensor response.[40] When the reductive gas molecules approach ZnO NWs, the released electrons increase the electron concentration in the NWs, so that the depletion layer becomes thinner (**Figure 6(a)**). A thinner depletion layer then leads to higher conductivity of the wires. The change in current or conductance can be monitored by electrical instruments. A similar mechanism is at play for the detection of oxidative gas using ZnO NW-based FET sensors. Once adsorbed on the surface of ZnO NWs, the oxygen atom of oxidative

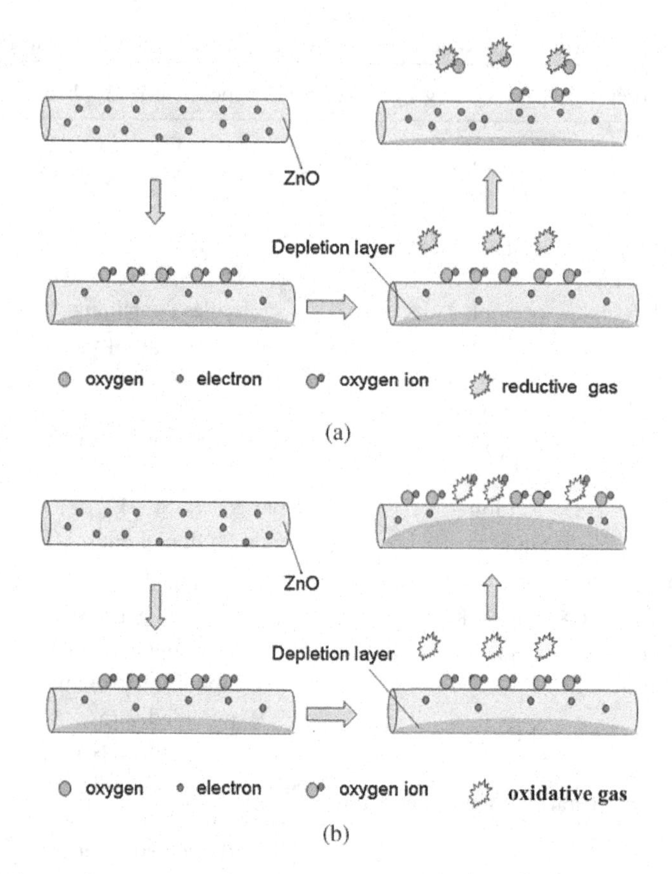


**Figure 6.** Scheme showing detection by ZnO-based sensors of (**a**) reductive gases and (**b**) oxidative gases.[40] Copyright 2011 Elsevier BV.

gas molecules extracts electrons from wires. In this case, the lowered electron density results in a thicker depletion layer in the wires (**Figure 6(b)**), leading to a lower conductance of ZnO NWs and thus a measurable decrease in the FET output current.[40]

## 4.1. *The measurement of the sensors*

In the measurement of NW FET sensors, analyte-semiconductor or analyte-receptor interaction is transduced into an electrical output signal. A change in the signal is usually measured using constant bias conditions, in which the gate voltage ($V_G$) and the source-drain voltage ($V_{DS}$) are set to facilitate current flow within the semiconductor NWs. The approach of an analyte toward the sensor is converted to a response in source-drain current, $\Delta I_{DS}$. Compared to simple chem-resistors, an important advantage of FET sensors is their ability to modulate the electrical signal and provide a more dynamic sensing window with a built-in signal amplifier. Multi-parameters can be obtained from a FET sensor, which means the analytes can be monitored *via* variation in numerous parameters, including conductivity, threshold voltage ($V_T$), and mobility ($\mu$), so that more information about the analytes can be obtained from the sensor response.[41]

Two testing methods are used widely to transduce the analyte event into an electrical signal: real-time detection and steady-state measurement.[15] Consider a biosensor integrated with a fluid channel, as an example of real-time detection: a buffer solution is injected into the device, and a baseline is established by recording the starting output current through the NW FET. The bio-analyte solution is then added into the device, resulting in a change in conductivity or output current. However, it is sometimes difficult to directly detect analytes in real-time, due to the screening effect of bio-analyte solution with high ionic strength (e.g., untreated serum with a short Debye length). In this case, a steady-state measurement method can be used, in which the sample is prepared in an aqueous buffer solution with very weak ionic strength, and then added onto the device.[15]

Multiple analytes and contaminants are often mixed in detections, such as volatile organic compounds (VOCs), or bio-analytes contained in blood or uric samples. To overcome this problem, sensor arrays using multiple types of NWs, or NWs modified with diverse receptors, have been developed as powerful tools for complicated detections, as well as to improve the performance of sensors by conferring better sensitivity and selectivity. As examples, **Figure** 7(**a**) shows the detection of 12 common VOCs by nine different sensors based on SiNWs modified with silane molecules containing various end groups and chain lengths,[29] and **Figure 7(b)** illustrates the detection of multiple cancer markers with SiNW FET arrays.[11] The SiNW FET array is composed of three independent devices containing different antibodies that are specific to the three different cancer markers. Real-time detection shows that each of the three cancer markers can only induce signals to the specific SiNW FET modified with the relevant cognate antibodies[7] — this demonstrates that the sensors are highly selective.

Calibration of such sensors for practical applications is challenging because of device-to-device variation in sensor parameters and in analytical performance of NW sensors (e.g., sensitivity and signal-to-noise ratio). One successful strategy to overcome that challenge has been applying the appropriate calibration model to sensors, including the Langmuir adsorption model and optimized calibration mode among others.[42] An optically-gated sensor has also been reported to exhibit self-calibration; in this case, the challenge of

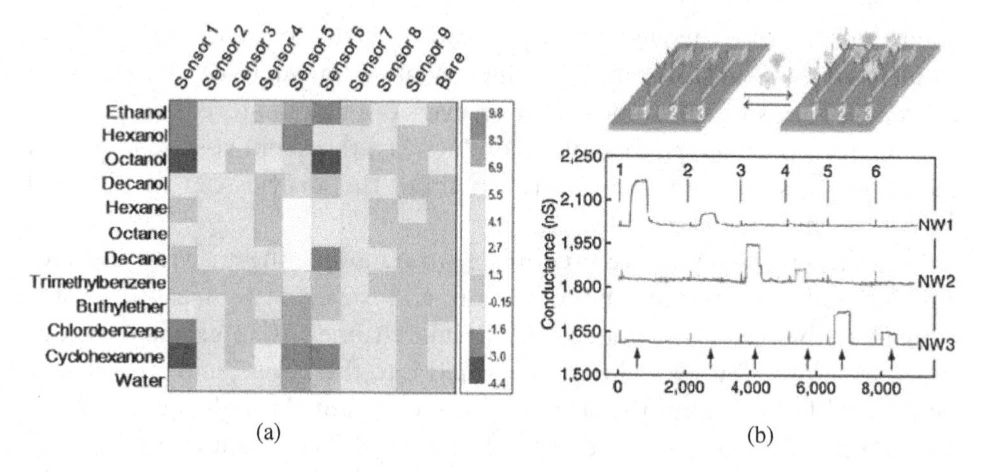


**Figure 7.** Detection behavior of NW FET arrays exposed to complicated samples: (**a**) multiple VOCs.[29] Copyright 2013 American Chemical Society (ACS) and (**b**) mixed cancer markers.[11] Copyright 2005 Macmillan Publishers Limited.

device-to-device variation was overcome by measuring the ratio of the current under dark *versus* light conditions, $[I_{sd}^{light}]/[I_{sd}^{dark}]$, leading to the measurement of absolute pH.[43]

## 5. Key Parameters for Performance of Nanowire FET Sensors

### 5.1. *Sensitivity*

The sensitivity of a sensor can be defined as the ratio between the amplitude of a sensor's signal and the original amplitude (i.e., the change in amplitude that was detected). The signal might be resistance, current, voltage, conductance, or some other measure.[40] Much efforts have been devoted to improve the sensitivity of NW FET sensors. For example, sensors with a higher density of analyte binding sites or a higher surface-to-volume ratio usually exhibit higher sensitivity. Other strategies have included optimization of the alignment of surface probing molecules, and the design of novel device configurations such as the Wheatstone bridge structure.[44,45] Sensitivity also depends on the signal-to-noise ratio, so reducing noise in the output signal can also effectively enhance the sensitivity of NW FET sensors.

### 5.2. *Selectivity*

The selectivity of a sensor indicates its specificity in responding only to desired target objects, compared to other objects that may be present. The selectivity of pure inorganic NWs is poor because many kinds of analytes can change the

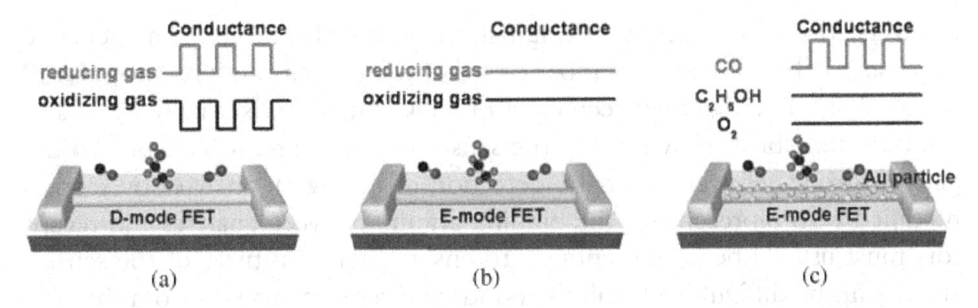


**Figure 8.** The design concept for the "one lock to one key" sensor configuration. (**a**) Conventional "D-mode" FETs for the detection of both reductive and oxidative gases. (**b**) Deep "E-mode" FETs with very large positive threshold voltage yield no response to any target gas. (**c**) Combing the deep "E-mode" FETs as a sensing platform, metal NPs are decorated onto the NW channel surface to introduce the gas specific selectivity and sensitivity to the particular target gas.[46] Copyright 2013 American Chemical Society (ACS).

state of the NW surface and thus generate similar responses.[40] NWs modified with functional groups exhibit better selectivity, especially when biosensors are coated with antigens or DNA. Nonetheless, sensor arrays are still considered to be the best strategy for achieving optimum selectivity (**Figure** 7).

Notably, X. Zou *et al.* demonstrated a "one key to one lock" hybrid sensor for precise sensing of specific target gases, using Magnesium (Mg)-doped $In_2O_3$ NW arrays decorated with various discrete metal NPs (**Figure 8**).[46] For a "D-mode" FET, both oxidizing and reducing gases can change the channel carrier concentration, by respectively withdrawing electrons from NWs or donating electrons to them, and can thus modulate the FET conductance (**Figure 8(a)**).[46] However, only reducing gases can be detected by n-type "E-mode" FET; oxidizing gases are not detected because they decrease the carrier concentration, shift the threshold voltage to a more positive direction, and yield a nil output current at the zero gate bias. By the same reasoning, no gases can be detected by the deep "E-mode" FET, which has a larger positive threshold voltage. (**Figure 8(b)**).[46] Yet the deep "E-mode" FET can serve as a perfect platform for decorating various metal NPs on NW surfaces, to introduce gas selectivity to sensors and improve their sensitivity (**Figure 8(c)**).[46] This is because metal NPs can respond to specific gas molecules through different catalytic reactions, e.g., Au NPs respond specifically to CO due to the low activation energy barrier.

## 5.3. *Reusability*

The reusability of a chemical/biological sensor is evaluated by the degree to which bonded analytes can be removed after sensory tests, and how easy it is to reverse the sensor back to its original status. It is also exhibited as a change in sensing behavior after a number of switches between the 'ON' and 'OFF' states.[7] The connection between gases or VOCs and NWs is usually weak, and can be readily broken by heating the sensor or applying a reverse bias to the gate electrode. However, for the detection of biological analytes by sensors modified with bioreceptors, the binding affinity between analytes and receptors must usually be strong enough to ensure high sensitivity of the sensor. It can thus be difficult to break the bonds and remove analytes from the NW surface after detection.[7] Various techniques have been developed to solve this problem. To break antigen and antibody interactions, for example, a technique called GSH/GST-tag has been developed.[47] Glutathione S-transferase (GST) and glutathione (GSH) molecules exhibit reversible association and dissociation interactions.[47] Sensors can be modified first by GSH, and then linked to GST-fused proteins, which are then used to detect specific target proteins. Once sensing measurements are completed, the GST-fused proteins can be removed easily by washing the sensors with GSH solution. This method has been adapted to fabricate reusable SiNW FET biosensors.[7,47] Numerous other performances including device stability, response speed, recovery time, and detection limit are also key parameters that influence the quality of NW FET sensors. Some external factors including temperature, humidity and chemical contaminations can also greatly affect sensor performance. With ZnO NW FET gas sensors, three steps are involved in the sensing process: adsorption, desorption, and activity of the oxygen ions. The activity of oxygen ions is a function of temperature. As temperature increases, increased ion activity leads to higher sensor sensitivity, but sensitivity actually declines once the temperature rises above what is optimum.[40]

## 6. Conclusion

FET sensors based on semiconducting NWs have many demonstrated advantages over conventional sensors. Detection of a broad range of chemical and biological objects has been achieved by employing diverse semiconductor materials and surface functional groups. The sensitivity of FET sensors can be significantly enhanced thanks to the high surface-to-volume ratio of NWs. The selectivity of FET sensors has been improved drastically by modifying functional groups on NW surfaces. Application of such sensors has succeeded in both vapor and liquid environments, using real-time detection or "steady-state" measurement.

Advancement in the field of chemical/biological sensors based on NW FETs will focus on further improvement of sensor performance, and on low-cost large-scale fabrication of sensors. The former can be achieved by introducing complex sensor arrays, while the latter will likely be realized by novel techniques such as roll-to-roll printing and self-assembly techniques.

## References

1. Chen PC, Shen G, Zhou C. Chemical sensors and electronic noses based on 1D metal oxide nanostructures. *IEEE Transac Nanotechnol* 2008;7:668–682.
2. Ramgir NS, Yang Y, Zacharias M. Nanowire-based sensors. *Small* 2010;6:1705–1722.
3. Barth S, Hernandez-Ramirez F, Holmes JD, Romano-Rodriguez A. Synthesis and applications of one-dimensional semiconductors. *Prog Mater Sci* 2010;55:563–627.
4. Chen Y, Liang D, Gao XPA, Alexander JID. Sensing and energy harvesting of fluidic flow by InAs nanowires. *Nano Lett* 2013;13:3953–3957.
5. Sarkar J, Khan G, Basumallick A. Nanowires: properties, applications and synthesis *via* porous anodic aluminium oxide template. *Bullet Mater Sci* 2007;30:271–290.
6. Hobbs RG, Petkov N, Holmes JD. Semiconductor nanowire fabrication by bottom-up and top-down paradigms. *Chem Mater* 2012;24:1975–1991.
7. Chen KI, Li BR, Chen YT. Silicon nanowire field-effect transistor-based biosensors for biomedical diagnosis and cellular recording investigation. *Nano Today* 2011;6:131–154.
8. Roy S, Gao Z. Nanostructure-based electrical biosensors. *Nano Today* 2009;4:318–334.
9. Baranowska-Korczyc A, Fronc K, Klopotowski L, Reszka A, Sobczak K, Paszkowicz W, Dybko K, Dluzewski P, Kowalski BJ, Elbaum D. Light- and environment-sensitive electrospun ZnO nanofibers. *RSC Adv* 2013;3:5656–5662.
10. Pachauri V, Kern K, Balasubramanian K. Field-effect-based chemical sensing using nanowire-nanoparticle hybrids: the ion-sensitive metal-semiconductor field-effect transistor. *Appl Phys Lett* 2013;102.
11. Zheng G, Patolsky F, Cui Y, Wang WU, Lieber CM. Multiplexed electrical detection of cancer markers with nanowire sensor arrays. *Nat Biotech* 2005;23:1294–1301.
12. Stern E, Wagner R, Sigworth FJ, Breaker R, Fahmy TM, Reed MA. Importance of the Debye screening length on nanowire field effect transistor sensors. *Nano Lett* 2007;7:3405–3409.
13. Li C, Curreli M, Lin H, Lei B, Ishikawa FN, Datar R, Cote RJ, Thompson ME, Zhou C. Complementary detection of prostate-specific antigen using In203 nanowires and carbon nanotubes. *J Am Chem Soc* 2005;127:12484–12485.
14. Curreli M, Zhang R, Ishikawa FN, Chang H-K. Real-time, label-free detection of biological entities using nanowire-based FETs. *IEEE Transac Nanotechnol* 2008;7:651–667.
15. Zhang G-J, Ning Y. Silicon nanowire biosensor and its applications in disease diagnostics: a review. *Anal Chim Acta* 2012;749:1–15.
16. Duan X, Rajan NK, Izadi MH, Reed MA. Complementary metal oxide semiconductor-compatible silicon nanowire biofield-effect transistors as affinity biosensors. *Nanomed* 2013;8:1839–1851.
17. Cui Y, Wei Q, Park H, Lieber CM. Nanowire nanosensors for highly sensitive and selective detection of biological and chemical species. *Science* 2001;293:1289–1292.
18. Lin T-W, Hsieh P-J, Lin C-L, Fang Y-Y, Yang J-X, Tsai C-C, Chiang P-L, Pan C-Y, Chen Y-T. Label-free detection of protein-protein interactions using a calmodulin-modified nanowire transistor. *Proceedings of the National Academy of Sciences of the United States of America* 2010;107:1047–1052.

19. Kim A, Ah CS, Yu HY, Yang J-H, Baek I-B, Ahn C-G, Park CW, Jun MS, Lee S. Ultrasensitive, label-free, and real-time immunodetection using silicon field-effect transistors. *Appl Phys Lett* 2007;91.
20. Chua JH, Chee R-E, Agarwal A, Wong SM, Zhang G-J. Label-free electrical detection of cardiac biomarker with complementary metal-oxide semiconductor-compatible silicon nanowire sensor arrays. *Anal Chem* 2009;81:6266–6271.
21. Zhang G-J, Luo ZHH, Huang MJ, Ang JJ, Kang TG, Ji H. An integrated chip for rapid, sensitive, and multiplexed detection of cardiac biomarkers from fingerprick blood. *Biosens Bioelectron* 2011;28:459–463.
22. Hahm JI, Lieber CM. Direct ultrasensitive electrical detection of DNA and DNA sequence variations using nanowire nanosensors. *Nano Lett* 2003;4:51–54.
23. Bunimovich YL, Shin YS, Yeo W-S, Amori M, Kwong G, Heath JR. Quantitative real-time measurements of DNA hybridization with alkylated non-oxidized silicon nanowires in electrolyte solution. *J Am Chem Soc* 2006;128:16323–16331.
24. Zhang G-J, Chua JH, Chee R-E, Agarwal A, Wong SM. Label-free direct detection of miRNAs with silicon nanowire biosensors. *Biosens Bioelectron* 2009;24:2504–2508.
25. Choi J-H, Kim H, Choi J-H, Choi J-W, Oh B-K. Signal enhancement of silicon nanowire-based biosensor for detection of matrix metalloproteinase-2 using DNA-Au nanoparticle complexes. *ACS Appl Mater Interfaces* 2013;5:12023–12028.
26. Pearton SJ, Ren F, Wang Y-L, Chu BH, Chen KH, Chang CY, Lim W, Lin J, Norton DP. Recent advances in wide bandgap semiconductor biological and gas sensors. *Prog Mater Sci* 2010;55:1–59.
27. Shaymurat T, Tang Q, Tong Y, Dong L, Liu Y. Gas dielectric transistor of CuPc single crystalline nanowire for $SO_2$ detection down to sub-ppm levels at room temperature. *Adv Mater* 2013;25:2269–2273.
28. Li C, Zhang C, Fobelets K, Zheng J, Xue C, Zuo Y, Cheng B, Wang Q. Impact of ammonia on the electrical properties of p-type Si nanowire arrays. *J Appl Phys* 2013;114.
29. Ermanok R, Assad O, Zigelboim K, Wang B, Haick H. Discriminative power of chemically sensitive silicon nanowire field effect transistors to volatile organic compounds. *ACS Appl Mater Interfaces* 2013;5:11172–11183.
30. Wang B, Haick H. Effect of chain length on the sensing of volatile organic compounds by means of silicon nanowires. *ACS Appl Mater Interfaces* 2013;5:5748–5756.
31. Wang B, Haick H. Effect of functional groups on the sensing properties of silicon nanowires toward volatile compounds. *ACS Appl Mater Interfaces* 2013;5:2289–2299.
32. Li H-H, Yang C-E, Kei C-C, Su C-Y, Dai W-S, Tseng J-K, Yang P-Y, Chou J-C, Cheng H-C. Coaxial-structured ZnO/silicon nanowires extended-gate field-effect transistor as pH sensor. *Thin Solid Films* 2013;529:173–176.
33. Wipf M, Stoop RL, Tarasov A, Bedner K, Fu W, Wright IA, Martin CJ, Constable EC, Calame M, Schonenberger C. Selective sodium sensing with gold-coated silicon nanowire field-effect transistors in a differential setup. *ACS Nano* 2013;7:5978–5983.
34. Zhuo M, Chen Y, Sun J, Zhang H, Guo D, Zhang H, Li Q, Wang T, Wan Q. Humidity sensing properties of a single Sb doped $SnO_2$ nanowire field effect transistor. *Sens Actuat B: Chem* 2013;186:78–83.
35. Huang Y-W, Wu C-S, Chuang C-K, Pang S-T, Pan T-M, Yang Y-S, Ko F-H. Real-time and label-free detection of the prostate-specific antigen in human serum by a polycrystalline silicon nanowire field-effect transistor biosensor. *Anal Chem* 2013;85:7912–7918.

36. Lin T-Y, Li B-R, Tsai S-T, Chen C-W, Chen C-H, Chen Y-T, Pan C-Y. Improved silicon nanowire field-effect transistors for fast protein-protein interaction screening. *Lab Chip* 2013;13:676–684.
37. Chen W-Y, Chen H-C, Yang Y-S, Huang C-J, Chan HW-H, Hu W-P. Improved DNA detection by utilizing electrically neutral DNA probe in field-effect transistor measurements as evidenced by surface plasmon resonance imaging. *Biosens Bioelectron* 2013;41:795–801.
38. Zhang G-J, Huang MJ, Ang JJ, Yao Q, Ning Y. Label-free detection of carbohydrate-protein interactions using nanoscale field-effect transistor biosensors. *Anal Chem* 2013;85:4392–4397.
39. Liu X, Lin P, Yan X, Kang Z, Zhao Y, Lei Y, Li C, Du H, Zhang Y. Enzyme-coated single ZnO nanowire FET biosensor for detection of uric acid. *Sens Actuat B: Chem* 2013;176:22–27.
40. Wei A, Pan L, Huang W. Recent progress in the ZnO nanostructure-based sensors. *Mater Sci Eng B* 2011;176:1409–1421.
41. Roberts ME, Sokolov AN, Bao Z. Material and device considerations for organic thin-film transistor sensors. *J Mater Chem* 2009;19:3351–3363.
42. Justino CIL, Rocha-Santos TAP, Cardoso S, Duarte AC. Strategies for enhancing the analytical performance of nanomaterial-based sensors. *TrAC Trends Anal Chem* 2013;47:27–36.
43. Peretz-Soroka H, Pevzner A, Davidi G, Naddaka V, Tirosh R, Flaxer E, Patolsky F. Optically-gated self-calibrating nanosensors: monitoring pH and metabolic activity of living cells. *Nano Lett* 2013;13:3157–3168.
44. Chu C-J, Yeh C-S, Liao C-K, Tsai L-C, Huang C-M, Lin H-Y, Shyue J-J, Chen Y-T, Chen C-D. Improving nanowire sensing capability by electrical field alignment of surface probing molecules. *Nano Lett* 2013;13:2564–2569.
45. Maedler C, Erramilli S, House LJ, Hong MK, Mohanty P. Tunable nanowire Wheatstone bridge for improved sensitivity in molecular recognition. *Appl Phys Lett* 2013;102.
46. Zou X, Wang J, Liu X, Wang C, Jiang Y, Wang Y, Xiao X, Ho JC, Li J, Jiang C, Fang Y, Liu W, Liao L. Rational design of sub-parts per million specific gas sensors array based on metal nanoparticles decorated nanowire enhancement-mode transistors. *Nano Lett* 2013;13:3287–3292.
47. Lin S-P, Pan C-Y, Tseng K-C, Lin M-C, Chen C-D, Tsai C-C, Yu S-H, Sun Y-C, Lin T-W, Chen Y-T. A reversible surface functionalized nanowire transistor to study protein–protein interactions. *Nano Today* 2009;4:235–243.

# Index

Printed in the United States
By Bookmasters

Printed in the United States
By Bookmasters